Writing in the
Health Professions

THE ALLYN AND BACON SERIES IN TECHNICAL COMMUNICATION

Series Editor: Sam Dragga, Texas Tech University

Thomas T. Barker
Writing Software Documentation: A Task-Oriented Approach, Second Edition

Carol M. Barnum
Usability Testing and Research

Deborah S. Bosley
Global Contexts: Case Studies in International Technical Communication

Melody Bowdon and Blake Scott
Service-Learning in Technical and Professional Communication

R. Stanley Dicks
Management Principles and Practices for Technical Communicators

Paul Dombrowski
Ethics in Technical Communication

David Farkas and Jean Farkas
Principles of Web Design

Laura J. Gurak
Oral Presentations for Technical Communication

Sandra W. Harner and Tom G. Zimmerman
Technical Marketing Communication

Barbara A. Heifferon
Writing in the Health Professions

TyAnna K. Herrington
A Legal Primer for the Digital Age

Richard Johnson-Sheehan
Writing Proposals: Rhetoric for Managing Change

Dan Jones
Technical Writing Style

Charles Kostelnick and David D. Roberts
Designing Visual Language: Strategies for Professional Communicators

Victoria M. Mikelonis, Signe T. Betsinger, and Constance Kampf
Grant Seeking in an Electronic Age

Ann M. Penrose and Steven B. Katz
Writing in the Sciences: Exploring Conventions of Scientific Discourse, Second Edition

Carolyn Rude
Technical Editing, Third Edition

Gerald J. Savage and Dale L. Sullivan
Writing a Professional Life: Stories of Technical Communicators On and Off the Job

Writing in the Health Professions

Barbara A. Heifferon

Clemson University

PEARSON
Longman

New York   San Francisco   Boston
London   Toronto   Sydney   Tokyo   Singapore   Madrid
Mexico City   Munich   Paris   Cape Town   Hong Kong   Montreal

To Lizzie and Leah

Vice President and Publisher: Eben W. Ludlow
Senior Supplements Editor: Donna Campion
Senior Marketing Manager: Wendy Albert
Production Manager: Donna DeBenedictis
Project Coordination, Text Design, and Electronic Page Makeup: Elm Street Publishing Services, Inc.
Cover Designer/Manager: Wendy Ann Fredericks
Manufacturing Buyer: Roy L. Pickering, Jr.
Printer and Binder: Hamilton Printing Co.
Cover Printer: The Lehigh Press, Inc.

For permission to use copyrighted material, grateful acknowledgment is made to the copyright holders on pp. 309–10 which are hereby made part of this copyright page.

Library of Congress Cataloging-in-Publication Data

Heifferon, Barbara.
    Writing in the health professions / Barbara A. Heifferon.
        p. ; cm. — (The Allyn & Bacon series in technical communication)
    Includes bibliographical references and index.
    ISBN 0-321-10527-3
    1. Medical writing. I. Title. II. Series: Allyn and Bacon series in technical communication.
    [DNLM: 1. Writing. 2. Documentation—methods. 3. Health Occupations. 4. Medical
    Records. WZ 345 H465w 2005]
R119.H397 2005
808'.06661—dc22

                                                                    2004019813

Please visit us at http://www.ablongman.com

ISBN 0-321-10527-3

    3 4 5 6 7 8 9 10—HT—07 06

CONTENTS

CHAPTER 3

Document Design Principles and Project Management   65

PART II GENRES IN THE HEALTH PROFESSIONS

CHAPTER 4 Medical Diagnostic Practices and Charting   91

CHAPTER

8   Grants, Proposals, and Government Documents   186

PART III   MEDIA AND CULTURES

CHAPTER 9   Multicultural and International Medical Writing   205

CHAPTER 10   Presenting Written Materials Visually   229

CHAPTER

11 Electronic Medical Writing   258

APPENDIX

A Common Greek and Latin Roots, Prefixes, and Suffixes   289

APPENDIX

B Frequently Used Visual Design Terms   297

APPENDIX

C Frequently Used Clinical Abbreviations   303

FOREWORD
by the Series Editor

The Allyn and Bacon Series in Technical Communication is designed to meet the continuing education needs of professional technical communicators, both those who desire to upgrade or update their own communication abilities as well as those who train or supervise writers, editors, and artists within their organization. This series also serves the growing number of students enrolled in undergraduate and graduate programs in technical communication. Such programs offer a wide variety of courses beyond the introductory technical writing course—advanced courses for which fully satisfactory and appropriately focused textbooks have often been impossible to locate.

The chief characteristic of the books in this series is their consistent effort to integrate theory and practice. The books offer both research-based and experienced-based instruction, describing not only what to do and how to do it but explaining why. The instructors who teach advanced courses and the students who enroll in these courses are looking for more than rigid rules and ad hoc guidelines. They want books that demonstrate theoretical sophistication and a solid foundation in the research of the field as well as pragmatic advice and perceptive applications. Instructors and students will also find these books filled with activities and assignments adaptable to the classroom and to the self-guided learning processes of professional technical communicators.

To operate effectively in the field of technical communication, today's technical communicators require extensive training in the creation, analysis, and design of information for both domestic and international audiences, for both paper and electronic environments. The books in the Allyn and Bacon Series address those subjects that are most frequently taught at the undergraduate and graduate levels as a direct response to both the educational needs of students and the practical demands of business and industry. Additional books will be developed for the series in order to satisfy or anticipate changes in writing technologies, academic curricula, and the profession of technical communication.

SAM DRAGGA
Texas Tech University

PREFACE

Writing in the Health Professions is a text about the language of medicine and its various shapes and forms in writing. *Writing in the Health Professions* looks at a wide range of health professionals' writing needs and includes a strong orientation toward visual and electronic forms of communication, with sections on visual communication, document design, Web sites, and multimedia productions— all now clearly necessary in today's high-tech medical and health writing tasks.

Do people in the health professions need to be able to write? Yes. Do the health professions need medical writers who are not necessarily health professionals per se? Yes, both are true. Whether one is a health professional or a professional writer, medicine and health provide special contexts for communication. Jennie Dautermann suggests, "[B]ecause hospital staff members [and other health professionals] must deal with rapid changes in medical technology, written communication in hospitals often functions in a context where split-second decisions may mean the life or death of acutely ill patients."[1]

Writing in the Health Professions addresses this situated writing, that is, writing found within not only specific workplaces such as hospitals, but also clinics, HMOs, health insurance companies, and other health settings. It is a predominately practical and applied text, rather than one that is theoretical and/or historical. The audience for this book is not just composed of students wanting to enter health fields or undertake studies in medical rhetoric or health communication, but also doctors, nurses, paramedics, health insurance practitioners, medical technicians and, most particularly, document designers and multimedia authors (formerly known as technical writers), who write a variety of health-oriented and medical documents.

Writing in the Health Professions also attempts to address and fill the gap in situated medical writing practice, but broadly. While I could not cover every single genre necessary for such a wide audience, I do cover many of those genres not covered before in texts about writing for or in the health professions.

A text that addresses the multiple forms of writing in and for the health professions is long overdue. In our field, there are a few texts that address physicians' and nurses' journal article writing practices, several texts that address medical language historically, and a few health writers' handbooks for general writing practices. In addition to starting to fill the gap in medical and health writing, this text can and should initiate research about writing in and for the health professions.

---

[1] Dautermann, Jennie. *Writing at Good Hope: A Study of Negotiated Composition in a Community of Nurses.* Greenwich, CT: Ablex, 1997.

Research areas abound in this field; because this text is broad-based, all the various genres and methodologies covered here need more in-depth research. *Writing in the Health Professions* can function as the jumping-off point for further inquiry.

Because I've been a health practitioner as well as an academic and medical writer, I am particularly enthusiastic about writing a text that addresses the concerns of the healthcare professions and those who write for them. In particular, I draw upon my medical experiences as a cardiopulmonary and special procedures technician to feature numerous examples and genres not usually covered in medical writing texts. But more importantly perhaps, I draw upon my experiences as a writing teacher. I am committed to providing a strong rhetorical basis for health communication practitioners as well as the most up-to-date writing and technical communication practices. This emphasis can be found throughout the text.

How Is This Text Organized?

Part I of *Writing in the Health Professions,* Considerations in Writing for Health and Medical Audiences, includes attention to audiences, ethics, and document design. Part II, Genres in the Health Professions, is based on the wide range of genres in which health professionals write. Part III, Media and Cultures, covers multicultural and international audiences, electronic and visual media within health and medicine.

Part I: Considerations in Writing for Health and Medical Audiences includes the basics for approaching writing in the professional health workplace. Chapter 1, Audience Analysis and Context Analysis, covers rhetorical strategies to analyze how to write for the variety of audiences one finds in the health workplace as well as for the various medical contexts. Chapter 2, Ethics in Medical Writing, discusses ethical considerations particular to health and medical writing. And Chapter 3, Document Design Principles and Project Management, examines the general visual communication and electronic communication practices now necessary in our profession and project management techniques to smooth the way for writers to more effectively manage their writing projects.

In **Chapter 1: Audience Analysis and Context Analysis,** I cover the material in audience analysis that many technical communication texts also cover, but the analyses are aimed toward particular audiences that medical writers address. I also discuss the type of writing and audience analysis done in traditional science writing, particularly those applications appropriate for health audiences. This chapter sets up methodologies for both students and practitioners. Such analyses should be done at the beginning of any writing and communication task or project so that the message reaches its audience(s) most effectively.

Because of what is at stake for writers and communicators in medicine, you will want to include **Chapter 2: Ethics in Medical Writing** in whatever table of contents option you choose for reading this text. Although most technical communication textbooks mention ethics, many do not give enough emphasis to this area. Paul Dombrowski's *Ethics in Technical Communication,* also in this series, includes some examples of medical ethics. Because ethics in writing also includes legal issues, you will learn some of the legal considerations you need to know to meet the needs of professional health writing at your particular site.

Chapter 3: Document Design Principles and Project Management focuses on basic document design principles and acquaints the reader with essential vocabulary. You will discover some of the theory and general design practices that underlie basic design decisions. You will also learn how to manage your documentation projects so that your time and resources are used effectively. In Chapters 10 and 11 in Part III, we follow up in practice and in more detail on what is outlined in general in this chapter on visual theory and vocabulary.

Part II: Genres in the Health Professions includes many of the genres in and by which health professionals communicate. If you look over the genres in this part, it will help you determine the most relevant chapters for your workplace. If you are a student as yet uncertain about the health professions for which you will be writing, a familiarity with them all will be helpful. Chapter 4, Medical Diagnostic Practices and Charting, covers the diagnostic practices that initiate and provide content for patient charts in hospitals and doctors' offices. Chapter 5, Medical Forms and Reports, discusses internal reports such as encounter forms and procedural reports with examples of each, plus forms that patients fill out. Chapter 6, Health Education Materials, examines documents such as handbooks, newsletters, brochures, instruction sheets, posters, flyers, shower tags, bookmarks, and Web sites aimed at patients and consumers. Chapter 7, Public Health Campaigns, analyzes the types of problems suitable for campaigns, ethical considerations, audience segmentation, media choices, campaign proposals, and assessments. And Chapter 8, Grants, Proposals, and Government Documents, considers grant writing in medicine, the language of government agencies and insurance companies, documentation standards, pharmaceutical documents, and medical sciences research reports.

In the first part of **Chapter 4: Medical Diagnostic Practices and Charting,** we move from the basics of how to design for health and medical audiences to the documents generated within medical workplaces themselves. First, we look specifically at the various diagnostic practices that generate the first reports or content for a medical record (a.k.a., patient chart). The writing of patient histories and physicals (HXP) also falls within this chapter's purview. A number of health professionals such as paramedics, physician assistants, nurse practitioners, nurses, as well as doctors, now are responsible for this important part of the diagnosis and assessment of patients.

Charting also is covered in this chapter because it anchors the written documentation of the individual patient's visit, whether that patient is being seen in a doctor's office, clinical laboratory, or hospital. Also, we look at an example of computer programs available for medical record keeping.

Chapter 5: Medical Forms and Reports covers internal reports such as encounter forms, progress notes, and procedural reports with examples of each, plus forms that patients fill in. You will learn both how to detect the specific histories and conventions of each document and how to prioritize the necessary information for each form.

In **Chapter 6: Health Education Materials,** we look at materials that are often directed at individual patients and other types of consumers. Such documents include handbooks, newsletters, brochures, instruction sheets, posters, flyers, shower tags, bookmarks, and Web sites aimed at patients and consumers or non-professionals. This chapter covers both traditional and on-line materials that medical writers develop for targeting groups of health information users. You will have a chance to apply the audience analysis learned in Chapter 1 and some of the design vocabulary developed in Chapter 3 to various genres represented in this chapter.

Chapter 7: Public Health Campaigns addresses large communication campaigns targeting mass audiences as opposed to materials targeting individuals and smaller groups. Methodologies are featured here for the developers/writers of those campaigns in information agencies, large hospital systems, nonprofit organizations, and government programs. This chapter discusses the research process, the design plan, project management, and outcome assessment or evaluation of large health campaigns. Prior to reading this chapter, you need to read Chapter 1 in this text on audience analysis, as this component is key to designing a good health campaign.

And **Chapter 8: Grants, Proposals, and Government Documents** covers writing proposals and grants in the medical arena, since public relations professionals, practitioners, and researchers often write proposals and grants as an essential part of their work. In addition, this section looks at writing for the government and health insurance companies, as well as those forms and documents (in general) that almost all health professionals use in their various healthcare settings.

Part III: Media and Cultures draws your attention to multicultural and international aspects of writing in the health field and the important new software and hardware developments within modern medical settings. In the future we will see dynamic and ever more digitized delivery systems of information. We are all aware that even within one's own culture, many other subcultures exist. If we plan to write effectively for the various audiences that now occur in most medical communications, we need to be sensitive to an increasingly multicultural and global orientation. Thus the three chapters in this last part cover the following information: **Chapter 9, Multicultural and International Medical Writing,** describes processes for analyzing cultures and resources for intercultural design, both in diverse settings with shared languages and in international settings with other languages; Chapter 10, Presenting Written Materials Visually, looks at applications of design language and theory developed in Chapter 3 with more hands-on information and oral presentation skills; and Chapter 11, Electronic Medical Writing, discusses storyboarding, scripting, Web site analyses and design issues, audio and video designs, PowerPoint, and interactive CD-ROMs.

Although the preliminary work on audience analysis can be found in Chapter 1, Chapter 9: Multicultural and International Medical Writing goes into more detail by drawing specific attention to various cultural groups found in our country as well as those outside our geographical borders. Other cultural groups examined within the United States will include African American, Latino, Asian, and

Native American. Most importantly, this chapter offers methods for analyzing culture that will help you know what to look for as you design for both multicultural and international health audiences.

Chapter 10: Presenting Written Materials Visually addresses the oral and visual presentation of material, because written texts often migrate to other forms. Also, much written material now is heavily dependent on graphics. Just as the design terms and principles were introduced for document design in Chapter 3 and Appendix B, we provide specific terminology for visual design and discuss key elements. Sometimes as health writers within large hospital systems, you will be asked to write presentations for doctors to accompany their PowerPoint or other multimedia presentations. Thus, although we do not discuss oral presentations at great length, we give you some reminders and a blueprint for integrating the oral and visual components of a presentation.

Chapter 11: Electronic Medical Writing looks at the many electronic forms of communication now used to reach audiences. They are increasingly apparent in the healthcare fields. Writing in electronic formats, however, is quite different from writing for more traditional formats. Therefore, this chapter will be one of the larger chapters in the book. Subheadings include:

- Web sites
- Videos/Audios
- Multimedia writing

At the end of Part III are three appendixes that list the various sources that we have pointed you to throughout the text. **Appendix A: Common Greek and Latin Roots, Prefixes, and Suffixes** gives you a beginning vocabulary that helps you to understand medical language better. **Appendix B: Frequently Used Visual Design Terms** gives you a vocabulary to use as you work on more visually oriented documents, whether they are designed and produced in-house or outsourced. And **Appendix C: Frequently Used Clinical Abbreviations** provides the most common medical abbreviations you will find and/or need to use in your documents.

Please refer to the Contents and Alternative Contents in the Instructor's Manual for the various ways you can use and read this book. Depending on whether you are a student or a practicing health professional or both, you will find alternatives for reading this text helpful. Medical writing can be daunting at first, but I hope you will come to discover that it is a rich, interesting, and exciting area to learn and to explore.

Instructor's Manual

The Instructor's Manual that accompanies *Writing in the Health Professions* provides the instructor with helpful background material and extends some of the concepts by means of examples. The discussion questions and exercises are already included in the text itself. In addition, you will find more references if you wish to read further and extend your knowledge in particular areas. Most important, the Instructor's Manual also contains a brief but important review of process writing for those students whose background in composition is weak or nonexistent.

Acknowledgments

There would never be enough pages in this book to thank the many people to whom I am indebted, so I will have to thank those who are and have been the most instrumental in my bringing this text to fruition. I could not open my eyes every morning without my daughters, Elizabeth Dee and Leah Adrianne. They mean the world to me. Also, I could not have written this book without my best friend and lifesaver, Judy Melton, who has been more than inspirational on a daily basis. In addition, Stuart Brown has been my mentor on this project from its conception to its birth. It was he who put this project on the radar screen and knew it was achievable. Mary De Shazer provided feedback at a crucial time in the process and let me use her as a sounding board for my ideas. Christi Conti, my graduate assistant in spring and summer of 2004, proved herself invaluable when it came to assembling the permissions for the book. My own physician, David Irvine, provided not only sound advice and good counsel, but he also saved my life in 2002. The reviews were valuable in the revision of the text, and for their help, I am grateful to Ellen Barton of Wayne State University, Amy Koerber of Texas Tech University, and Elizabeth W. Staton at University of Colorado's Health Sciences Center.

I would also like to thank my extended family, other friends, and colleagues who lend me their ears, eyes, and hearts on a regular basis, especially Mac Wood. Lastly I'd like to thank my dean, Janice C. Schach, and my college and university, the College of Architecture, Arts and Humanities at Clemson University, for their support and for a research grant in the spring of 2003, which allowed me to commit my time and energy to the writing of this book.

BARBARA A. HEIFFERON

Audience Analysis and Context Analysis

 Overview

This chapter discusses audience analysis and context analysis, topics that many technical communication texts cover. However, we aim these analyses toward particular audiences that you, as a health professional or a student going into a health profession, will address. An understanding of audience and context, including attention to gender, race, age, class, attitudes, and culture, will enable you to more effectively accomplish your document purposes. We use the terms *audience analysis* and *context analysis* as ways to talk about analyses that are broad enough to translate across the various disciplines, both science and humanities, that you as medical writers come from. We also address the type of writing and audience analysis (or lack thereof) done in traditional science writing, some of which is and some of which is not appropriate for health audiences.

In this chapter, you will learn the answers to the following questions about audience and context analysis:

- How do I begin to analyze?
- How do I determine the context or rhetorical situation?
- How do I determine audience needs?
- How do I choose research tools?
- What is usability testing?

This chapter sets up methods that both students and practitioners can use to analyze various audiences and rhetorical situations before beginning writing projects in order to reach the relevant audiences and thus be most effective. In addition, we include here a number of research tools or methodologies that will ensure that you are meeting audience needs and that will reveal deficiencies in documents or in health campaigns to help you revise. Some of the techniques we

1

cover here are library and Internet research, field research, surveys, interviews, focus groups, and usability strategies. These tools can be used at numerous points in the development of documents, manuals, and campaigns—not just at the beginning or end of development, but along the way as well.

How Do I Begin to Analyze?

Once you have determined the purpose of your document, you will next begin to move toward an analysis of both context and audience. If you are acquainted with process writing, you will remember that each time we design a new document or write on a chart or speak to another person in our workplace, we are in a rhetorical situation. Any situation that demands some type of communication is in fact rhetorical. As health communicators, in order not to fail at our message to our audience, we need to analyze the situation. Another way to view the rhetorical situation is as the context for writing/speaking/visually representing our communication task. "Context" should not be confused with "content." Context is the bigger picture surrounding our particular communication assignment. In this chapter I'm using a particular project my students and I developed as an example of how important researching context is to correctly targeting your audience. If we assume, for example, that African American women don't come in for free breast screenings because they are embarrassed about examinations, we have missed some of the more important reasons for their absence: transportation to and from campus, parking, and fear that "free" healthcare is a euphemism for some experimental drug therapy or forced sterilization. Had we not done our research, we would not have targeted these particular problems to solve before we developed our outreach materials.

To begin your analysis, develop a list of questions about both the context and audience for your specific project. If you don't know the answers to questions, begin your research with those unanswered questions in mind. You can research through the Internet, library, via phone calls, emails, and interviews . . . whatever it takes to determine the answers. The more research you complete at the beginning of the actual design process, the less revision you'll have to do and the less chance of missing your target audience.

How Do I Determine the Context or Rhetorical Situation?

Geography and Environment

Geography, whether your context is international, national, regional, or local, is a key consideration. Determining geography is a two-sided approach. First, where is your audience located; and second, where are you developing this document? Look at the country, region, city, town, or rural area in which your audience is located. A small hospital with few facilities for an extensive health promotion campaign will need a more scaled-down campaign than a major university teaching hospital. Working on documents for a large, international pharmaceutical company that will finance state-of-the-art health campaign materials offers a very different context.

You need to ask the same questions about your target audience. Will you be writing a manual for installing a new telemetry system or designing the instructions for nurses on its use in an intensive care unit (ICU)? Will an audience in a small rural hospital in the Midwest have different needs than an audience in downtown Chicago? The geography or the setting of your audience may predetermine some of your context.

Think of geography the same way you might the setting of a play or novel. What role, for example, does the audience you are writing to play in the community? Is the hospital a major employer in the area? Does it compete with other hospitals for top billing? Do certain types of patients go to one hospital as opposed to another? Or is it perhaps a health insurance company that is number two in the nation and, like Avis Rental Car once did, needs to convince consumers that it tries harder?

Environment also includes access. What difficulties do patients face in order to access health facilities? Is there ease of access or are there problems in accessing certain healthcare sites? Access was a crucial consideration in the healthcare project mentioned above in which we needed to persuade African American women from a public housing site to come to a nursing and wellness center on a university campus for free breast and cervical cancer screening. Because we discovered that one of the problems was that there was no parking available on campus for these patients, part of the project design involved transportation: one team of students redesigned a bus schedule (Exhibit 1.1) and advertised the free bus service that went directly to the wellness center as a way to increase participation in the program. Because of interviews and surveys with the target audiences, we were able to identify needs we had not anticipated.

History

What is the history—both of existing documents and the situation you are addressing? Do you need to send a memo that clearly lays out procedures that two different departments both say is the other department's responsibility? For instance, has the radiology department at your regional health center historically refused to process medical records intrinsic to the heart catheterization procedure that takes place in the cardiac diagnostic department? Will you need to structure a document that persuades radiology to include these records in their storage space?

All situations have some type of history. Having a history does not mean that it is necessarily negative or that it will cause problems; some histories are quite benign. But before you begin your document project, you need to think through its historical context.

We often refer to the oral history of a particular place or setting as **folklore**. Folklore is made up of stories passed from employee to employee, especially

Purple Route (D)   Northeast Residential

Service from C-1 Lot or P-1 Lot on request only

All times are departures except those outlined in red

Clemson-Central Library
Clemson-Central Senior's Center
Heatherwood Place Apartments
Ingles
Hunters Glen Apartments
Issaqueena Village Apartments
Heritage Apartments
Ridgecrest Apartments
Littlejohn Community Center
Winn-Dixie
Tillman Place Apartments
Redfern Health Center
East Library Circle
Thornhill Village Laundry
Cherry @ McMillan
P-1 Parking Lot

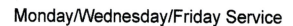

Monday/Wednesday/Friday Service

Clem-Central Senior Center	7:18	8:25	8:28	9:30	9:33	10:25	10:28	11:40	11:43	12:45	12:48	1:50	1:53	2:55	2:58	4:00	4:01	5:25
Apartments on Issaqueena Trl	7:20	8:19	8:30	9:24	9:35	10:29	10:40	11:34	11:45	12:39	12:50	1:44	1:55	2:49	3:00	3:54	4:03	5:19
Ingles	7:25	8:16	8:35	9:21	9:40	10:26	10:45	11:31	11:50	12:36	12:55	1:41	2:00	2:46	3:05	3:51	4:06	5:16
Littlejohn CC	7:28	8:13	8:38	9:18	9:43	10:23	10:48	11:28	11:53	12:33	12:58	1:38	2:03	2:43	3:08	3:48	4:11	5:13
Winn-Dixie	7:30	8:11	8:40	9:16	9:45	10:21	10:50	11:26	11:55	12:31	1:00	1:36	2:05	2:41	3:10	3:46	4:13	5:11
Tillman Place	7:34	8:06	8:44	9:11	9:49	10:16	10:54	11:21	11:59	12:26	1:04	1:31	2:09	2:36	3:14	3:41	4:17	5:06
Thornhill Village	7:43	8:02	8:48	9:07	9:53	10:12	10:58	11:17	12:03	12:22	1:08	1:27	2:13	2:32	3:18	3:37	4:21	5:02
East Library	7:46	8:00	8:51	9:05	9:56	10:10	11:01	11:15	12:06	12:20	1:11	1:25	2:15	2:30	3:21	3:35	4:24	5:00

Tuesday/Thursday Service

Clem-Central Senior Center	7:18	8:25	8:53	9:55	10:23	11:25	11:53	12:55	1:23	2:25	2:23	3:55	4:01	5:25
Apartments on Issaqueena Trl	7:20	8:19	8:55	9:49	10:25	11:19	11:55	12:49	1:25	2:19	2:25	3:49	4:03	5:19
Ingles	7:25	8:16	9:00	9:46	10:30	11:16	12:00	12:46	1:30	2:16	2:30	3:46	4:06	5:16
Littlejohn CC	7:28	8:13	9:03	9:43	10:33	11:13	12:03	12:43	1:33	2:13	2:33	3:43	4:11	5:13
Winn-Dixie	7:30	8:11	9:05	9:41	10:35	11:11	12:05	12:41	1:35	2:11	2:35	3:41	4:13	5:11
Tillman Place	7:34	8:06	9:09	9:36	10:39	11:06	12:09	12:36	1:39	2:06	2:39	3:36	4:17	5:06
Thornhill Village	7:43	8:02	9:13	9:32	10:43	11:02	12:13	12:32	1:43	2:02	2:43	3:32	4:21	5:02
East Library	7:46	8:00	9:16	9:30	10:46	11:00	12:16	12:30	1:46	2:00	2:46	3:30	4:24	5:00

Exhibit 1.1 Bus Schedule

from people who have been on site for a number of years to new hires just join-ing the work community. It is important to know the folklore of an organization in order to more fully understand its history and especially the history of atti-tudes that may have developed over a number of years. Many histories, how-ever, include a "we've always done it this way" attitude. You need to be aware of such attitudes. When a document goes into the feedback loop and cycles back onto your desk with comments, some of the comments may be growing out of a historical context. You also need to know that histories of workplaces and dis-course communities may include folklore that is not factual. Often there are group myths that most have believed about a situation, another department, the administration, or other employees or fields that are considered outsiders (as op-posed to being a member of your work group). In one clinical laboratory where I worked, the folklore included the belief that the pathologist on call would not answer if our lab called but would only respond to a medical director's call. When I was supervising the lab and needed a consult with the pathologist, I called him and found that he was very approachable and willing to do whatever we needed. Somehow the idea had gotten started, and our communication with pathology had suffered for years based on a myth. You need to develop an ana-lytical stance that helps you separate the myth from the fact when you work on projects. At the same time, knowing even a nonfactual group attitude can be helpful as you work to address people at their own level of understanding, what-ever that might be.

Some organizations, especially larger ones, may have you develop docu-ments or communication systems for distant and faceless audiences. You may not be able to easily determine for whom you are writing. There are a number of ap-proaches possible. If time is a major constraint, you may need to rely on your ex-perience in the field. However, usually time permits at least a phone call or email and a chance to ask questions about what has been tried in the past, or what the person who is *in situ* knows about the specific context or audience.

Ideally, one would be able to travel to the site to ask the kinds of questions and do the necessary field research to understand the context. (See "Field Re-search" later in this chapter.) But if both time and money make this difficult, then the Internet and library resources are invaluable. Many databases now exist (in-cluding archival ones) that help us find out a great deal of specific information. Demographics, especially national census data, are available through the Internet. For example, see http://www.census.gov for U.S. Census Bureau data and http://www.census.ac.uk/ for the United Kingdom Census Information Gate-way. Some universities have also made more searchable databases that feature helpful categories such as U.S. counties, population estimates, Equal Employment Opportunity (EEO) files, and school districts. They also include geographic area profiles for their particular areas. Access to other good government documents and data can be found at the more general site: Fedworld Information Network (www.fedworld.gov), a site that also links to the U.S. Government Information Locator Service, a database for U.S. government documents.

To give you an example of how important geography and history are, con-sider a recent public health campaign in South Carolina. A relatively small but well-funded hospital decided to target the population of their geographic area (a small but growing town in the Southeast and the towns in its adjoining counties) in an effort to reduce the number of inappropriate emergency room visits. Hospital administrators had heard of a campaign in Boise, Idaho, that had successfully addressed this issue. A major part of the campaign included handing out a 371-page book that listed healthcare problems, detailed prevention strategies, and care that could be given at home, and included a list of symptoms that would signal the need for trained professionals. I was called in as a possible consultant for this project, which was well underway. When I asked what questions had been asked and what research had been done up-front, I discovered that very little had been done prior to the implementation of the campaign, but over two million dollars had already been sunk into the project. It was immediately obvious that this lack of research could prove problematic. The book would be printed and distributed by the thousands in the entire multiple-county area, and one copy would be delivered to every resident. The demographics in the Boise, Idaho, area, however, are different from those in rural South Carolina. Education levels and literacy rates in the Boise population are quite high. The small town and rural area in South Carolina that the campaign was targeting had educational levels and literacy levels that were low. Early assessments of this campaign showed very little success. The target population was not one that spent much time reading. The television set was on constantly in the homes when people were not at work, and a book was not the first choice for consultation in any type of health situation (or perceived emergency).

Fortunately, after some months, other components were added to the book distribution. A hotline was added that included a nurse-on-call 24 hours per day to answer questions. The hotline has been much more successful than the book. The history, by the way, also was revealing. The audience in the Boise, Idaho, area was historically Mormon with a strong emphasis on both education and family. In short, the history of family reading time and children being encouraged to read rather than spend time on television and computer games was more firmly established than in this particular region of the Southeast. Millions of dollars could have been saved by studying the situation first, especially a simple and quick comparison of the demographics of the two different geographic regions. An understanding of history could have revealed very important information about the potential success of such an endeavor.

How Do I Determine Audience Needs?

One of the most effective ways to determine the needs of your audience is to ask. Below you'll find various research strategies you can use to determine your audience needs. In health writing, it's very important to shift the emphasis toward more involvement of the audience early in the design process. When this process has been applied in various rhetorical situations, the documents have been much more successful in reaching their target audiences. By doing interviews in the Creekwood housing project with the women we were targeting and also giving them surveys to fill out later, we found out what their real issues were. We involved them in determining what kinds of health information would be most interesting and helpful to them. They suggested a health fair on the topic of breast screenings at the nearby community center, which was well attended. Our clients also talked about the difficulties in transportation and in understanding the poorly designed bus schedule. Frankly, I couldn't decipher it either. As mentioned above, the students and I redesigned that bus schedule to make it easier to understand (see Exhibit 1.1).

Planning

I cannot emphasize planning too much as you begin to write. When you develop materials for health audiences, you should first submit a proposal that includes an audience analysis. A very brief example of such an analysis done in an undergraduate class working on a project directed at food stamp recipients can be seen in Exhibit 1.2. The larger the campaign and the more health risks and funds involved, the larger and more thorough the analysis must be. A good demographic analysis up-front saves patient health and lives as well as costs and time. In the Boise project, before applying it to the rural South, a thorough demographic analysis would have saved much time, energy, and money for the hospital system involved. Such an analysis (paired with geographical and historical considerations) will include information about media access, local vs. global audience, gender, race, class, education, culture, age, and attitudes.

Media Access

One of the first considerations is to choose a medium that works best for your audience. If you are designing a set of instructions for a nurse in the ICU, will it be more effective to communicate on-line or with a hard-copy version or both? Ask yourself, what access does your audience have to computers, what software does your audience use, and how experienced are they in using those programs? Often we assume that everyone has access to computers, and all professionals are equally savvy in using them. This assumption is wrong. Cindy Selfe and others have written about the "Digital Divide" that separates the haves from the have-nots (see Moran and Selfe 1999). This issue is sometimes a class issue and sometimes also a race issue, but almost always an educational and cultural one. Impoverished people do not have the access to computers that people in higher income brackets enjoy. Even those who have the means to afford computers are not always interested in learning how to use them fully or on more than a very basic level. Even the most intelligent and qualified nurse or physician or other health professional can have a job that limits their use of computers, although more and more people in health professions are now using computers and sophisticated software. A key task is to identify the medium (media) most available to your audience for your document projects.

Audience Analysis

The main goal of our project is to increase the participation of eligible South Carolinians in the Food Stamp Program (FSP). We are going to produce a public service announcement (PSA) to educate the audience about the program. We intend to encourage people in need of food-aid that, if eligible, the FSP is available and easy to sign up for. The information presented here was collected from other sources: a 1999 fiscal report put out by the United States Department of Agriculture (USDA) Office of Analysis, Nutrition, and Evaluation; demographic statistics of food-aid recipients from America's Second Harvest, a national network of hunger relief organizations; the Food and Nutrition Services' website; and statistics from the Department of Human Services.

Gender:
There is a significant difference in the gender of people receiving food-aid. America's Second Harvest reports that of those adults receiving emergency food-aid, 62% are female and 38% are male. The statistics provided from government sources have food stamp participants to be 67–75% female.

Race:
The USDA reports that in 1999, 40.5% of the FSP participants were white, 36.1% were African-American, 18.1% were Hispanic, 3.1% were Asian, and 1.5% were Native American.

America's Second Harvest has similar statistics: 43.8% of their food-aid clients are white, 35.3% are African-American, 15.0% are Hispanic, 1.9% are Asian, and 2.1% are Native American.

An important observation is that whites comprise 72.6% of the U.S. population and only 41% of food stamp participants, while African-Americans, though only 12.1% of the total U.S. population, comprise 34% of food stamp participants.

Class:
The USDA reports that 89% of households in the FSP live in poverty and that 35% live on half what is considered poverty. In order to be eligible for the FSP, Federal guidelines do require the gross monthly income of the household to be 130 percent or less of the Federal poverty guidelines. The Department of Agriculture reports that the average gross monthly income per food stamp household is $584.

Educational Level:
America's Second Harvest reports that among its clients, 36% have a high school diploma or equivalent, 40% have not completed high school, and only 5% of all have attended college or received a college degree.

Exhibit 1.2 Audience Analysis

Attitude Toward Project:
There was no information to be found about the client's attitudes toward the FSP, but we believe that most people who are eligible for food stamps feel apprehensive because of the stigma concerning participation in the FSP. We believe that given the proper information, they will see that the program is for their own benefit and nothing to be ashamed of.

Age:
The USDA and other federal sources report that children (age 0–17) comprise just over half of food stamp participants (52.8%) and the elderly (age 60+) make up 8.2% [adults (age 18–59), 39.0%].

America's Second Harvest reports that their participants are 15.9% elderly, 46.3% adults, and 37.8% children. One has to take into account that the America's Second Harvest network aims to serve a disproportionate number of those two vulnerable groups, namely, children and the elderly.

Household Composition:
In 1999, 92% of all benefits from the FSP went to the 75.8% of food stamp households with children or elderly persons.

According to the Department of Human Services, 55.7% of the FSP households included children. Of these households that include children, 68.5% were headed by a single parent, most of which were women.

Of the 20.1% of food stamp households that included the elderly, 78.5% of them were single-person households.

Also, the USDA reports that a substantial proportion of food stamp households contained disabled persons (26.5%).

The average household size is 2.2 persons for those not containing children, while households with children average 3.3 persons.

The study in the *Journal of Nutrition Education* found that married couples are less likely to use food stamps, while having children in a household increases the probability of using food stamps.

Food Stamp Participation:
Average monthly participation in the Food Stamp Program has declined from 27.5 million in 1994 to 18.2 million in 1999, a 33.8% decline. Still, the Food Stamp Program remains the largest domestic food and nutrition assistance program administered by the U.S. Department of Agriculture.

(continued)

Exhibit 1.2 *(continued)*

In South Carolina, 127,000 households received food stamps in 1999, accounting for 1.6% of the households receiving food stamps nationwide. South Carolina households remain certified for the program for an average period of 13.2 months— only Arkansas, Connecticut, and South Dakota have longer certification periods.

One study found that 35% of the nationwide decline in FSP participation can be attributed to the strong economy, while 12% is the result of program reform and political variables. The same study shows that 29% of the decline was associated with a decrease in the number of households with annual income below 130% of the poverty line. Fifty-five percent was associated with a decline in the proportion of low-income households who participate in the Food Stamp Program and may be attributed to economic and/or program changes.

Attitudes (toward the ease of receiving food stamps):
To be eligible to receive food stamps, most families must meet specific income guidelines. For example, items such as the number of people in a family, income from earnings, shelter costs, and childcare costs determine a family's food stamp eligibility. Most often, individuals who are eligible for food stamps do not apply. The USDA finds that many times, former welfare recipients believe that their earnings disqualify them from receiving food stamps. A prior study cites the lack of communication between caseworkers and clients as a reason why individuals are not aware of food stamp eligibility. The USDA identified several practices within many social services that deter food stamp applications. This includes untimely application processing, lack of bilingual staff, and confusion over rules and regulations affecting eligibility. The most common reason for not applying seems to be the application procedures. The average application takes five hours of client time and at least two trips to the Food Stamp Program (FSP) office. In addition, lack of flexible FSP office hours and locations deter many food stamp applicants.

Exhibit 1.2 *(continued)*

Local vs. Global Audience

As you choose what medium to use for your message, you will next want to consider if the project is directed at a fairly specific audience or a more generalized one. Will you need to design materials that target local audiences or multiple audiences? Once you have gathered information for the medium and the focus, you can then move into the many other issues that feed into an audience analysis. A local audience is usually fairly small and confined to a particular geographical setting or within a certain group; it may involve in-house work specific to a particular organization. A global audience implies one made up of individuals on the national, multinational, or international level.

In some ways the local focus for your media development is easier, because you have a certain range within which you design. That does mean, though, that you must pay close attention to the specific needs of the audience. Global work is sometimes harder, because no one project can please everyone, and if you work across cultures, your project may offend someone, either because it is translated incorrectly, or a color or graphic offends or is incomprehensible to another culture. Both local and global deliverables can be field-tested, but in the larger, global project, even field-testing or usability testing is difficult at best for the variety of audiences they will reach.

Gender

Basic census data can provide gender information, but when designing for an intended audience, your health materials may need to target a specific gender. We can all think of certain health issues germane to women and to men separately. In approaching a health problem, smoking for example, a medical writer might need to develop separate appeals for different genders. Advertisers do this all the time. We can look at the recent targeting of young females by the tobacco industries. The marketers have discovered specific and savvy ways to appeal to young women, and the statistics indicate that their campaigns have been successful. Antismoking campaigns would be smart to appeal to women smokers in similar ways to get them to *stop* smoking.

At the same time, we also want to avoid stereotyping our audience, whether we're talking about gender, race, or economic status. If we operate from a distorted or imbalanced view of our audience, we can offend the very audience we need to reach. For example, if we stereotype or essentialize[1] women, we might think that to use an emotional appeal would be the best way to go. However, this approach stereotypes all women as emotional, which they are not. Some men are more emotional than some women. Instead, we must examine the rhetorical situation more closely. If we were creating a brochure advertising a seminar for women physicians or health administrators, we would likely use more logical and ethical appeals.[2]

Race

Many of the same considerations that apply to gender apply to race: avoid stereotyping, study your audience, and let them educate you in appropriate ways to reach them. For example, scientific writing acts as a gatekeeper or insider discourse, allowing those that are highly educated in science or biomedicine in and keeping those not familiar with the technical language out. Language is not neutral or transparent but carries certain dominant cultural values about class and race. Also, medical studies and the way they are set up can have hidden assumptions. For decades, all cardiac research was done on men. Only lately have cardiology researchers discovered that women have very different symptoms during heart attacks and also react to palliative treatment very differently from men. The assumption was that both genders react the same way to cardiac events. Therefore, we need to be aware of hidden assumptions we make and embed within our studies, the texts we write, and the kind of language we use.

Often race and culture blur together, especially since race itself is a social construct (although of course races exist, "scientifically" the races do not break down in the same way we have constructed them). Some biologists believe that we should classify humans according to blood types or other criteria. We often attribute characteristics to race that in fact may apply to culture, class, or genes (other than racial markers) instead.

Often characteristics blur and no clear lines can be drawn in our analyses, but as I mentioned above, involving the audience in the invention, planning, and design process for health materials helps. For example, in one particular graduate seminar in bilingual design, the executive director of a nursing and wellness center gave me a particular problem: to design health diagnostic materials that would bridge the gap between English-speaking nursing and nurse practitioner students and Spanish-speaking Mexicano farm workers. This nursing center had a well-equipped mobile van that visited farms and orchards to provide primary care as part of the national Migrant Health Project. When workers came in for healthcare, the nursing students were unable to triage and prepare people for further treatment in the van because of the language problem. After learning about audience analysis, the communication students began to interview the Mexicano farm workers and other Latinos who were working in the area, asking them what would most help make access to healthcare easier. One of the workers we interviewed came up with ideas for documents that students could design and nursing students could use to triage. One Mexicana joined our class on a regular basis to provide translation help and to lead further discussion of cultural issues.

One issue we discovered, thanks to our audience's help, was that modesty is of particular concern to Mexicana women. Therefore, the seminar students designed charts that women could point to rather than having to point to their bodies, a transgression of their cultural norms. Students designed a diagnostic book that used graphics more than words to illustrate various health problems; this design, too, came from the audience rather than being imposed on them. These documents are now being field-tested for further revisions.

Class

Class is sometimes particularly difficult for Americans to discuss, because we operate under a cultural myth that we live in a classless society. However, this myth is exactly that—a fantasy—and although class structure in the United States is not completely rigid and does allow for some mobility, we tend not to see class issues because of our cultural training. Various ways of writing and communicating, however, reflect different classes, and our planning of documents needs to be able to cross these lines or appeal to classes in ways that make the documents accessible. Just as other boundaries blur in the taxonomy set up in this chapter between race and culture, boundaries also blur between class and educational levels. Often middle- and upper-class children receive the most education and are thus able to maintain their class status, while children in lower economic brackets tend to have fewer opportunities and fewer class norms that motivate them to see themselves as college students, for example.

As I pointed out earlier, scientific writing often privileges upper class and highly educated individuals. If we are writing for poverty-stricken white families with malnourished children, obviously using scientific writing will make those messages inaccessible. The writer must walk a thin line to avoid stereotyping or condescending to groups because of perceived educational differences. This means that you have to be realistic about your audience's educational level so that you can design documents that are understandable without being perceived as talking down to people. Having relatively simple messages that don't preach to people or belittle them is important. For example, one nutritional brochure my students and I developed for low literacy levels played on the idea that we recognized that our target audience consisted of good and caring mothers who wanted to provide the most nutritious meals for their children at an affordable cost. To have criticized junk food choices some single mothers feed their children would have offended the very people we wanted to reach and would have been based on a negative assumption regarding their interests. The most helpful attitude takes into account class differences and sees them as strengths. The writers/designers who get to know their audiences will be able to more effectively design documents that will be readable and comprehensible to the people they are trying to reach.

Education

Educational levels can vary according to regions and states within the United States. Again, good data exist in on-line census databases. Some regions and particularly some states have better educational systems than others; thus researching the educational levels of your target audience is always wise. If a writer within an educational and training facility of a regional medical facility is asked to design a PowerPoint presentation for a physician's talk at a conference, it will be useful to know if that conference is for other physicians with the same specialty or general physicians. Perhaps the PowerPoint is meant to address nurses and other health professionals as well. Therefore, within highly educated groups of individuals, too, various levels of expertise and specialized knowledge exist. Therefore, the writer/designer needs to know the educational levels even within highly educated groups in order to effectively reach the target audience.

Culture

We've discussed culture where it has overlapped with other categories within an audience analysis. Cultural norms and the various languages within different cultures are of prime importance when it comes to designing writing projects. There are cultures within relationships and families; there are cultures within workplaces and organizations; there are cultures within states and countries.

How does one look at cultures? In technical and professional communication and in medical communication, too, we borrow methodologies from anthropologists and look at the artifacts of a particular culture. We use ethnographic tools (not the same as doing a complete and exhaustive ethnography), such as the following questions, to arrive at a better understanding of a particular culture:

- What values do you find in the culture's or organization's official medical documents, such as mission statements, annual reports, marketing materials, Web sites, etc.? Are the values that you see in the materials for the outside world the same as those in the culture's or organization's internal world? For example, in the United States we emphasize a free enterprise system of healthcare in our global advertisements. Would everyone in the United States agree with this form of healthcare?
- What relationships are revealed in the media about a culture? (See Chamber of Commerce literature, etc.) Are these relationships expressed in a health organization's documents? Are these relationships reflected in actual interactions? Within a hospital, for example, the Web site may promote the "family" nature of the workers in the hospital and indicate that both skilled workers and professionals are on a level playing field, yet within the hospital a traditional hierarchy may be a more accurate depiction.
- What kinds of folklore about bodies and our health do we find in the culture or organization? What kind of values do they reflect and how are those reflected in our health practices?
- What kinds of hierarchy exist within the health organization? How is communication structured and what hierarchy (or lack of hierarchy) does this reflect? For example, what access do you as a medical writer for a large HMO have to public officials or top administration? What access does your audience have?
- Look at hospital or other health agency organizational charts to see who reports to whom. Who gets to initiate communication? To whom and how must others respond?

Pay especially close attention to language issues. How familiar is your audience with medical terms? In addition, you will have studied the geography or environment. As a writer in a health organization or profession, you should practice your anthropological skills by examining your workplace, and then use these same skills to analyze your audience and its environment for the documents you design, whether these are internal or external documentation projects.

- What kind of language is appropriate in the setting for which you are designing? Remember that there are many cultures within one culture. For example, the same word in one region may not be used in another region of the same culture. Often idioms are used in common parlance to describe and denote certain diseases. These idioms can vary drastically from culture to culture.
- What terms do employees use to address themselves and each other? Employees are often sensitive about how they are addressed, whether they are "medical associates," "EKG techs," or "cardiac professionals." Also, in many hospitals it is customary to address nurse-colleagues by their last names, while physicians are called Dr. XXXXXX.

I'll discuss more language issues in the multicultural chapter (Chapter 9).

Age

Age is another important issue to consider as we target our audience. Again, good databases catalogue this information if we are designing for large groups. Different age groups need different media and have different design issues. The most difficult task is to design for all ages. In many ways this goal is impossible, although we are often asked to achieve it in our work.

Elderly patients and consumers have specific design issues in terms of audio and visual needs. We will talk about that more in later chapters, but we will touch briefly on it here. Disability activists refer to those with no obvious disabilities as TABs. TAB stands for *temporarily able bodied*. They argue that if we live long enough, we will all be disabled. The aging process itself produces characteristics that our culture sees as disabilities. Therefore, design issues such as size of text, color (for visibility and contrast), speed (in PowerPoint slides or multimedia presentations), loudness (in audio tapes or CDs), and graphics that older people can identify with are just a few of the design considerations to be made.

Middle-aged people have different health concerns as well as different lifestyles and patterns. The same is true of teens and children. Sometimes we use readability tests to determine age and ability levels in documents. If you are designing for a particular age group, unless you are very much in touch with or are a person of the age you are targeting, it's best to research the particular needs of that age group and to conduct readability tests, usability tests, and focus groups to see if you are on target. Such time investment early in the process means not having to redo expensive media because it fails to reach the age group it was intended to reach.

Attitudes

We have touched briefly on attitudes in the discussion of history and culture. Also, we must consider **assumptions**. Assumptions and attitudes are closely related. If you can get at what your audience will assume, then you can get at many of their attitudes. Unless you understand the attitudes of the audience you are trying to reach, you can fall short of your goals. You must anticipate what people already believe about what you are trying to tell them. The following exemplifies this strategy.

A graduate student recently completed her Master's project on conflict resolution for a high school with the highest violence rate among high schools in the surrounding area. She developed an interactive, multimedia CD in Hyper Studio that taught ninth graders effective, nonviolent body stances and communication techniques to keep verbal confrontations from being physically violent ones. Before she began the design project, she spent time on site getting to know the ninth graders. She then surveyed them to determine their attitudes toward violence and the cultural assumptions of their rural area. In addition, she looked at the history of settlement in the area. She found out that the area had been settled by predominately English and Irish settlers, and at one point in the early 1800s, very few middle-aged males had both of their eyes intact because of blood feuds (literally an eye-for-an-eye) in this particular area. The surveys revealed that today's parents from the same area expected their teens to fight (physically) to defend their families and personal honor if either were verbally challenged. The graduate student used this information to design her project, finding ways to address the issue of honor by solving challenges in certain ways rather than backing down or fighting. Ordinarily, we could have assumed parents would *not* want their teenaged children to be in physical fights, but that assumption at this particular location was wrong. To design with that assumption in mind would have made the entire project ineffective.

How Do I Choose Research Tools?

Humanities students usually are familiar with library research, whereas science students usually are familiar with laboratory research. Health professionals may be familiar with either or both of these. Often trained in social science research methodology, a combination of humanities and science research methods, healthcare students and professionals have a wealth of investigative tools available to them. Below you will learn various research strategies germane to the kinds of writing that health professionals do.

Library Research

Let's begin with library research, making the assumption that most of you have some familiarity with library research. Both public and university libraries usually have an information center, an information line, and a reference desk where you can call or email for a quick answer to your search for a specific fact or for help in conducting a search for specific information. If good demographic, historical, and other contextual studies already exist, that would be the place to begin your research. According to Mary Sue MacNealy, "[T]here are four issues to consider: type of research on which the source is based, place of publication, principal author, and recency of publication" (MacNealy 1999, 16). Timeliness is a factor in online sources, too. If a Web site has not been updated lately, that may be a sign that information there is not current or that the site is not being actively maintained.

Your first job is to determine if the studies you are examining are historical, theoretical, rhetorical, or empirical. When you examine research reported by others, you want to be aware of good research methodology so that you are not reading skewed analyses based on poor data. Is research that is not based on good designs and analyses published? Unfortunately, yes. You therefore need a good working knowledge of empirical research as grounding for good document design, both for studies you might conduct as well as those you read. You can quickly determine if a research report is empirical by looking at the methodology. Except for historical and theoretical work, you will find evidence of empirical methods. There may be empirical studies embedded in historical and theoretical research, so be aware that not all empirical research is reported out in the IM-RAD[3] format, because some disciplines (especially humanities disciplines) do not report data in this scientific framework. A writer with little time will still need to be able to scan reports and quickly determine if the data is worth basing documents on or not. A good understanding of empirical research is even valuable in patient and other interviews you might conduct, as well as in surveys, usability tests, and focus groups. Writers need to be able to interpret data. If, for example, a researcher gives you data to publish or digitize, you need to be able to ask relevant questions and make sure that the data you report is not missing critical features for which other professionals will hold you accountable. Also, it bears repeating that working with patients' health and lives puts a special onus on us as researchers.

MacNealy also discusses research in document design itself; we will talk more about those works in a later chapter. She recommends Felker's *Document Design: A Review of Relevant Literature,* which reports empirical findings in human factors, psycholinguistics, and typography, areas important not just in audience analyses but rich in the kinds of issues important to writers in the health professions. Karen Schriver's much more recent review of empirical research in designing documents is also valuable.

In addition to being aware of the research methods your data is based on, you need to be aware of the place of publication. You will want to be aware of the scholarly weight or credibility various health journals or medical scientists carry. As a member of a health profession, you will soon learn what sources are credible sources within your particular field. The same criteria apply to on-line sources. Is the Web site a personal site or is it connected with a well-known and prestigious organization? Does it look scholarly or professional as befitting a medical site or does it look like someone's personal home page? Another question to ask about on-line sources is: is the Web site well constructed? Sloppy design and construction may equal sloppy thinking and writing. However, a well-designed Web site does not necessarily mean the content is valid. No matter what the medium, the writing has to be accurate and documented to be credible.

You also need to be aware of the medical author's credibility, just as you are aware of the doctor's or medical researcher's credibility within your work site. Citation frequency is one way to measure an author's credibility, but make sure this author is not being cited for flawed or controversial studies. You can also consult supervisors or experts with more experience in the field to check sources. The same holds true for Internet sources:

- Is authorship for the site clearly established, and are there links to a brief bio or vita?
- Does the Web author link to other credible Web sites?
- Does the publication indicate an awareness of other literature published in the field (either on-line or in print)?
- Are there cited materials and are citations correctly attributed?

Also, especially in the health field, the latest and most recent publication builds credibility for your documents. However, do be aware that medical studies are released that have not stood the test of time. Often the public thinks that research studies are prematurely released, because "facts" on diet, medication, etc., seem to vary from week to week. Therefore, you need to keep in mind that you are working with and trying to persuade an already skeptical lay audience. Be willing to use the expertise of others.

Internet Research

First, let's discuss Internet research in general, and then we will review Internet research specific to the health professions. Again, my assumption is that both health professionals and students will already have at least an intermediate or average knowledge of the Internet. Much of your research can be done via the Web if you are conducting contextual and demographic analysis of large groups of people. Above I've already mentioned several Universal Resource Locators (URLs) that would be useful for demographic research. Although it's difficult to write about specific search engines and databases because of quickly changing technologies, here is some brief information that is helpful to research. In Chapter 11 you will learn more about how to write medical copy for the Web and how to design for other electronic health media.

Another important source for your general research is databases. For general research the largest database is DIALOG information services, which provides access to articles, statistics, news, conference proceedings, and other data in medical, scientific, and technical documents from academic and trade journals, magazines, and over 100 newspapers (Markel 2001, 148). You can also consult PsycINFO, which houses some 1,000 journals going back to 1967, and UNCOVER, with a variety of journal tables of contents from recent entries. The Humanities Index and the Social Sciences Index are useful, too. MacNealy recommends Elizabeth Smith's article in a 1996 *Technical Communication Quarterly* for electronic databases useful to technical and professional communicators. More and more CDs and DVDs are available with full texts from such places as *Fulltext Sources Online* (Medford, NJ: Information Today—http://www.infotoday.com), which comes out with a directory twice a year listing full texts that are on-line (Markel 2001, 148).

There are a few specific databases that are particularly helpful to the writer in the health professions: Grateful Med (see text below), PubMed, and MEDLINE. (Also see Chapter 6 for patient-oriented Web sites.)

Grateful Med was and PubMed still is a search engine for medical databases, and PubMed, "a service of the National Library of Medicine, provides access to over 11 million citations from MEDLINE and additional life science journals. PubMed includes links to many sites providing full text articles and other related resources" (http://www.ncbi.nlm.nih.gov/PubMed/).

Grateful Med, also a feature of the NLM, had functioned as a search engine for numerous databases. According to recent news on the Web site: "Internet Grateful Med (IGM) will be retired in phases in the coming months. All databases searched by IGM are static, and have not been updated since December 2000 . . . The new NLM Gateway is replacing IGM. The Gateway does simultaneous searching in multiple NLM databases" (http://igm.nlm.nih.gov/).

The NLM has many databases available to medical writers, but the most widely used is MEDLINE with journal articles published since 1966. Other more specific databases are also available such as: BIOETHICSLINE, AIDSLINE, AIDS-DRUGS, AIDSTRIALS, ChemID, DIRLINE, HealthSTAR (in health services, technology, and administration), HISTLINE (history of medicine, not histology), HSRPROJ, OLDMEDLINE, and POPLINE (Gastel 1998, 26).

Many public Web-based databases are also popular with consumers, such as WebMD. This type of Web site can lead you to discussion groups, bulletin boards, and chatrooms that help you learn about particular audiences.

Both Web sites and databases require a working knowledge of Boolean operators to conduct key word searches. Consult the search tips or advance search features to learn how to use the specific Boolean operators for the search engine or database you are using. For more sources use the asterisk * to increase the number of hits; for example cardi* will turn up cardiac, cardiology, cardiologist, cardiogram, cardiography, and other related terms. Another way to increase source options is to use the word *or*, as in *cardiogram or cardiography*. On the other hand, to decrease options and be more specific, you can use *and* or *not*: cardiac *and* cardiogram, cardiogram *not* cardiac. Further, if you are looking for a specific article on electrocardiograms written in 1998, you could specify: "cardiograms—1998."

On-line discussion rooms and chat rooms are excellent for quick answers to specific queries and to be in touch with particular audiences. With all the medical support groups on-line now, you can have instant access to people who, for example, suffer from a chronic ailment you may be researching. Do be aware certain types of people may be more likely to be on-line and somewhat public about their illnesses, whereas others in your audience may be silent. Also, some population groups have less access to computers as we mentioned above when we discussed the "Digital Divide." Bulletin boards and/or Usenet newsgroups are also sources for information. You may be able to become part of a professional medical discussion group as well. Again, be aware that some groups will carry more credibility than others.

Field Research

First, we need to define what we mean by field research for professional health writers and communicators. Obviously, we are not talking about conducting major studies in areas of health, but rather gathering information that is helpful to your writing projects. By field research we are referring to information gathered outside of libraries and the Internet or World Wide Web. We are referring to gathering information from primary sources, particularly people in both public and private spaces (for example, on the street or in someone's hospital room). We think of field research in writing for the health professions as an ethnographic method of gathering information through observation (observer) and/or interacting with the audience (participant/observer). In addition, some usability tests can also be conducted in the field, so field research blurs the boundaries between such methods as thick description[4] found in ethnography and methods such as surveys, interviews, focus groups, and usability testing. We will discuss field research as an ethnographic tool and a type of usability test, and then give more specific information about surveys, interviews, focus groups, and laboratory-based usability tests below. Even though we suggest this form of data gathering is an ethnographic tool, do not confuse it with a real ethnography, which like its precursor in anthropology is usually an intense study of several years within a specific culture or cultural group and includes a thorough description of the setting or physical surroundings in addition to much other thick description.

In order to observe and record accurate data about a group of people and the context in which they find themselves, you will want, also, to triangulate your data. To triangulate means to take multiple observations and/or use more than one observer and/or different forms or sources of data collections (interviews with several people in your target audience, their on-line comments in a chat room, etc.) Why is it important to triangulate data? Suppose you are researching how people in hospital admissions react to the admissions procedure in order to streamline the communication flow. You will want to observe a number of different patients with different intake personnel and at different times of day. Suppose you only observed one intake employee, a clerk who had an unusual way of complicating the process by filling out duplicate information on another form rather than copying it via a copy machine? If you only observed that individual and then wrote a memo suggesting that the form be duplicated, and everyone else was doing it that way, your credibility as a communication specialist would be compromised. In addition, you would waste your own and others' time and resources. You might have a colleague also record the same interviews you are watching; other observers may note different features about the interview or confirm what you are observing.

This kind of research also reveals good information about the context in which the rhetorical situation you are researching is taking place. For example, if the situation described above was in a crowded hospital lobby in which the distractions were multiple, the noise levels loud, and the chairs uncomfortable and too far from the admissions desks or partial cubicles, you might not find a communication flow that was flawed because of human habits, but one that was taking longer because incoming patients could not focus well or hear the admissions staff members very well.

Your analysis of the data you collect should align itself with the reasons you conducted research on your audience and context in the first place. Your analysis will consist of categories of findings depending both on what you are seeking to find out and what you observe. In the example above, perhaps the researcher could have divided the intake procedure into various segments of the data that needed to be filled out. Thus she could have noted that one of the employees did one part of the procedure twice. The ways of setting up and analyzing a field study are as multiple as the fields to be studied. Although this method is usually not a quick study, a busy medical writer will develop strategies for finding the information she wants to find even in contexts that usually take longer to study.

Surveys

Surveys are also an excellent way to determine the needs of your audience. They are especially useful in determining the demographic statistics (ages, health problems, etc.) and the attitudes, preferences, and beliefs of your target audience. Make sure you have a very clear purpose in mind before you begin your survey, just as you determine purpose and audience before any communication project.

Most of us have participated in numerous hard copy and on-line surveys, so it is a familiar genre for us. Some of us have also participated in face-to-face and telephone surveys. There are advantages and disadvantages to each of the means of surveying audiences. If you send out a hard-copy survey, you may be able to avoid some of the bias that you encounter (or bring with you) in the face-to-face surveys. Also, because you are simply mailing them out, passing them out to a group of people, or leaving them in a place where you hope people would fill them out and send them in, you have much less of a time commitment involved in the effort of gathering information. Also, depending on your budget, hard-copy surveys can be less expensive (especially in paying for personnel to administer them).

The disadvantages to hard-copy surveys include low return rates, sometimes an unclear sampling, and difficulty in asking follow-up questions or clarifying confusing questions for your readers. The statistics on returns of mailed-out surveys are quite discouraging. Although some researchers will accept low rates of return (under 50 percent), other researchers claim that such a low rate makes the data you do collect unreliable. According to two social scientists, typical rates can fall between 20 and 40 percent (Frankfort-Nachmias and Nachmias 1992). However, if you are conducting research in which any feedback on attitudes is helpful, you may be willing to accept the lower rates. Do bear in mind that a low rate of return, if you chose to target a representative sample of the larger target audience, can skew your results. For example, if your representative sample includes a 50–50 balance for gender and your return rate reflects a 30–70 gender breakdown, then your survey can be problematic. It depends, of course, upon the purpose of the survey.

Also, if there are no opportunities to meet with the person you are surveying, you will not be able to answer any questions the person may have about the survey or ask any follow-up questions as you can do in a face-to-face or telephone survey. On-line surveys, however, can include an email address, and sometimes your audience will ask you questions that you can answer; survey takers may be willing to answer follow-ups on-line as well. Such responses will depend on your audience and how you set up the on-line survey.

On-line surveys also have advantages and disadvantages. When you send a survey via email or put it on the Web, the time for delivery is almost instantaneous, and with email delivery you know very quickly whether or not the survey has been received. Also, your respondent can then respond at her convenience. In addition, Chou suggests that this format is best for international audiences, although the Digital Divide (as C. Selfe would also point out) presents problems both in this country and internationally (Chou 1997, 197–207).

In surveys and in interviews, especially those that ask about health and illness in addition to other personal questions, writers must be able to construct sensitive questions that do not offend the survey takers. In person-to-person surveys, especially within hospital and clinic settings where patients are already under stress, uncomfortable, and in an unfamiliar setting, it's necessary to take great care in asking questions. The rule of thumb is to ask the most sensitive and invasive questions last (if indeed they need to be asked). Build up trust with the individual you are questioning and determine before you ask questions whether or not your respondent is likely to feel comfortable answering such questions.

Exercise sensitivity in relation to issues such as gender, race, class, and culture. Although "political correctness" is often maligned and satirized, it was originally intended to correct age-old insensitivity and blatant discrimination. Be aware of your audience, what offends the members of it, and how best to approach different groups of people; show sensitivity and awareness of their needs. For example, my technical writing students could not understand why a Power-Point nutrition guide they composed for a group of state highway workers was offensive. The students were featuring different fast-food chains and suggesting healthier choices among the lunches offered there. On a Taco Bell slide, they had chosen to use a Chihuahua dog. I pointed out that Latino audiences had been offended by the use of this dog as a representative of Mexico and its food. These kinds of issues need to be considered in constructing survey questions, too. Audiences can be uncomfortable in discussing their income. Often respondents are more comfortable checking a box that lists income ranges rather than revealing how much they make per year.

For formal surveys, you need to pay close attention to sampling techniques. There are many types of samples with specific guidelines depending on the type of questions you are asking and quantitative measurements you have planned. Therefore, it's best to consult a research methodology book or experts within the workplace to set up a formal and quantitative study, unless you can claim such expertise. Here are a few issues to consider. If your surveys require detailed statistical analyses, it's best to consult statisticians (unless you happen to be one). For a medical writer, a working knowledge of statistical analyses is useful depending on the type of writing your work requires; in pharmaceutical writing, for example, it would be mandatory. Remember to always build in more respondents than you actually need because of attrition rates. This rule is true for both qualitative and quantitative surveys. Obviously the larger the sample, the more accurate any quantitative data you gather will be.

Interviews

Like surveys, interviews are also an important way to assess your audience's attitudes. For more in-depth information, especially when troubleshooting a document that has not worked or a public health campaign, it is useful to conduct personal interviews. Interviewing is more of a qualitative research approach germane to case studies, ethnographies, feminist research, and some usability tests. Your sample is typically small because of the amount of time and personnel it takes to conduct face-to-face interviews. If you record the interview (easier to capture all the data), you must ask the respondent for permission first. If the information will be kept confidential and/or anonymous, you must also let your interviewee know that in advance. Any patient information will be confidential because of the 1996 HIPAA Act. (See the next chapter for more about this important legislation.) Here, too, sensitivity is key to a good interview as well as good question construction. One hint: prepare more questions than you have time to ask in anticipation of a very quiet or brief discourse style on the part of your participant. While surveys can be fill-in-the-blank, check-the-box, multiple-choice, open-ended, or a hybrid of these formats, interviews tend to be more open-ended. You might need some demographic information up-front, but avoid yes-no questions as they quickly shut down information unless you are in an emergency situation and need answers fast.

Of course all healthcare involves interviewing patients on the part of the physicians, nurses, and technicians who care for them. Those interviews are typically formatted in terms of information that is fairly routine in order to make accurate diagnoses or elicit health histories. Given the constraints of managed care, it is difficult to take the kind of time that a good interview demands, so the patient interview shrinks to a very terse and often tense discursive practice. This situation, of course, is counter to eliciting good information, but often healthcare providers have to work within the constraints that are in place. You should be aware, too, that often patients tell you what is really bothering them or ask the questions they are most worried about at the end of the interview.

Focus Groups

Another useful audience research methodology in the health professions is focus groups. Marketing professionals are probably the most adept at this kind of research, but it is now playing a useful role in audience analyses outside of marketing. Marketers use them to get feedback on products. Many have found that focus groups are useful in getting feedback on various documents, on public health campaigns, on new videos developed for patient education, etc. Unlike interviews, which rely on the interaction between the two people involved, focus groups are dependent on group dynamics and group interactions to yield the most information for your audience analysis.

Focus groups work best when you get people to talk to each other in response to specific questions you ask about certain topics. Your goal as a researcher and facilitator is to make people feel comfortable quickly. You want to set up an environment that stimulates informal conversation yet keeps your respondents on task. You should not control the input, but you want people to interact in ways that inform your feedback. Participants should be free to discuss, challenge each other in nonhostile ways, change their minds, and form common opinions with each other. For example, if you give them three information booklets on diabetes and the participants start out suggesting that one format is better than another, and then change their minds and promote, instead, a booklet they didn't choose first, that's okay. Their interchanges form a thick description (like in larger ethnographies) that you can tap into for information, either formally by coding your responses or informally by drawing preliminary conclusions based on the responses that you recorded.

You can record, videotape, or use a person as recorder to collect the data. When you recruit people for the groups, don't approach them formally. Instead you want to talk about the focus group as a discussion rather than as research. To select participants, you can use traditional sampling techniques to recruit people, but sometimes convenience sampling is called for depending on your research design. If you are developing a video for elderly people's eye care, your sampling criteria will, of course, match a certain demographic group. If you are testing general audience response to new materials promoting your medical clinic, you want a good cross section of people, perhaps people in a mall setting, for example, for your convenience sample.

You should ask someone else to moderate and/or record the group session. The best facilitators of focus groups are people with excellent social skills and good rapport with groups; someone who can make people feel comfortable from the very beginning by socializing as participants arrive is an effective moderator. You do not want to build up resistance to information sharing or alienate the participants. The goal is to establish a balance between creating a positive atmosphere and keeping people on task and getting work done. It is also best to start with a list of simple but effective rules that set boundaries, but still be open and flexible. You can do this via handouts, verbal instructions, PowerPoint slides, posters, etc. You want your questions to reflect the openness of the focus-group format, so open-ended questions are the best to ask. Discussion prompts are helpful, too, but don't ask "why" questions, because participants often feel defensive with that kind of question (MacNealy 1999, 189).

As moderator, you will want to keep dominating participants in check and draw out the more reticent ones. However, you need to be able to do this skillfully so that you are not creating resistance or hostility. Usually, the time frame for a focus group is not more than an hour or two or you risk wearing out the goodwill and participation of your audience members. Make sure you have made the amount of time required clear up-front, and remember that less time is often more productive than more (one-hour is ideal). It is especially important to know how much information you want covered in the time period, so that you can be effective in moving conversation along and truncating diversions and time-wasting after the first segment (approximately one-sixth of the total time) of group bonding is completed. Make sure audience members feel that their input is valued, and, if you are working on a project costing millions of dollars (such as the text-centered public health campaign based on the successful Boise model above), you might want to offer a monetary incentive or other reward for participation.

For data analysis, you can code the information or take overall impressions. Ideally, you will conduct several different focus groups to determine whether the information from several groups is the same and can be used for your purposes. If you do only one focus group, you run the risk of your information being idiosyncratic. Multiple focus groups can take place back-to-back within the same day if you are recording information during the sessions.

Usability Testing

Once a health document(s) or computer program has been developed, we have still another form of analysis that can be applied: usability testing. There are several forms of usability testing, and different definitions exist as well. Most experts recognize at least two forms: field-testing and laboratory-based usability testing. Field research (see above) and field-testing are not the same, although they share similarities because of their contexts. Field-testing is valuable both during and after document development. Increasingly, industries are relying on this form of feedback and audience analysis in their design process, if they have both the time and funds to support it. You can actually use field-testing and usability testing at any stage in the document/software development process, but you do need some form of the deliverable (an early draft, for example) in order to get useful feedback from potential users.

One example of such testing includes several bilingual health diagnostic documents developed in one of my classes. These documents have been field-tested in five clinics around the state. Although the drafts of these projects look complete, we consider them beta phases in a longer design project. We received extensive feedback on the documents, and a graduate student built his Master's project on revising these documents based on the results of our field-testing. In this example, the clinics, mobile healthcare units, and Latino school nurses' offices used these deliverables in their clinical settings; they told us how well the documents worked and what changes should be made to make the documents even more effective.

A crucial aspect of usability testing is to determine where and how your project will be used. Because our diagnostic book was designed for use in clinical settings, and because this design is also used with Latino farm workers who follow the harvests, testers needed to conduct these particular usability tests in the field. To bring field workers and patients into a laboratory setting (usability testing facility) would be intimidating and invasive for our audience (see Chapter 2 on Medical Ethics). Laboratory testing, as we will soon discuss, involves videotaping and a setting that would not be comfortable for our audience members or, in some cases, even possible for them to access given their work schedule. Our documents were originally designed for a mobile unit visiting migrant farm worker camps from 4–10 p.m., because workers basically work long hours, literally from dawn to dusk. Also, issues such as immigration status could make them less than sanguine about being videotaped and/or audiotaped.

Other audiences are suited for a more laboratory-like setting or a usability testing facility. If you want to test a new interactive software program for charting, for example, you might bring medical records clerks or nurses (depending on who uses this program) into the laboratory to see how easily they access the program and enter data. This audience works in a clinical setting and with computers on a regular basis, and if their schedules and time permits—also sometimes a problem—they could most likely be persuaded to participate.

Usability testing, especially if you have to rent the facility, may be an expensive proposition. Again, whether to do this type of research is dependent on a number of factors. First, consider the cost of the overall project. Instituting a new computerized charting system could be very costly, and there are a number of different types. It may be worth designing and conducting your own usability test within an office on site with a computer and note-taker, a hybrid form of usability testing located between field testing and laboratory testing. If you have to spend large amounts of time retraining employees and large amounts of money on equipment and/or software, it would be worth your time to spend some money and time investigating the programs you are considering.

What Is Usability Testing?

See Ann Duin (1993), J. Dumas and Janice Redish (1993), J. Rubin (1994), Jacob Nielsen (1999), and others for more in-depth information on usability testing, as the overview below is not intended to provide the know-how for an actual test.

The following suggestions apply to formal usability testing (lab-based and hybrid) but are flexible, particularly in usability tests conducted in the field. Usability testing can sound somewhat arcane or be difficult to describe, so here is a brief outline of the steps of a typical formal or lab-based test to give you a general idea of what usability testing is. Consult usability texts for the complete instructions. Preparing for a usability test, the practitioner would:

- Determine the purpose of his or her test (whether field or lab or a hybrid of the two), and be specific about outlining the tasks he or she wanted the user to do, especially if the plan called for a quantitative measurement process.
- Determine the audience for the study. There may be more than one audience. For example, in an emergency room there may be a software program that enables the nurse to screen patients in the waiting room via a laptop computer. Therefore, both patients and staff are the audience. Because the staff member would be using the computer and assisting patients verbally with information, the ER nursing staff would be the primary audience.
- Decide where and how to test the users and investigate the feasibility and appropriateness (to purpose and audience) of the testing environment.
- Decide how to select participants and what sampling technique to use in the selection process. Most lab-based usability studies test 5–10 participants. Field-testing is more variable, with sometimes many more participants. Also determine incentives: are funds available or other incentives to enlist users?
- Develop a plan or a proposal with the most specific detail that could be gathered. Include timelines and budgets, too.

- Develop and gather usability test materials/documents. The practitioner would need the following documents for a formal usability test:
  - Legal permission forms.
  - Script for introducing users to the test (must be written so it does not vary from user to user). It tells users the purpose for the test without being so specific as to predetermine and bias answers.
  - Survey questions for the test (sometimes pre- and post-test surveys).
  - Instructions for the user to follow for the test.
  - Means for recording data (can be notebook, audiotape, videotape, another note taker in addition to the tester).
  - Incentives (if used).
- Select participants based on predeveloped criteria.
- Schedule participants so that there is time in between each person without having them wait for the tester to finish with the last participant, and without an opportunity for participants to brief each other about the test.
- Be responsible for checking out equipment and setting up a room (unless a natural setting is called for by the study).
- Conduct a pilot test to make sure the usability study works well.
- Conduct the usability test.
- Follow these steps after greeting the participant (flexible):
  - Explain the test using an introduction or orientation script (see above).
  - Have participant sign the permission form.
  - Administer the first survey (pre- or background survey).
  - Make sure the participant understands the procedure.
  - Ask the user to begin to follow the instructions. (Often formal testing includes think-aloud protocols in which the users voice all of their thoughts. They will also require some prompting if they fall silent.)
  - Record the data after testing voice levels (if audio- and videotaping).
  - Give the participant the incentive if one is used and thank the person.
- Analyze the usability test data:
  - In the plan, decide up front what form of qualitative and/or quantitative measures to use.
  - If coding the data, have another person read through the data according to the codes to assure inter-rater reliability.
  - Report the feedback needed to revise the documents or other media.

Field-testing can run the gamut from highly formal (rare outside a laboratory setting) to very informal. My technical writing students prepared survey questions to field-test documents developed in an earlier semester's class. They each were asked to interview five people via convenience sampling at local areas where low-income families received services and shopped. In many cases, the students discovered that it was easier to be flexible even in using the survey questions because the people they interviewed volunteered feedback in varied and casual ways. Thus students recorded comments and then wrote a thick description of their encounters immediately afterwards. Although the tests were very informal, we had spent several weeks preparing for them, and students were able to record and document important and useful data for our revisions.[5]

We have not yet discussed the ethical dimensions of conducting research with patients and health professionals. This topic is addressed in the next chapter.

Summary

In this chapter you have learned that both audience and context analyses are key to developing good health writing techniques. Nothing wastes more time and resources than designing expensive and elaborate health campaigns that incorrectly target their audiences and fail miserably, some even causing a backlash or other negative and unintended results.

You learned how to begin the process of analysis by developing lists of questions about the context of your communication task, questions that ask about geography and environment. For your audience analysis, you will ask questions about demographic aspects: media access, local vs. global approaches, gender, race, class, education, culture, age, and attitudes. Under each of the categories, ways of going about determining and finding this information were featured.

In addition, research methodologies were discussed, so that you have many choices for collecting the data, depending on your time and other resources. We discussed library research, Internet research, field-testing, surveys, interviews, focus groups, and usability testing. Each method carries its own advantages and disadvantages. Also, as a researcher, you do need to be aware of how credible the information sites you are studying are, especially on-line sites. There are several means of determining credibility as pointed out in the above section on Internet research.

Once you have this information gathered and analyzed, you're ready to start planning your communication effort. The first part of the plan, before you get too far into a new project, is to weigh the ethics of what you plan to do in the given context and with the particular audience. Ethics are an important key in health communication and medical rhetoric. You want to be sure that you parse the ethics of your particular goals and the means you use to get there.

Discussion Questions

1. Why is it important to ascertain the purpose of your particular writing or other communication task before you begin? Isn't this rather obvious? Discuss what could go wrong if you don't agree on the purpose of the effort up-front. Why does the idea of a clear purpose help make sure your communication is on target? Can you think of health communications that haven't worked well? Was it because they weren't correctly targeted to the real purposes of the project?

2. Why is an audience analysis important to your particular writing or other communication task? Can you think of examples in which audiences were not properly targeted and ones that were? Imagine some possible health education

projects you might do and discuss how you would determine your audience's attitudes toward what you want them to learn.

3. Can you find ways to involve audience members in the planning and designing of communication tasks? How would you involve an x-ray technician in the revision of a technical manual for new x-ray equipment? What could this person tell you about your writing task? Could you involve elderly patients in a nutrition program aimed at them? How would this be possible? What problems might you encounter? Think of other situations in which audience involvement would be possible and others in which audience involvement would be difficult.

4. Discuss the different factors of audience analysis and see if you can take an example and apply the various factors such as race, gender, education, class attitudes, etc. Can you think of ways different demographics will change the details of your communications?

5. Discuss the various research methodologies. When would you use each? When would you not use each? What experience do you have with the methods? Share the experiences with class members.

6. What are the differences between field-testing and usability testing? What are the advantages and disadvantages of each?

Exercises

1. Before you answer the questions listed in the text, write down the information you would like to know about your audience. Think of everything that you don't know now that you think will be most important to find out before you compose your health document or project.

2. Using the census databases on the Internet featured in this chapter, research the city or county in which you were born and fill in as many aspects of an audience analysis that you can with these databases.

3. Based on your analysis above, what other information do you need to find out? Make a table in which you list that information in one column, where you will find that information in another column, and then where you actually find the information. This matrix can act as a model for your future audience analyses.

4. Taking your audience analysis, what contextual information do you still need? Using various methods, find the information you need and add that to your matrix. (Experiment with different ways of displaying the information so that it is most useful to you.)

5. Form a team with your classmates. Choose an audience and contextual analysis that one of you has developed in one of the above exercises. Now, after you have chosen a specific health communication project as a team, each of you should take a different research method and see how it would or would not be appropriate to your project. Then choose one or more of the research possibilities based on your findings and present your ideas to the rest of the class.

6. Find several health education and information Internet sites (like WebMD, etc.). Do screen captures of the sites, print them out and see whether or not

these sites would have benefited from some (or more) usability testing. How are they reaching their audience(s)? Do you think they did a good job of analyzing the context and their audience? Find some more local sites to compare to the better-known sites to get good comparisons and write these up in reports.

Works Cited

Chou, C. 1997. "Computer Networks in Communication Survey Research." *IEEE Transactions on Professional Communications* 40: 197–207.

Duin, Ann. 1993. "Test Drive-Techniques for Evaluating the Usability of Documents." *Techniques for Technical Communicators.* Eds. C. Barnum and S. Carliner. New York: Macmillan: 306–335.

Dumas, J. and Janice Redish. 1993. *A Practical Guide to Usability Testing.* Norwood, N.J.: Ablex.

Fedworld Information Network. www.fedworld.gov. U.S. Department of Commerce. July 21, 2000.

Felker, D., Editor. 1980. *Document Design: A Review of Relevant Literature.* Washington, D.C.: American Institutes for Research.

Frankfort-Nachmias, C., and D. Nachmias. 1992. *Research Methods in the Social Sciences,* 4th ed. New York: St. Martins.

Gastel, Barbara, M.D. 1998. *Health Writer's Handbook.* Ames, Iowa: Iowa State University Press.

Gould, E., and Stephen Doheny-Farina. 1998. "Studying Usability in the Field: Qualitative Research Techniques for Technical Communicators." *Effective Documentation: What We Have Learned from Research.* Ed. Stephen Doheny-Farina. Cambridge, MA: MIT Press.

Grateful Med. National Library of Medicine of the National Institutes of Health. 6/30/2000. July 17, 2000. http://igm.nlm.nih.gov/

MacNealy, Mary Sue. 1999. *Strategies for Empirical Research in Writing.* Needham Heights, MA: Allyn Bacon.

Markel, Mike. 2001. *Technical Communication,* 6th ed. New York: Bedford/St.Martins.

Moran, Charles, and Cynthia L. Selfe. 1999. "Teaching English across the Technology/Wealth Gap." *English Journal* 88, no 6: 48–55.

Nielsen, Jakob. 1999. *Designing Web Usability: The Practice of Simplicity.* Indianapolis: New Riders Publishing.

PubMed Web Site. National Library of Medicine of the National Institutes of Health. 7/1/2000. July 18, 2000. http://www.ncbi.nlm.nih.gov/PubMed/

Rubin, J. 1994. *Handbook of Usability Testing: How to Plan, Design, and Conduct Effective Tests.* New York: Wiley.

Schriver, Karen. 1993. "Quality in Document Design: Issues and Controversies." *Technical Communication* 40: 239–257.

Selfe, C., and R. Selfe. 1994. "The Politics of the Interface: Power and Its Exercise in Electronic Contact Zones." *CCC* 45, no. 4: 480–504.

Smith, Elizabeth. 1996. "Electronic Databases for Technical and Professional Communication Research." *Technical Communication Quarterly* 5: 365–385.

United Kingdom Census Information Gateway. Economic and Social Research Council. Maintained by David Martin. 5/5/2000. July 17, 2000. http://www.census.ac.uk/

U.S. Census Bureau. United States Government Census Bureau. 6/1/2000. July 20, 2000. http://www.census.gov

U.S. Government Information Locator Service. Superintendent of Documents—Government Printing Office. 7/1/2000. July 20, 2000. http://www.access.gpo.gov/su_docs/gils/

Endnotes

1. "Essentialize" means that women (or other gender or race) are labeled as being *essentially* nurturing or *essentially* less aggressive than men. In other words, instead of looking at these traits as being a result of cultural conditioning, they are thought to be an embedded quality in every woman. *Essentializing* puts women into a restrictive box formed by definitions. If a woman, then, is not nurturing or passive, she is criticized or not accepted and blamed by the culture for not having the traits the culture thinks are innate to women.

2. Logical and ethical appeals refer to rhetorical strategies that include statistics, logical reasoning, and credibility. See rhetoric texts for more information.

3. IMRAD is the standard form for scientific reports. The acronym stands for: Introduction, Methodology, Results of the research, Analysis of the results, and Discussion.

4. *Thick description* is an anthropological term used to describe note taking in an ethnographic research setting. The observer (or participant/observer) records everything she or he observes in great detail, whether or not it will figure into the final report. The idea is that more of the full context will be revealed, and a fuller picture will be painted for those who read the study. The details or thick descriptions flesh out what could otherwise be an acontextualized study in which many of the variables are missing.

5. See Gould and Doheney-Farina (1998) for their take on field-testing.

Ethics in Medical Writing

Overview

Because of what is at stake for writers and communicators in medicine, no medical text can neglect careful attention to ethics. Although most technical communication textbooks mention ethics in general, many do not pay enough attention to this area. In this technical writing series, we have an excellent book on ethics for technical communicators: Paul Dombrowski's *Ethics in Technical Communication.* Larger health facilities usually have a vetting process that oversees the documents produced by its communicators, but smaller organizations do not. Regardless of the institution's ethical and legal representatives, you as a medical writer are of course still expected to have a strong and well-developed perspective of the ethics in health and medicine.

This chapter moves deductively from the larger, more global issues and examples of medical ethics to the American Medical Writers Association Code of Ethics. From there the focus shifts to very specific copyright and other legal issues within medical writing.

In this chapter you will find answers to the following questions:

- Why do I need to be aware of ethics?
- How does ethics affect medical writing?
- What are other examples of ethical situations?
- What is AMWA's Code of Ethics?
- What copyright issues concern health writers?
- How do I obtain permissions?
- What do I need to know about duplicate publication?
- What do I need to know about copyright law?
- What digital copyright issues must I know?
- What are on-line ethics and etiquette?

- How do I write for differently abled audiences?
- How do I write for multicultural audiences?
- How do I protect patients' rights?
- What is HIPAA?
- How was HIPAA phased in?
- What are Patient Bills of Rights?

Why Do I Need to Be Aware of Ethics?

Because we work with texts and other media that deliver health and life-and-death information, rather than fiction or opinion, our communicating tasks carry much responsibility. The U.S. justice system, for example, has ruled that instruction booklets or manuals can be defective, and if a person using them suffers some injury or problem, the writing professionals can be held liable and sued (Markel 2001, 21). In addition to text-based communication such as medical equipment manuals, postoperative instruction sheets, and chronic disease management booklets, medical communicators must often represent certain kinds of data graphically. Because statistics can be manipulated and easily used to misrepresent certain findings, we must be aware of possible problems with designing Web sites, representing data, and especially reporting out statistics.

These are some of the potential ethical dilemmas you could encounter:

- If you are writing descriptions of a product (such as a new prescription cream to grow hair), you may be asked to exaggerate the claim that hair will be restored to increase sales or to obscure research findings that indicate the drug if contacted by a pregnant woman could cause problems in her unborn child.
- If you have incomplete data about the debts and losses of a health organization, MaxHealth, you may be asked to rush publication of the budget in order to make the situation look better to its stockholders, who are scheduled to vote for new officers. By obscuring the actual financial "health" of the organization, you may be helping to elect or reelect a group of officers that are not ethical or truthful to their own stockholders.
- If you are asked to develop a Web site for your HMO or hospital system, you might be tempted to copy the html code from the site of your competitor hospital system and tweak it slightly to avoid the time and cost of obtaining copyright permission from the owner/designer. You may also *want* to make your Web sites look alike in order to confuse viewers looking for the other health system.
- If you are asked to help market certain charting software for doctors' offices, you might be tempted to understate the many hours it takes to understand and learn the new software. You may also be asked to overstate its advantages, such as speed, ease of use, and cost effectiveness.

- If you are designing certain documents or Web sites, it might be easier and faster not to subject every part of it to the type of accessibility analysis necessary for differently abled people to be able to use your media.

These are all examples of scenarios you could encounter as a medical writer, and they are all ethically problematic or patently unethical.

How Does Ethics Affect Medical Writing?

Yes, if we work in medical settings, we must be aware of the controversies and sensitivities of audiences who may have moral and religious reasons for opposing a particular form of research; stem cell research, for example, has been a big focus of attention recently. How we report research from this and other research areas can be crucial. That rhetorical situation, like all such situations, will be dependent on the audience(s) for which we are writing. Because stem cell research is dependent on "growing" cells in an "in vitro embryo" that is slated for extinction, many politicians, legislators, and members of the public have reacted strongly to the ethics of such research. Researchers and physicians saw potential help for patients with Parkinson's disease, heart disease, diabetes, Alzheimer's, and spinal cord injury (Zitner 2001, 1A). However, religious leaders and legislators voiced loud opposition; the Pope called stem cell research "evil" (McQuillen 2001, 1A). Columnists lined up on both sides, both for and against the research and the government funding of such research (William Raspberry, Suzanne Fields, Ellen Goodman, Rabbi Marc Howard Wilson, and Cal Thomas, to name a few). Physician journals such as the *Medical Ethics Advisor* did the same. As a medical communicator and writer, you cannot afford *not* to know what issues are currently the subject of heated debate. If you were asked to prepare a funding proposal for your researcher to a foundation or a speech to a community group, not knowing what context this form of research had been and has been a part of could result in some disastrous assumptions and poorly persuasive texts. This example illustrates where contextual analysis and ethics come together (see contextual analysis in the previous chapter). Ethics, then, becomes a part of the context you must not only carry yourself as writer/designer, but also as an important part of the context for your audience.

What Are Other Examples of Ethical Situations?

When I worked for a number of years in a heart catheterization laboratory, it was sometimes painfully clear that for the radiologists on staff, the main criterion in the studies was to get the very best possible X rays. While this desire usually yielded the best diagnostic route for the patient, there were some times when the patient's well-being was not the prime consideration. For example, an anomaly might be discovered that radiologists are more interested in capturing on film, studying, and writing about the patient rather than letting him get off the hard, cold table in a stress-inducing, equipment-filled room and return to supportive family and friends. The ethic here is a form of expediency because the goal is a perfect specimen or film shot in which the patient is objectified. As health professionals and ethical writers, we have to ensure that human beings never cease to be "subjects" in our minds, so that they do not become objects instead.

Another more recent example is one that affects the health of millions of people worldwide and has been the subject of much investigative reporting, some whistle-blowing, and other media exposure: the tobacco industry's role in profiting from a product that results in slow, agonizing deaths. As can happen in technical memos, the humanity and health of the consumer is lost. But the kinds of technical reports and research data from the tobacco industry went beyond techno-speak to a deliberate attempt to disguise facts and downplay the effect of smoking on the human body.

In 1958 the American tobacco industry formed a group called the Tobacco Industry Research Committee (TIRC) that began to challenge scientific reports that reported a link between cigarette smoking and lung cancer. The group crafted documents that attacked earlier documents released by medical scientists and researchers studying the linkage (Dombrowski 2000, 159). For the last forty some years, medical researchers, private and public groups of concerned citizens, and the government itself have battled with the tobacco industry to take some responsibility for its encouragement of an addiction that harms almost every person coming into contact with cigarette smoke. Only since 1997 have some strides finally been made in the battle of documents . . . now legal ones as well. Up until recent times, the tobacco industry never lost a court case, but over the last few years it has paid out some 350 billion dollars in settlements. Unfortunately because the well-paid lawyers always manage to avoid final judgments by settling these huge lawsuits out of court, no legal precedent has been established that can then be cited to create a watershed effect for victims of lung cancers and other diseases caused by or made worse by smoking (Dombrowski 2000, 161).

Through careful manipulation of wording of various reports, documents, memos, and other responses to scientific findings, the tobacco industry has sidestepped obvious responsibility. In the following memo we hear a typical manipulation of technical, medical language that can serve as an illustration of an ethical dilemma for health writers who may find themselves having to write or refute such documents:

> You will remember that it can be shown experimentally that smoke, or smoke condensate, has certain undesirable effects on animals, including tumor production on the backs of mice. At R. & D. E. [research, development and engineering] we cannot disregard these results, although in themselves they do not prove any connection between smoking and human disease. It is our belief that the cigarette of the future must have reduced biological activity, and when I speak of biological activity I mean those adverse effects such as tumor production on mice. (Dombrowski 2000, 171)

The language in this memo disguises the growth and causes of cancerous tumors by referring to them as "biological activity." We have very little sense of the human agent here, and we are given little sense of what "biological activity" is. If we heard this term fleshed out to its full meaning, we would hear about the true horror of the disease process of lung cancer; we would hear about lesions and erosion of tissue, pain, and loss of ability to have full lung function. This language obscures the facts, hiding meanings from the readers: an unethical use of terminology in a medical research report.

As problematic as this terminology obscuring the harm of cigarette smoking is, even more horrendous is "the Phillip Morris report on the 'positive effects' of cigarette smoking in the Czech Republic, which concludes that cigarette smoking is a good thing because of the 'health-care cost savings due to early mortality' " (http://nosmoking.org/july01/07-27-01-2.html).[1] We can hope that such a memo would be beyond our capability for insensitivity.

The above examples illustrate why it is important to understand what it means to communicate ethically, rather than to obscure certain realities through jargon, objectify human beings, or manipulate the truth for political motives. More frequently the ethical situations involve less dramatic abuses, but it's important to be able to have sorted out the ethical implications of your work before you are in such situations. It is easier and more comfortable to assume that everyone in the health professions is ethical and is interested only in the patients' welfare. The health industry can be exactly that: an *industry*, a business. Often huge profits are at stake. Do not be surprised if the bottom line is profit instead of patient welfare in some cases and in some institutions. As an ethical medical writer, therefore, it's particularly important to be on your toes. One group of medical writers that has written its own code of ethics is AMWA.

What Is AMWA's Code of Ethics?

AMWA (http://www.amwa.org) is a nonprofit association founded in 1940 for professional medical communicators. In addition to regular meetings and conferences, the organization also sponsors workshops, special seminars, and a quarterly journal. The association also "provides a forum for the exchange of ideas and the improvement of professional writing skills" (Fraser 1992, 181).

The American chapter also has European and Canadian sisters. The European Medical Writers Association (http://www.netlink.co.uk/users/emwa) sponsors a newsletter and a spring conference in different locations throughout Europe. EMWA's membership includes a wider group of medical writers that includes "academics and professionals working in-house or freelance for pharmaceutical and medical communications companies and research institutes, or in the wider field of journalism" (Fraser 1992, 181).

AMWA's Canadian sister at http://www.amwa-canada.virtuo.ca/ is also an active and useful organization for those of us writing in and for the health fields.

On the facing page is the excellent set of guidelines developed by AMWA in 1974 and revised several times since that time. The principles listed, although developed for a more specialized group of writers who tend to mostly write research and academic articles in conjunction with physicians and biomedical

American Medical Writers Association Code of Ethics

Preamble
The American Medical Writers Association (AMWA) is an educational organization that promotes advances and challenges in biomedical communications by recommending principles of conduct for its members. These principles take into account the important role of biomedical communicators in writing, editing, and developing materials in various media and the potential of the products of their efforts to inform, educate, and influence audiences. To uphold the dignity and honor of their profession and of AMWA, biomedical communicators should accept the ethical principles and engage only in activities that bring credit to their profession, to AMWA, and to themselves.

Principle 1
Biomedical communicators should recognize and observe statutes and regulations pertaining to the materials they write, edit, or otherwise develop.

Principle 2
Biomedical communicators should apply objectivity, scientific accuracy and rigor, and fair balance while conveying pertinent information in all media.

Principle 3
Biomedical communicators should write, edit, or participate in the development of information that meets the highest professional standards, whether or not such materials come under the purview of any regulatory agency. They should attempt to prevent the perpetuation of incorrect information. Biomedical communicators should accept an assignment only when working in collaboration with a qualified specialist in the area, or when they are adequately prepared to undertake the assignment by training, experience, or ongoing study.

Principle 4
Biomedical communicators should work only under conditions or terms that allow proper application of their judgment and skills. They should refuse to participate in assignments that require unethical or questionable practices.

Principle 5
Biomedical communicators should expand and perfect their professional knowledge and communications skills.

Principle 6
Biomedical communicators should respect the confidential nature of materials provided to them. They should not divulge, without appropriate permission, any confidential patent, proprietary, or patient information.

(continued)

Exhibit 2.1 American Medical Writers Association Code of Ethics

---

Principle 7
Biomedical communicators should expect and accept fair and reasonable remuneration and acknowledgment for their services. They should honor the terms of any contract or agreements into which they enter.

Principle 8
Biomedical communicators should consider their membership in AMWA an honor and a trust. They should conduct themselves accordingly in their professional interactions.

Original: Eric W. Martin, PhD 1973
First Revision: June 1989
Second revision: April 1994
Reprinted with permission: Minick 1994

Exhibit 2.1 *(continued)*

---

scientists, are nevertheless easily applicable for us as health communicators in a variety of workplaces.

Now that we've moved from general ethical situations in medical communication to the more specific situations we medical writers find ourselves in, let's examine the following issues inherent in writing in and for the health professions: copyright issues, shareware, the data protection act, informed consent, ethical research agreements, patients' bill of rights, and living wills.

What Copyright Issues Concern Health Writers?

Because so much of our work is now digital as well as on hard copy, we will discuss copyright in both types of media as well as include a sample letter requesting copyright permission.

As writers we are often in a somewhat privileged position in that much of what we write and publish, whether through desktop publication or through outsourcing, will be assumed by readers to be both ethical and legal. Readers usually *assume* we are acting ethically; this assumption gives us an even greater responsibility to *be* ethical. Although we cannot cover all situations and contingencies here, we can certainly offer some guidelines and ensure that as health writers we are aware of both ethical and legal aspects within different genres of medical writing.

Much of the writing we do within our workplaces will not require any special permission, because it comes within the purview and under the oversight of our larger organization. Often we are given internal manuals and other materials to revise that have no attributions and have been written by staff members who held our jobs at a previous time. Such use of others' work is legal under the label *work for hire*. The revisions and original texts we compose for our organization belong to the organization and not to us.

We do need to be careful not to put our organization in jeopardy (as well as our own jobs, ethics, and reputations) by being less than knowledgeable about copyright issues especially in the development of external materials. There are situations in which we highlight the name of our institution and others in which we cannot make reference to our place of work without permission. In a sense we represent our institution at all times and therefore must be mindful of its reputation as well as our own. We must ask our own supervisor or manager for permission, and, if that person does not have the authority to grant it, we must either discuss the publication with someone further up the line or request that our immediate supervisor do so for us. Keep a paper trail of all conversations and meetings especially regarding legal or ethical issues in case there is a question of copyrights and permissions after the materials have been published and distributed.

However, loyalty to one's institution, also an ethical stance, does not mean that one blindly follows an institution's lead if the institution itself is acting unethically. Although whistle blowing is not an easy route for any person to take, when necessary it calls attention to unethical practices within the healthcare industry and can save patients' lives and enhance patients' quality of life.

One of the ways that unethical medical situations come to light is through investigative reporting. If you are a health writer for a major newspaper or news service, you most likely enjoy reporting the positive aspects of medicine, the ones that offer hope and cures to people suffering from chronic and acute illnesses. As Gastel comments, "Many of us enter health writing partly because we have high regard for medical research and healthcare and enjoy reporting favorable news such as medical advances. However, medicine also has a sorrier side, and documenting and publicizing problems in realms such as healthcare can help lead to their solution" (1998, 102).

In other words, there are times when good in-depth, investigative reporting is necessary for exposing unethical situations. We need only to remember the Tuskegee project in the not-so-distant past to be aware of some of the atrocities committed in the name of medical science.[2] Areas for possible investigation are: safety of pharmaceuticals and devices, environmental hazards – their danger to public health, competence of physicians and other health professionals, health insurance fraud, governmental health agency incompetence and bureaucratic nightmares, inappropriate medical association lobbying, hiding of research findings if it impacts company profits, etc. "At a recent conference on investigative reporting, one reporter told of exposing a fertility clinic that was illicitly giving women other women's eggs. Another recounted uncovering the case of a heart transplant program that was no longer doing transplants—and yet was still accepting patients. A third reporter told of investigating a doctor who was doing unnecessary medical procedures—and doing them incompetently" (Gastel 1998, 102). Unfortunately, opportunities for such reporting are very prevalent.

How Do I Obtain Permissions?

Writing for the health professions varies from singular activity to collaborative work, although most of it will involve collaboration and, at a minimum, feedback from others as we revise. If there is an authorial role (many promotional compositions, for example, carry no attribution), we must acknowledge our colleagues' input. In addition, we must ask permission to use their contributions rather than assume it. Your particular context will provide a set of guidelines for this type of situation, but it's good to be aware of the more formal routes—and the ethical ones. Larger institutions will carry written policy on copyright and other issues in their policies and procedures manuals, and you should request it and have it on hand. Smaller organizations may rely on oral tradition or folklore rather than formal written policies. If, on the other hand, you are expected to publish some work on your own and you do not name the institution (and in this case you might also not want to list it in your biographical information as author), you do have the freedom to publish such material under the moniker of literary freedom granted to academics and others.

Because the preponderance of texts about medical writing address journal articles and writing for publication, we will only touch briefly on the ethical issues involved in that rhetorical situation.

We should also point out that if you need to use material from biomedical journals, fair use rules apply; in short, material may be reproduced for educational, not-for-profit purposes without regard for copyright. Some of those journals (over 500) also follow a set of guidelines developed in 1997 called *Uniform Requirements for Manuscripts Submitted to Biomedical Journals,* which addresses ethical issues for medical writers who publish in journals or who assist other health professionals in publishing in journals (Fraser 1992, 185). These guidelines carry instructions to the authors, not to the editors of the publications. The advantage for writers is the consistency of requirements over a number of journals, so that one need not reformat articles so frequently. The journal editors also have the responsibility of letting writers know "in their instructions to authors that their requirements are in accordance with the *Uniform Requirements for Manuscripts Submitted to Biomedical Journals* and to cite a published version" (Fraser 1992, 186).

What Do I Need to Know About Duplicate Publication?

Redundant or duplicate publication means that a published work overlaps substantially with one already published (Fraser 1992, 186). Unless you state your article is being reprinted, you cannot publish the same document without attribution and obtaining permission from the editor of your earlier publication. Whether you are publishing electronically or in print-based media, you cannot send the same article to more than one journal at one time. Only after the article is rejected by one journal can you submit it to another. You can still present the article or parts of it at a conference or other professional gathering, but if the article is under consideration at a journal, it is wise to mention that fact. This system is in place to protect intellectual property rights. If you or someone else has created a text (including visual, graphic, and electronic texts) that's your work that needs to be protected and credited. In addition, medical researchers protect publications that report their discoveries and breakthroughs.

For example, biomedical scientists compete to publish their discoveries first and to get the substantial material and social capital that follows on the heels of such a discovery (see Crick and Watson's DNA and the French/American HIV-virus discovery controversies). Be careful that the rush to publish does not lead you to abandon your ethics during the process.

If an article is in the publication process at a journal and a journalist wants to publish a press release based on the work, check with the journal or publishing house (Fraser 1992, 187). An exception to this ethical policy could occur if an early release would be for the good of a number of people, such as in a public health emergency.

In addition, Fraser suggests

> [w]hen submitting a paper, the author should always make a full statement to the editor about all submissions and previous reports that might be regarded as redundant or duplicate publication of the same or very similar work. The author should alert the editor if the work includes subjects about which a previous report has been published. Any such work should be referred to and referenced in the new paper. Copies of such material should be included with the submitted paper to help the editor decide to handle the matter. (187)

Such careful and ethical attention to publication of articles will keep you and your health professional in the good graces of the important journals in the field.

You can publish the same article in another language if all of the following conditions are met:

1. The authors have received approval from the editors of both journals; the editor concerned with secondary publication must have a photocopy, reprint, or manuscript of the primary version.
2. The priority of the primary publication is respected by a publication interval of at least one week (unless specifically negotiated otherwise by both editors).
3. The paper for secondary publication is intended for a different group of readers; an abbreviated version could be sufficient.
4. The secondary version faithfully reflects the data and interpretations of the primary version.
5. A footnote on the title page of the secondary version informs readers, peers, and documenting agencies that the paper has been published in whole or in part and states the primary reference. A suitable footnote might read: This article is based on a study first reported in the [title of journal, with full reference]. Permission for such secondary publication should be free of charge. (Fraser 1992, 187)

Some health writers write within or for government agencies such as the National Institute of Health or a state department of health and often have different guidelines for publishing:

> For some groups, such as those working directly for government departments, it is often a contractual requirement that the employer's permission be obtained prior to any publishing activity, whether or not the publication identifies the employer or the author's place of work, or contains information derived from working for that employer. Such a severe constraint on the publishing freedom of a professional is rather inhibiting but must be complied with. In case of doubt, it is important to ask whether or not formal permission is needed. If it is, a request must be made in writing and a written reply must be obtained as must a copy of the contractual statement which states that permission is needed. (Cormack and Benton 1994, 6)

Differences in copyright issues and procedures also vary depending on whether you are publishing via your own desktop or outsourcing to a professional publisher. Sometimes the publisher also checks submitted materials (Cormack and Benton 1994, 6). But as author, you should be responsible at all times for your own honesty and integrity in order to establish trust with your audience, showing an awareness of the importance of *ethos* or credibility as a professional communicator.[3] If you are working with other professionals' data, don't be afraid to question their use of sources and follow up on any questions about their research findings that you find suspicious or not fully reported. In the long run, you are doing yourself, your colleagues, and your organization a favor. Especially in medicine, where an error in writing up a treatment regimen or research data can result in serious harm to or the deaths of patients, every situation requires special vigilance and maximum effort in accuracy and checking of facts and sources.

Sometimes in order to help explain a situation as ethical or unethical and to make ethical choices yourself, it's useful to place the situation within a context and a history. In other words, we don't always understand how certain choices came about, and we can be more assured of where the ethics lie if we are willing to dig deeper. Gastel, for example, claims that

> Good health writing often looks beyond the purely biomedical, putting its subject in broader context. Check reference and other sections of libraries for works containing material on historical, social, economic, ethical, and other aspects of medicine. Examples of such works include *The Cambridge World History of Human Disease* (Kiple 1993) and the *Encyclopedia of Bioethics* (Reich 1995). (Gastel 1998, Chapter 2)

What Do I Need to Know About Copyright Law?

Written in 1976, the Copyright Act "protects the author of any published or unpublished work—such as printed material, software, and photographs—whether the author is an individual or a corporation. The author is entitled to profit from the sale and distribution of the work in exchange for making the work accessible"

(Markel 2001, 24). For more information, check with the U.S. Copyright Office (http://lcweb.loc.gov/copyright).

In addition to publishing a text in another language or working with in-house texts, within certain contexts you can use materials—usually texts, not graphics—without permission under guidelines called "fair use":

- Research
- Scholarship
- Teaching
- Journalistic reporting
- Criticism or commentary

But in order to use the materials as "fair use," you should consider the following four factors:

- Is the material to be used for profit? (It's much easier to consider "fair use" for nonprofit purposes.)
- Is the material to be used for the public good? (Often in medical settings, this criterion is met.)
- Will the original author of the material lose profits as a result of your use of the materials?
- How much of the work are you using? (Usually the rule-of-thumb is to use 10 percent of the material or less.)

If you have any questions about copyright, fair use, or other such issues, you should consult your supervisor and/or legal counsel.

Keep all permissions on file and accessible; you can print emails if permission is obtained that way, too. Do seek counsel from your organization or legal staff if you are concerned that an email in print may not be full protection for you. Some situations require signatures on print-based material for legal purposes.

Sample Letter Asking for Permission

Edward J. Huth suggests that copyright or permission request letters be fairly straightforward and comply with the usual polite forms of address and discourse as well as spell out in exact detail what quoted material would be used and in what context. As you might expect, context here is everything. It would be unethical to quote someone in a critique, for example, without giving someone a chance to decline first, especially in for-profit situations. In Exhibit 2.2, the physicians' group is using an editor's comments about HMOs in a brochure recruiting patients for that particular group.

In academia, of course, other rules apply as scholars critique each other with regularity; within journal articles permission is not solicited, but citation is required of course. The scholar does not stand to profit monetarily (although social capital would be a whole different ballgame), and the context within academia allows for this practice as part of its normal procedures.

William W. Parmley, MD, FACC
Editor-in-Chief, Editorial Office
Journal of the American College of Cardiology
415 Judah Street
San Francisco, CA 94122

Dear Dr. Parmley:

I am requesting permission to quote a full paragraph and a partial paragraph from your editorial published in the September 1996 edition of *Journal of the American College of Cardiology*. The quotation is to be used in a brochure my physicians' group is preparing for patients experiencing problems with receiving optimal care within Health Maintenance Organizations (HMOs).

The citation for the quoted material would be:
Parmley, William W., MD, FACC, Editor-in-Chief. "Anecdotes in Medicine: Do They Have Value?" *Journal of the American College of Cardiology*. 28.3 (September 1996): 795.

The full paragraph to be quoted begins with "A man in his late 50s, self-referred" and is paragraph three (3) on page 795. The partial paragraph follows immediately and is paragraph four (4) on page 795. It begins with "This single experience is representative of the kinds of experiences that we are having with the current health care system." I would continue with that paragraph and end with the sentence: "To withhold care from the very patient who will benefit the most is a principle that turns medicine wrong side up."

Acknowledgment of the permission to use this material will be given in the brochure. You may wish to specify the form of acknowledgement.

Thank you for considering this request,

Barbara Heifferon, PhD
Health Communication Department
Clemson Regional Physicians Group
778 Crepe Myrtle Drive
Clemson, SC 29678

Exhibit 2.2 Sample Letter Requesting Permission

Illustrations

Just as text is copyrighted, so are illustrations, photos, and other graphic images that are published. When you write for permission to use these copyrighted images, you will often be required to pay a fee. This requirement seems to be consistent for most graphics. My students got permission to use a copy of an original painting for free in exchange for crediting the National Farm Workers Association (see Exhibit 2.3). This graphic showing farm workers in the fields by a Latino painter was appropriate for a bilingual health diagnostic guide developed by graduate students in a seminar on bilingual document design.

Often if you are illustrating a document you are designing, whether a print-based or electronic one, you can find good illustrations in medical texts and journals. Again you must write for permission and often must pay a copyright fee. Students often have better chances of not having to pay fees, because their work is within an educational context, so they are not selling their documents or books for profits. Also, "some sources, such as U.S. government publications and *The Sourcebook of Medical Illustration* (Cull 1990), do not bear such restrictions. Keeping alert for illustration ideas early in an information search and placing the ideas in a file can save effort later" (Gastel 1998, 13).

Logos, Trademarks, and Registered Trademarks

Logos (plural of logo, not to be confused with the rhetorical concept of *logos*), trademarks, and registered trademarks are like copyrights on products or companies. As an ethical writer in the health professions, you must be able to recognize when these various marks protect certain property and when to use the symbol system that recognizes this ownership.

Often larger hospital systems, physician groups, and health insurance companies have logos, which are individual symbols made up of various different letters and designs to signify a commercial entity or an organization. Within academia many of our universities have a logo for the university that is a registered trademark. In addition, strict laws are enforced governing their use. You have to have permission to use the logo, and it needs to be used appropriately. If your organization is large enough and you are designing materials to be sent outside the workplace, most likely you will have a style guide or a policy and procedures manual dictating the terms of its usage. You have to realize, too, that you can't change and manipulate the logo, especially if it has been registered. For many situations in larger organizations, you'll have preprinted stationery and forms with the logos already on them. If you are good at design and at a small health facility, you may be asked to come up with a logo. Research other logos so that you do not inadvertently make one identical or too similar to ones that are already used by other entities.

In addition to logos, you need to be aware of trademarks. Markel defines trademark as a "word, phrase, name, or symbol that is identified with a company. The company simply uses the ^{TM} symbol after the company name to claim the design or device as a trademark" (Markel 2001, 26). Trademarks often help organizations fight for their rights to certain designs and products because they are recognized by the state justice system as symbols belonging to certain companies. If you do documentation for a medical equipment company, you will have to be aware of when and where to incorporate their symbols into your work.

Registered trademarks are similar to trademarks themselves but permit companies to claim ownership in federal and international courts as well. This process

Exhibit 2.3 Cover Art: National Farmworkers Association

Source: Heifferon et al. 2002

can take several years but it also gives increased protection to the organizations. Markel defines *registered trademark* as: a "word, phrase, name, or symbol that the company has registered with the U.S. Patent and Trademark Office" (26). The symbol ® is used after the company's name.

As an ethical medical writer you will need to protect the logo, trademark, and/or registered trademark of your company, organization, or client. The documents you prepare, whether print-based or on-line, will need to correctly represent and protect its or his/her rights. You can provide protection in the following ways:

- Always include the trademark or registered trademark symbol next to the design and put an asterisk. Then next to the asterisk at the bottom of the page or end of the document, mention that "Blank is the registered trademark of Blank company."
- Don't use a trademark name as a verb; for example, "Before you start logging the prescriptions into the computer, xerox them so that you do not lose the original information." (Xerox is a trademark used when referring to a xerographic copier.)
- Don't use the trademark to show possession or as a plural. You cannot say Cokes are damaging to your health or that Coke's acid levels increase the chances of kidney stones.

What Digital Copyright Issues Must I Know?

In the following section we will move away from strictly paper-based copyright and discuss issues within electronic media. We will also include a discussion of software usage, as that area certainly has ethical implications for medical writers, too.

Software, as we know, is copyrighted, and individual copies or site licenses for sharing software programs are required. Medical writers/communicators are expected to follow the same guidelines here as any ethical technical writer. Within academia, at least at my university, regular memos are sent that suggest officials can visit your university office and ask you to turn on your computer and show them the software on that machine. You are expected to have the documentation to back up any programs you have installed on your university computer, if the software is not provided at your site.

Shareware

One exception to this usual practice of licensing software for individuals or sites is a genre called *Shareware*. Shareware programs are software programs distributed for free, but you usually pay for shipping and handling. The companies distributing these products allow you to use them for a limited amount of time (30–90 days), and then you must register and pay for the software to continue to use it. Some companies market new software in this way, getting user feedback and banking on a certain percentage of those users to register and buy their products. Often computer magazines carry ads (and sometimes CDs with the programs) for shareware; you can also find it on the Internet. Although you can make copies for your friends, those individuals will be subject to the same rules as you are (Brunard 1997, 56–57). While a number of software programs are available to share, few medical and health software programs are free of charge. You can download some examples for free to try out or show your facility, but usually medical software is protected and expensive.

Freeware

Be aware that there is another genre of software that *is* free and it is listed as being in the public domain. In other words, it's not copyrighted and does not require you to register and purchase it. Instead it is available to use for free and can be copied and distributed to others for their free use, too. Often the same companies distribute both shareware and public domain software (or freeware) (Brunard 1997, 57). It's not always clear in magazines which is which, so you might want to inquire if you aren't sure. Most of the companies have Web sites and list other contact information and capabilities.

Web Pages and Access Issues

Designing health-oriented Web pages presents still other ethical dilemmas for the health communicator. Many of us design pages by copying code from other pages to our own pages. We borrow graphics, fonts, and other visuals with little attention to copyright. In a professional setting (or a personal one, for that matter) such borrowing of code is unethical and illegal. The same copyright laws that cover print materials also cover digital texts and graphics. You cannot use text and graphics from others' Web sites even if the materials are not specifically copyrighted on the site.

A number of ways exist for us to get permission to use others' designs and give proper attribution, such as:

- Emailing the owner of a page and asking permission
- Sending a formal letter asking permission
- Paying for commercial packages and designs and listing the company/owner in citation information
- Citing the page and the copyright information after gaining permission

Although the information above seems quite straightforward, complicated exceptions do exist to the overall more simple guidelines. Also, if sites are listed or have graphics and other designs that are listed as shareware or freeware, those materials are then available for your use.

Because some changes can occur as Web pages are transmitted, especially depending on users' browsers, screen resolutions, and other settings and equipment, a number of people argue that Web site designs are not copyrighted. Also, users can greatly change the designs, unlike when they are printed in hard-copy format. But, in fact, the design of Web sites is still considered copyrighted because it is an original work created by a person (Markel 2001, 627).

Links and linking to others' pages present other legal and ethical conundrums. Some argue that it is illegal to build composite Web sites made up of others' sites only. Unless the site is covered under fair use guidelines (in other words, an educational site with no profit-making motive or ability), composite sites can be considered unethical and illegal if copyright permission has not been attained.

Are lists of links on someone's Web page protected by copyright? Again, the answer depends on the context and specific situation. If the links are listed on someone else's page and you simply copy them over, then you have violated copyright. But, if you link to others' sites and create your own original list, then copyright is usually not an issue. It never hurts to ask for permission to link to others' sites, especially because it signals them that they may experience increased traffic on their servers. Also, some linkage presents ethical dilemmas if someone with self-serving interests links to sites with an effort to gain clients or profits or to forward ideological or other political agendas.

Graphics, Music, and Videos

Graphics, like illustrations in print materials and like Web-page designs, are subject to copyright, too. Unless you have purchased a package or a design, you cannot use the design without paying copyright fees. Graphics, unless they are used for educational purposes, are considered original work. Whatever electronic format (or print format for that matter) you may have in mind for the graphic, you must be careful to obtain copyright permission.

The same rules apply of course to .wav and .midi files (and other electronic music files). Just as you pay royalties to embed popular songs in your Multimedia® Director, Authorware, or Dreamweaver programs, you must pay royalties to use theme songs in all of your productions. Be sure, also, if you are a project manager in a health firm, that you budget to pay for the various permissions that you might need for your projects.

Videos, too, require health writers and health organizations to pay royalties. Sometimes if you are preparing public service announcements or other campaigns for nonprofit groups, if you tell the studios what the video or music is used for, they will often give you some of the work you need gratis in exchange for an acknowledgment. Some companies gain tax write-offs in this manner, too, and may be more willing to go this route.

What Are On-line Ethics and Etiquette?

If as writers/communicators/composers we are using others' words from electronic or oral forums, we also need to be aware of ethical issues here. Because many of us use email frequently and skillfully, we often aren't aware of the ethics

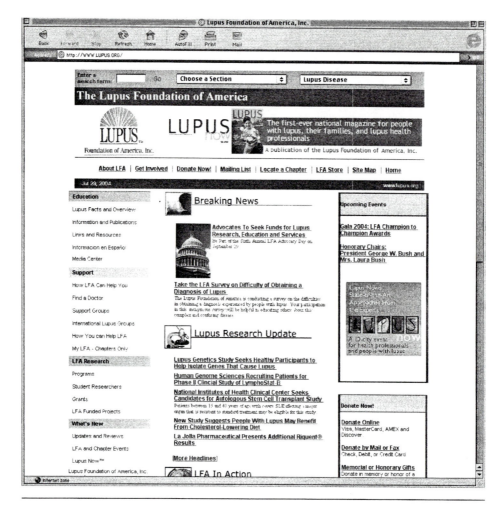

Exhibit 2.4 Web Site for Health Support Group

involved in this form of communication. For example, any time we forward someone's personal email (not so problematic in material already forwarded and dependent on this form of distribution), we should ask permission of the author. Often it is more a matter of personal privacy whether the person who authored the message wants it to be distributed to someone else. The same is true of discussion groups either real and physical or virtual. Before using someone's words, we need to make sure we have their permission.

As we mentioned in Chapter 1, often it is helpful to us as composers seeking feedback to try out ideas or solicit input from Internet discussion groups, chat rooms, or on-line support groups (see Exhibit 2.4). Before using such discussion material, every person whom you quote must have given permission.

Sometimes you can also find useful information in newsgroups maintained by commercial on-line services. Gastel recommends the Web site Deja News at http://www.reference.com, which

> lets you search for newsgroups on given topics and also provides access to recent postings from the groups. "There's a support group for everything on-line," one free-lance writer observes. Monitoring the messages can yield story ideas, alert you to patients' concerns, and aid in identifying people to interview. And posting messages of your own can lead you to information and sources. (Gastel 1998, 63)

If you are entering such a site with the intent to use material or asking questions and visiting the group or groups as a participant/observer, good ethics requires that you inform the group about your intent. In short, you need to announce your intentions and ask permission up-front, and then, if you indeed use such material, you should ask permission again from the individuals in the group whom you quote or paraphrase. Let them see the exact wording and phrasing that you use so that they can validate both your accuracy as well as your representation of their words. Because such groups are excellent resources for health writers, by treating the participants ethically, you will be able to revisit sites and keep your sources patent and trusting.

How Do I Write for Differently Abled Audiences?

Access in this case does not only mean the kind of access we writers can get to various designs, graphics, and texts. Access here also includes being aware of differently abled audiences and what kinds of access issues arise for them as we design our documents, Web pages, etc. In many ways the Internet has made materials more accessible to people who are differently abled. However, there are issues that all Web designers need to be aware of, but especially medical writers, because our audiences usually have a larger percentage of differently abled and elderly people than other specialties.

The courts ruled that the Americans with Disabilities Act (ADA) does cover Web sites and other visual texts, and there are a number of ways to make sure you are designing accessible sites (Markel 2001, 626). Also, see the World Wide Web Consortium's *Web Content Accessibility Guidelines (WAI)* http://www.w3.org/WAI/GL/, http://www.usablenet.com, and http://bobby.watchfire.com/bobby/html/en/ for guidelines. Bobby will check and assess your site for you for free. Some of the software manufacturers are paying increasing attention to accessibility. Macromedia Dreamweaver, for example, carries a download called the "508 Accessibility Suite," which you can download for free from http://www.macromedia.com/. You can visit the University of Maryland's Web site and sign up for the Disabilities Studies in Humanities (DIS-HUM) list for excellent feedback on issues regarding people with differing abilities.

Do consider the following categories of disabilities in your designs for the Web, PowerPoints, slides, videos, and audios.

Vision

For people who have difficulty seeing or must depend on text-to-speech converters:

- Provide text-only versions of every page you compose
- Use 12-point font or larger
- Use sounds to signal that a user has clicked a button or selected an icon on interactive sites
- Use pop-up screens on graphics or describe graphics in text

Hearing

For people with hearing difficulties (and remember a geriatric audience will have this in common):

- For sounds, also create visual cues and markers for button clicks and icon selections, such as flashing or color/pattern changes
- Record voices and music so that users can increase the volume and so that the sound is clear and not muddied
- Provide captions and textual cues for voice-overs, etc.
- Avoid sounds in the highest ranges, as hearing loss occurs here first and most commonly
- Use American Sign Language and subtitles in videos
- If you include a person talking in a video, make the face large enough for lip-reading

Digital Mobility

Try to avoid using the mouse (build in keystrokes instead) and make clickable areas large and easy to click on for people whose hand motions are not easily controlled (Markel 2001, 626).

Avoid making assumptions about what a particular audience can or can't do. For example, if you are working on a health information product for people who suffer from Parkinson's, research the disease thoroughly so that you have an understanding of accessibility issues before you plan your designs.

How Do I Write for Multicultural Audiences?

We talked about audience analysis in Chapter 1, and we've also devoted Chapter 9 to multicultural and international medical writing, but it is necessary to include a few reminders here in the ethics chapter, too. As mentioned above, it is important to know about copyright and other legal and ethical regulations within other countries if you plan to market materials in countries other than your own. As in any sensitive analysis of audience, in rhetorical situations within other countries, often ethical norms and principles can be quite different from your culture, and you need to be especially careful in making assumptions about another country's or cultural group's ethics.

Many situations can occur in preparing materials for other groups and other cultures such as preparing materials showing women doctors as well as male doctors in a culture in which women do not have the right to practice. What is your ethical duty here? To be fair to women and not include bias because another country is biased would make sense; however, you also then have to risk offending a male audience. On the other hand, to uphold sexism is not an ethical practice in this culture, and, if you are from this culture, it would seem unethical to support such practices.

Also, one needs to be careful in preparing health education materials that are so opposite from the ethics of another culture as to be totally rejected by that culture. For example, if you need to promote certain hygiene practices to prevent an epidemic that will cost many lives, you will need to put the greatest good for the greatest number of people first if you believe in utilitarianism as an approach to ethical behavior. (See Dombrowski's *Ethics in Technical Communication,* Chapter 3, for a discussion of the various types of ethical standpoints most recognized within the U.S. culture.)

How Do I Protect Patients' Rights?

This area has taken on special importance since the HIPAA Act of 1996. As Fraser says,

> Patients have a right to privacy that should not be infringed without informed consent. Identifying information should not be published in written descriptions, photographs, and pedigrees unless the information is essential for scientific purposes and the patient (or parent or guardian) gives written informed consent for publication. Informed consent for this purpose requires that the patient be shown the manuscript to be published. (Fraser 1992, 188)

Of course for those health writers who work with medical records and for those who have direct contact with patients, the patient has to be clearly identified. Even as a professional health worker and writer, though, we have certain confidentiality to uphold, whether it's speaking to another patient about a different patient on the floor or discussing details outside of the medical setting with unauthorized people.

For those rhetorical situations that do not call for this familiarity with patients, writers usually omit identifying details about patients if they aren't essential; however don't change relevant data to try to keep the patient from being identified. To be on the safe side, it's always advisable to seek and obtain informed consent and to gain that consent in only the most ethical ways. It's also advisable to mention that you have informed consent in a brief sentence or phrase within whatever communication product you have constructed. Be careful, too, using photographs of patients. As Fraser suggests, ". . . masking the eye region in photographs of patients is inadequate protection of anonymity" (188).

As an ethical medical communicator, you should have a working knowledge of patients' rights and what types of protections they have in various states and countries. In this country, you will find a strong system in place making sure patients' rights to privacy are enforced. Wherever you work, you should be aware of the legislation and social norms of your particular organization, state, and country as these rights vary somewhat from place to place.

What Is HIPAA?

HIPAA, the Health Insurance Portability and Accountability Act of 1996 (a.k.a. the Kennedy-Kassebaum Act), is the set of guidelines for the medical community in this country that protects our patients' rights.[4] The following information gives a basic outline of the federal regulations on the medical profession concerning confidentiality of patient information.

The federal government has been very specific in providing this set of guidelines, which must be followed when disclosing a patient's medical information.

The passage of HIPAA in 1996 is a visible representation of the response of the United States Congress to the will of the American people. In a poll conducted by Louis Harris and Associates in 1992 (4 years before the passage of HIPAA), it was found that 24 percent of healthcare leaders knew of violations concerning patient confidentiality. Although this is far from a majority, it is certainly an unacceptable amount of violations.

Also in a 1996 poll conducted by the same well-known pollsters, it was found that only 18 percent of those polled found the unauthorized use of confidential medical information as a proper practice for medical professionals.

The results of the polls conducted by Louis Harris and Associates give us clear indicators as to the reasons behind the mandate of the federal government expressed in HIPAA.

HIPAA is not only intended to deter the unauthorized use of patient information but also to create a more positive environment in which the medical profession operates by doing the following:

- Make health insurance more portable
- Reduce administrative costs
- Create medical savings accounts
- Regulate group health plans
- Combat fraud in medical reporting

All healthcare providers are subject to the rules and regulations of HIPAA. The ultimate privacy of the patient, since the passage of this act in 1996, has become much more of an acknowledged goal for which the medical community is expected to strive.

HIPAA is intended to set up a national standard for the treatment of "Protected Health Information" (PHI) and "Individually Identifiable Health Information" (IIHI). This act, by the very nature of federally legislated mandates, takes precedence over all state law (unless the state law is more stringent than the federal). Ann Mathias, Esq., a compliance officer for Magee Women's Hospital in

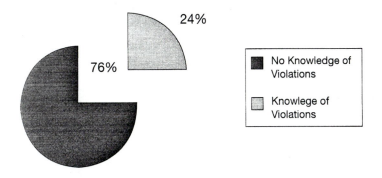

Exhibit 2.5 Healthcare Leaders with Knowledge of Violations

Pittsburgh, Pennsylvania, describes four key concepts that need to be regarded when considering the federal mandates on healthcare providers regarding the issue of privacy:

- Protected Health Information (PHI)
- Restrictions on use and disclosure
- The "Minimum Necessary Standard"
- Patient Rights

First, let us discuss HIPAA's implications concerning Protected Health Information. PHI is a subcategory of Individually Identifiable Health Information. IIHI is defined as "information collected from an individual that is created or received by a healthcare provider, health plan, public health authority, employer, life insurer, school or university, or healthcare clearinghouse and relates to the past, present, or future physical or mental health condition of an individual. . . " (Mathias). More specifically, PHI is defined as "Individually Identifiable Health Information" (IIHI) that is: transmitted electronically, maintained electronically, transmitted or maintained in any other form (i.e., paper or verbal), or anything that is in what we view as the medical record. In short, PHI is defined as any sort of medical report or form that concerns the patient's "past, present, or future physical or mental health condition. . . " (Mathias). Certainly the treatment and disclosure of any report filled out with patient information would fall subject to the mandates of HIPAA under the definitions of IIHI and PHI.

Second, let us discuss the "use and disclosure" of medical information as defined by Ann Mathias. There are certain instances in which authorization is not required for disclosure. These instances fall under the "Treatment, Payment, and Operations" (TPO) exemptions. Mathias states that healthcare providers "do not need patient authorization to use PHI for TPO—providing patient care, getting paid for patient care, and running the organization." Other exemptions for patient authorization to disclose PHI include "functions such as public health, national security, [and] law enforcement." In short, medical providers do not need authorization from a patient when their medical information is being used simply to conduct the everyday business of a healthcare provider or when an official government entity has asked for the information with concern for the public safety. These loopholes in disclosure have certainly created room for abuse of the system and contributed to the skepticism of the American public concerning the privacy of health information.

Although there are some exceptions to the disclosure of information contained in medical reports, more often than not the disclosure of patient information requires the authorization of the patient. According to Mathias, authorization is required when PHI will be used for the following purposes: "marketing, fundraising, [or] research." Legal authorization of medical information must be in writing and include all of the "necessary elements" that are intended for the specified use. Mathias also makes certain to note that the patient may revoke authorization at any time.

Finally, healthcare professionals must consider the "Minimum Necessary Standard" when taking into account the authorized disclosure of private medical information. According to Mathias, the "Minimum Necessary Standard" mandates that healthcare professionals "must make reasonable efforts to limit use or disclosure to the *minimum* amount of PHI *necessary* to accomplish the intended purpose of the use or disclosure." For example, if a physician is writing an article to be published in the *Journal of the American Medical Association* that details the current procedures used in the treatment of Alzheimer's disease, he or she may not obtain permission for the disclosure of information pertaining to the patient's past orthopedic difficulties. Information concerning orthopedic treatment would not be necessary for the academic discussion of current Alzheimer's treatment.

How Was HIPAA Phased In?

The Health Insurance Portability and Accountability Act of 1996 was far from a simple bill that was passed and immediately enacted by Congress. What was actually contained in the Bill that was passed in 1996 recognized the need for federal regulation in this area but did not issue specific, legal regulations. Congress actually provided for 36 months in which it was to pass specific regulations. If none were passed, the Health and Human Services (HHS) Department was then to be given six months to issue a proposed set of regulations. During this period the Department was to receive feedback from patient groups and the medical community. This feedback was to be taken into account and then the HHS was to issue final regulations.

HIPAA Timeline:

- *1996: The Passage of the Health Insurance Portability and Accountability Act.* Congress passed this act in response to the acknowledged need for a federal mandate to protect "Individually Identifiable Health Information." This act did not specifically create the federal regulations itself, but provided that if Congress failed to enact such regulations within 36 months, the Secretary of Health and Human Services was to create enforceable regulations concerning privacy standards within six months.

- *1999: Time is up!* The 36 months that Congress allowed itself expired without the passage of specific regulations.
- *November 3, 1999: Proposed HIPAA Privacy Regulations.* As mandated by Congress, the Health and Human Services Department issued their proposed Privacy Regulations. These regulations were only a proposal, were not enforceable, and were sent out for feedback.
- *December 28, 2000: Final HIPAA Privacy Regulations.* HHS received over 52,000 comments on their "proposed Privacy Regulations" and published the final (legally enforceable) Privacy Regulations.
- *April 14, 2001: Privacy Regulations Become Effective.* All applicable medical organizations are subject to the Privacy Regulations. (American Health Lawyers Association)

The privacy regulations finally came into effect, and none too late, because the American public's general feeling about the privacy of their medical information continued to be one of concern. For example, in a poll conducted by Gallup in November of 2000 it was found that the "privacy of their health information is very important" to 77 percent of Americans.

In the same poll it was found that 61 percent of Americans were "very concerned that their personal health information might be made available to others without their consent."

Today, even with these constraints in place the American public is still extremely skeptical of the attention paid to the privacy of their health information. We medical writers must begin to do our part in changing this popular opinion. As professionals, we must treat the private information of patients with special care, and, if possible, remind the health professionals we work with to do the same. The federal government has done its best to enact regulations to alleviate this skepticism, and it is now in the hands of the healthcare providers to gain the trust of the public by the ethical treatment of patients' private information.

What Are Patient Bills of Rights?

In addition to federal and state legislation, hospitals, clinics, and even some doctors' offices now have their own Patient Bill of Rights like the one featured on the front page of the University of Missouri Health Care system's Web site (http://muhealth.org/~patient/rights.shtml).

As medical writers who often work globally, we need to be aware that other countries, too, have acts protecting their patients. The United Kingdom has an excellent set of guidelines called the Data Protection Act, their version of HIPAA. For those of us who must often work with and within databases, we call your attention to the eight general principles of the act and suggest as ethical writers we follow them.

Data Protection Act

In the United Kingdom, if you keep details about other people on disk, you must register under the Data Protection Act. There are eight general principles behind the act. These are as follows:

- Personal data must be collected and used fairly without deceiving those concerned.
- Personal data must be registered under the Act.
- If you are going to pass on personal data to other people, you must state this when you register.
- You should not keep more personal data than you need.
- You have to try to make sure that personal data is accurate and up-to-date.
- You should not keep personal data for longer than you need to.
- If you hold personal data about somebody, they have the right to know what it is and to check that it is accurate.
- You have to try to make sure that personal data is not changed, destroyed or disclosed without authorization. (Fraser 1992, 82–83)

To register costs a reasonable fee, and you will need to fill out forms that require the following information:

- the personal data that you are recording
- what you use the data for
- where you got this information
- to whom you may want to disclose the information
- any overseas countries where you may want to send the information (Fraser 1992, 83)

In the United Kingdom you do not have to register information in every type of database you develop. In this country, and particularly in universities, faculty members and students must register any research they do with human subjects, even if it involves only data and communication as our work usually does. It's always better to register and cover oneself legally. At Clemson University for example, the Institutional Review Board requires such registration, and often the health communication work I do qualifies for waiver status. But it's always best to have the safeguards in place and to keep them renewed and up-to-date for each human subject project.

Living Wills

One of the most ethically charged discussions of patients' rights centers on how and when a patient would choose to die. Because this ethical discussion involves important documentation (living wills and Oregon's Death with Dignity Act), medical writers should have a working knowledge of this area. As a healthcare professional, you need to make sure those who have the power to make decisions about a patient are aware of these documents. They often save family members, significant others, and close friends the agony and painful decisions that otherwise attend the dying and deaths of their loved ones. A well-written document (these are state legal documents) can make ethical choices clear and less painful.

This type of document is getting more and more common in the United States. As the baby boomers age and require more healthcare, patients' rights will take center stage in this culture. As medical writers, we need to have an understanding of the impact that such documentation can have in life and death decisions and also be able to write and express these ethical choices as succinctly and clearly as possible. The more these situations can be spelled out, the more effective our ethics will be and the better care patients will receive.

In even a further (or the furthest) step in patients' rights, we now have one state, Oregon, which has enacted a Death with Dignity Act that was voted in by Oregon voters in 1994. In February of 1999, the *New England Journal of Medicine* published a report on the law's first-year results (Chin et al). Many predictions that detractors had warned about did not materialize. This measure has not yet gained the support or the acceptance that living wills have engendered, but again, as ethical medical writers, we need to be aware of the ethical conversations around us in order to design and write sensitive health materials. Being aware of the issues gives us a more accurate reading of the context for particular materials we develop.

Medical ethics will continue to play key roles in our culture and in other cultures as technology continues to move forward with genetic research, stem cell usage, and other developments that spark keen debate. Our awareness and sensitivity to the context can help our audience members clarify and understand the debates around them by translating issues into understandable prose for nonspecialists and lay audiences. By not obscuring implications of various technologies, as was done in the cigarette companies' reports, we can perform valuable public service and can provide ethical insights into the difficult decisions that all citizens need to be prepared to make.

Summary

In this chapter, you've covered a large territory of medical ethics, with the focus on your responsibilities to create ethical documentation and protect patient rights. If you weren't quite clear before on why you needed to be aware of medical ethics, you should be thoroughly convinced now of the importance of this area for writers and designers. You read a number of examples in which medical writers and designers were put in situations ranging from problematic to unethical and illegal. Medical facilities and corporations can be like other big businesses in which profit becomes the bottom line. In that case, you need to have a strong sense of your own personal ethics as well as the ethical responses of the medical writer and designer. To help facilitate your understanding of your obligations, the American Medical Writers Association Code of Ethics was included in this chapter to serve as a set of helpful guidelines.

Because copyright and intellectual property issues are so important in our litigious culture, you learned the parameters of copyrights for medical writers and designers, too. You learned about *work for hire, fair use,* and the *Uniform Require-ments for Manuscripts Submitted to Biomedical Journals,* so that you could under-stand the various contexts that impact copyright obligations. These contexts help determine, too, when and how to obtain permission in writing for external audi-ences. Besides texts, other print areas that require copyright knowledge include graphics, illustrations, photos, logos, trademarks, and registered trademarks.

Equally important today as information architects, we need to know copy-right and ethical issues in digital work. Those genres include but are not limited to: shareware, freeware, emails, listservs, chat rooms, on-line support groups, Web pages, CDs, digitized graphics, audios, and videos. Just because some prac-tices in Web work and other genres are common (such as copying code) doesn't mean that they are legal or ethical. This area, because of its relative newness, will continue to define both ethical and legal issues as the culture adapts and adopts further iterations of today's technical abilities. Thus all of our questions may not be answered for some time.

Another important question for us in this chapter was what should you know about designing for differently abled audiences. Because so many of our patients are differently abled, we have an ethical obligation to design taking their needs into account, whether they be in the areas of vision, hearing, or digital and other mobility. People who conduct research in the area of disability studies are keenly aware that there is a differently abled culture, too, so when we talk about multi-cultural audiences, we need to include that culture. Although we talked about multicultural issues briefly in Chapter 1 and will spend Chapter 9 on that in much more depth, we did consider that different cultures can have very different views on ethics from those of our own culture.

Lastly in this chapter, we talked about patients' rights and the legislation and documents that address this topic. Although this area was featured last, it is the single most important issue in medical ethics and for us as medical writers. In ad-dition to the HIPAA act, there are many other laws and documents that we are re-sponsible for in our work. We discussed living wills and Oregon's Death with Dignity Act as examples of the kinds of documents in which good medical writ-ing can truly have an impact on life-and-death decisions.

Discussion Questions

1. What do you think of when you hear the term *medical ethics?* Does everyone in your class think of medical ethics in the same way? How would you define *medical ethics?* Why? How do your classmates' definitions match yours? What differences are there? Can you come up with a class definition that you can all agree on?

2. Given the tobacco industry medical ethics example in the beginning of the chapter, what roles did and could the medical writers play? What are all their options? Once you come up with their options, discuss the various outcomes of each option. Which one would you choose and why?

3. Choose one or more medical ethics case. Using the *American Medical Writers As-sociation Code of Ethics,* which principles apply to your cases? Are there principles

left out that you would add? Which ones and why would you add them? How would you explain to your employer that you couldn't follow his or her orders precisely, because to do so would violate your code of ethics?

4. You were given a number of contexts within which to understand the role and importance of copyrights. What are the differences among the contexts? Is it clear when and why copyrights are important? Why is copyrighting both an ethical and a legal issue? Do you think it will become more or less important in these contexts in the future?

5. After reading the section on digital copyrights, discuss your experiences with designing Web pages, visiting chat rooms, being on listservs, etc. What issues have come up for you in your digital experiences? Is it important to heed copyrights in electronic media the same as we do in print media? Why or why not? Do you think that duplicating CDs or sharing software without paying royalties or purchasing copies is ethical or unethical? Should the laws regarding electronic copyrights be as strict as they are now or stricter?

6. In drafting living wills and in writing legislation for a Death with Dignity Act, how do clarity and accuracy become ethical responsibilities for the medical writer? Do such documents tend to lag behind popular opinion or lead the way? Are there more choices patients should have? What documentation could be designed to give patients more rights and more choices? Are there times when patients need to have fewer rights? If so, when?

Exercises

1. Find medical ethics cases other than those featured in this chapter. Design a PowerPoint® that outlines what medical writers' roles would be and what they could and could not have accomplished. If you think they should have blown the whistle on the organization, show what documents they might need to do so. What were their other alternatives? What would you choose and why? Make your PowerPoint® a visual argument for your position.

2. Divide the class into teams of two partners. Each team should find a different Web site advertising a medical product (pharmaceutical, machine, device, etc.). Prepare a memo to read to your class while you show the Web site (if your class has the appropriate computer setup) that answers the following questions: What copyrighting do you notice on the site? Are there logos or trademarks and where are they? Are there any medical or legal disclaimers? Are they frightening or reassuring to patients and why?

3. Do a brief report in which you compare the Web sites of the American Medical Writers' Association (AMWA), Europe's Medical Writers Association (EMWA), and AMWA–Canada Web sites. What can you tell about the three organizations? How are they similar and how are they different? What cultural differences do you see?

4. Examine several medical and health-oriented Web sites. Are they accessible for differently abled patients? Pull up the Bobby Web site and see if the Web sites have built-in accommodations for this audience or not. Visit other cultures' Web sites that address the same or similar topics. How do they make their Web

sites accessible? Are their strategies different or the same? How? Make a graphic display on a poster that shows the various Web sites and some of your analysis. Write the rest of the analysis up in a brief report to include with the visual you have created.

5. Interview a medical writer, a lawyer (or law student), and a philosopher (or philosophy graduate student) and ask them for their definitions of medical ethics. What experiences have they had with digital publishing and copyrights? What is their understanding of copyrighting from your standpoint (small health facility writer or large corporate health systems writer)?

6. Role-play how you would handle meeting with a family to discuss a living will document.

Works Cited

American Health Consultants. 1999. "Increase in Debates, State Bans Don't Change Central Fact: It's Coming." *Medical Ethics Advisor*, 15, no. 5 (May 1999): 49–60.

American Health Lawyers Association (AHLA). © 2003 AHLA. March 2, 2002. http://www.healthlawyers.org

American Medical Association Manual of Style, 9th ed. 1998. Baltimore: Williams and Wilkins.

American Medical Writers Association Web Site. Copyright 2003 AMWA. August 10, 2001. http://www.amwa.org

AMWA's Canadian Sister. 2003. August 7, 2003. http://www.amwa-canada.virtuo.ca/

Bobby 2002–2003. August 7, 2003. http://bobby.watchfire.com/bobby/html/en/

Brunard, Philip. 1997. *Writing for Health Professionals: A Manual for Writers,* 2nd ed. Boca Raton: Chapman & Hall.

Clemson University Institutional Review Board. IRB Waiver. 8/1/2003. August 7, 2003. http://www.clemson.edu/IPC/orcSite/orcIRB_New.htm

Chin, Arthur E., M.D., Katrina Hedberg, M.D., M.P.H., Grant K. Higginson, M.D., M.P.H., and David W. Fleming, M.D. 1999. "Legalized Physician-Assisted Suicide in Oregon— The First Year's Experience." *New England Journal of Medicine* 340, no. 7 (Feb. 18, 1999): 577–583.

Cormack, Desmond F. S., and David C. Benton. 1994. *Writing for Health Care Professions,* 2nd ed. Cambridge, MA: Blackwell Science, Inc.

Cull, Peter, Editor. 1990. *The Sourcebook of Medical Illustration: Over 900 Anatomical, Medical, and Scientific Illustrations Available for General Re-Use and Adaptation Free.* Boca Raton: CRC P-Parthenon.

Data Protection Act 1998. 1998. August 7, 2003. http://www.hmso.gov.uk/acts/acts1998/19980029.htm

Deja News. 2003. August 7, 2003. http://www.reference.com

DIS-HUM LISTSERV. University of Maryland. 7/31/2003. August 7, 2003. http://www.umd.edu

Dombrowski, Paul. 2000. *Ethics in Communication.* Needham Heights, MA: Allyn & Bacon.

EMWA. Europe's Medical Writers Association. 7/24/2003. August 7, 2003. (http://www.emwa.org)

Fraser, Jane. 1992. *How to Publish in Biomedicine: 500 Tips for Success.* London: Radcliffe Medical Group.

Gastel, Barbara, M.D. 1998. *Health Writer's Handbook,* 1st ed. Ames, Iowa: Iowa State University Press.

"Declaration of a Desire for a Natural Death." 1999. South Carolina State Government. Kay Barrett, Esq., Attorney. Clemson, SC. May.

Heifferon, Barbara, Jacob Barker, Melissa Tidevell, et al. *Diagnosti condo la salad de Trabajador.* Clemson, SC: Clemson, 2002.

Huth, Edward J., M.D. 1990. *How to Write and Publish Papers in the Medical Sciences,* 2nd ed. Baltimore: Lippincott, Williams & Wilkins.

————. *Writing and Publishing in Medicine,* 3rd ed. 1990. Baltimore: Lippincott, Williams & Wilkins.

Kiple, Kenneth F. 1993. *The Cambridge World History of Human Disease.* London: Cambridge University Press.

Macromedia Dreamweaver. "508 Accessibility Suite." Free download for access. August 7, 2003. http://www.macromedia.com

Markel, Mike. 2001. *Technical Communication,* 6th ed. New York: Bedford/St.Martins.

Mathias, Andrew. "Report on HIPAA." *Medical Reports: Analysis of Genres, Styles, and Language.* Andrew Mathias, Chris Welch, and James Galloway. Unpublished manuscript.

Mathias, Esq., Ann. Personal telephone interview by Andrew Mathias. April 20, 2003.

McQuillan, Laurence. "Pope Calls Stem Cell Research Evil." *The Greenville News.* July 19, 2001: 1A.

Minick, Phyllis, Editor. 1994. "The American Medical Writers Association's Code of Ethics." *Biomedical Communication: Selected AMWA Workshops—A Practical Guide for Writers, Editors, and Presenters of Health Science Information.* Bethesda, MD: American Medical Writers Association.

No Smoking. 7/27/01. August 23, 2002. (http://nosmoking.org/july01/07-27-01-2.html

Parmley, William W., MD, FACC, Editor-in-Chief. 1996. "Anecdotes in Medicine: Do They Have Value?" *Journal of the American College of Cardiology* 28, no. 3 (September): 795.

Patients' Bill of Rights (S. 1344). http://www.findarticles.com/cf_0/ m0BSC/29_8/ 55868006/01/article.jhtml.

Reich, Warren T. 1995. *Encyclopedia of Bioethics.* New York: MacMillan.

University of Missouri Health Center's Patient Bill of Rights. The front page of the University of Missouri Health Care system's Web site: (http://muhealth.org/~patient/ rights.shtml).

U.S. Copyright Office 8/1/01. August 31, 2001. (http://lcweb.loc.gov/copyright)

Watson, James D. 2001. *The Double Helix: A Personal Account of the Discovery of the Structure of DNA.* Carmichael, CA: Touchstone Books.

World Wide Web Consortium. *Web Content Accessibility Guidelines (WAI).* 7/29/2003. August 7, 2003. http://www.w3.org/WAI/GL/

Zitner, Aaron. 2001. "Senators Use Bible for Lesson on Life in Stem Cell Debate." *Los Angeles Times.* Reprinted in *The Greenville News.* July 19, 2001: 1A, 4A.

Endnotes

1. Thanks to Sam Dragga, series editor, for pointing this out. I'm quoting him here.
2. The Tuskegee project was initiated in the 1940s and 1950s as a longitudinal study of syphilis among African American men in Tuskegee, Alabama. Although there was a known cure for syphilis at the time, this cure was withheld from those with syphilis so the results of this horrible disease could be studied and documented. This study was

truly one of the worst horror stories in American medical history. Often, in African American populations, there is a reluctance to take advantage of healthcare opportunities offered by white health workers. When asked, many members of this population are well aware of this history, as well as the forced sterilization of African American women, also in the not-so-distant past.

3. If *ethos* is an unfamiliar word and concept for you, please consult a rhetoric text.

4. Thanks to Andrew Mathias for his research and written report on HIPAA in my spring 2003 Honors Technical Writing class. I did make some changes, but the research was done by Andrew.

CHAPTER

Document Design Principles and Project Management

 Overview

Another phrase for designing communication is information architecture. Consider yourself an architect of health information or a designer of health messages and medical texts, who needs to effectively construct a document in any medium or in multiple media. Because some of you will not have an extensive background in technical communication or experience in document design, this chapter will cover the principles of design briefly. In addition, you will find a design vocabulary you can use to communicate to in-house and outsourced designers and printers with whom you will interact.[1] This chapter also articulates some of the design practices that underlie the document design decisions you will need to make. In Chapter 10, "Presenting Written Materials Visually," and Chapter 11, "Electronic Medical Writing," in Part III, you will learn in more detail about the various media and how they specifically impact your health information architecture.

In addition, this chapter will cover project management and, briefly, an analysis of organizational communication flow problems; both are important issues, especially in larger health systems in which voluminous information flows in, is processed, and is then sent out or filed within medical archives.

This chapter will address the following questions:

- What should I do to ensure good design?
- What principles do I need for designing?
- How do I talk about visual document design?
- How do I manage large projects?
- How do I manage communication flow?

Once you are ready to consider designing the documents, you will have already considered the ethics of the project, done an audience analysis, determined the purpose of the documents, and understood good writing practices necessary for developing the content of documents. Just as it was in the previous chapters, process is also key to designing and managing communication projects. Two of the most important voices in document design and management in the field of technical and professional communication are those of Karen Schriver and JoAnn Hackos. Experts in visual design include Charles Kostelnick, Edward Tufte, John Berger, and others. However, here you will learn to tailor their input for writing in the health professions.

What Should I Do to Ensure Good Design?

Good document design, for example in some of the myriad of forms used for hospital patients, can greatly increase speed of access to healthcare options for patients, as well as ensure efficiency and accuracy in treatment. Good documents can help patients more accurately report their health issues, so that practitioners can focus more quickly on patients' primary problems and concerns.

There are several ways to achieve good documents in general:

1. Make readers or users want to use your document. Documents, too, have a sense of credibility. Messages that are accurate, well designed, and carefully crafted will make them credible to users.
2. Help readers or users prioritize the information. Especially in medicine, patients and staff need to understand how information is valued. In a well-designed document, it should be clear what information is most important—in a sense, good document design *triages* information, just as emergency room personnel, emergency medical technicians (EMTs), and paramedics must *triage* patients within a medical situation.
3. Make it easy for users to navigate through the information, whether on a Web site or within a printed document. Readers need to be able to follow a format logically. If they are guessing how to move through a document or site, the users will get frustrated and confused, adding more time and emotion to an already stressful situation. Medical staff members are usually so overworked and hurried, they need to be able to at least depend on documents and sites not to slow them down.
4. Make sure the story that one is telling in the document is the one the designer means to tell. As health professionals, the design and data in our medical reports must represent an accurate picture. Edward Tufte, a visual communication expert, suggests that we need to make sure the story we tell in the data, especially numerical data, is true to our understanding of the story as we first received it:

> When we reason about quantitative evidence, certain methods for displaying and analyzing data are better than others. Superior methods are more likely to produce truthful, credible, and precise findings. The difference between an excellent analysis and a faulty one can sometimes have momentous consequences. (1997, 27)

What Principles Do I Need for Designing?

In order to be a good document designer in the health professions, you need to communicate using at least some of the design and document terminology professional writers, communicators, and designers use, in order to work well with them. If you have had a technical writing class or an art or design class, you will be familiar with some of these terms. If you don't have a background in this area, and many health professionals do not, you need to be aware of the following concerns in good document design:

- Layout (or grid layout)
- Headings or headers
- Margins and/or white spaces
- Typography

These elements are deceptively simple, but very important. Any one of the elements, if not used appropriately, can make your document unreadable.

Larger organizations will have their own style sheets that predetermine your logos, style choices, and layouts. You can set up your own workstation to reflect these styles, and your documents will remain consistent with the corporate image or brand that is represented in all the documents. You may also be asked to develop a style sheet for a particular department or organization. The following considerations will be especially important for you as a health communicator.

Because we want to start with the overall appearance of the document, we will first discuss layout.

Layout (or Grid Layout)[2]

Layout is best approached by thinking of the document or documents as elements on a surface. First you determine the media, then the orientation. Once you know the size of the documents you will produce, you can begin to think in terms of placing the various texts and/or images on the page. On Web pages, you can

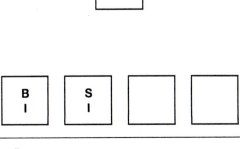

Exhibit 3.1 Storyboarding

develop a template for subsequent pages, so that your overall layout or look of the page remains consistent.

In Web design, we use *storyboarding* as a way to sketch out the layout of the pages of a Web site. In print design, you can also *storyboard* or use a sketch to lay out the components that need to go in the document. If you start with this step, you will have a better design and ultimately save more time than if you have to spend time redoing the site later.

If you think of a large CPR poster, for example, you know that certain steps will have to fit onto that poster in order to have the full information represented. You may want to take paper and cut it into multiple pieces of uniform size according to the number of steps you will need to fit on the poster. Then you will have a sense of the size of each graphic and text block.

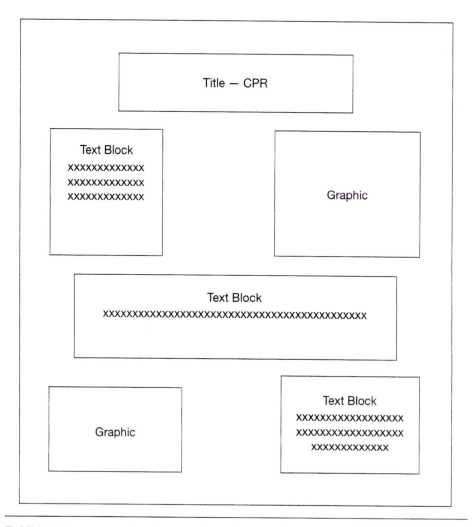

Exhibit 3.2 Arranging Graphics and Text Blocks in Layout

Exercises for Back Pain
(Arial – san serif 22 pt)

White space (3 spaces)

Flexion Exercises (Arial – 16 pt)

White space (2 spaces)

Curl-Ups (Arial – 12 pt)
xxxxxxxxxxxxxxxxxxxxxxxxxxxxxxxxxxxx
xxxxxxxxxxxxxxx (Text – Times Roman or any serif font 10 pt)
Graphic
White space (1 space)

Knee-to-Chest (Arial – 12 pt)
xxxxxxxxxxxxxxxxxxxxxxxxxxxxxxxxxxxx
xxxxxxxxxxxxxxx (Text – Times Roman or any serif font 10 pt)
Graphic
White space (2 spaces)

Extension Exercises (Arial – 16 pt)

White space (2 spaces)

Backward Bend (Arial – 12 pt)
xxxxxxxxxxxxxxxxxxxxxxxxxxxxxxxxxxxx
xxxxxxxxxxxxxxx (Text – Times Roman or any serif font 10 pt)
Graphic
White space (1 space)

Press-Ups (Arial – 12 pt)
xxxxxxxxxxxxxxxxxxxxxxxxxxxxxxxxxxxx
xxxxxxxxxxxxxxx (Text – Times Roman or any serif font 10 pt)
Graphic
White space (1 space)

Shoulder Lifts (Arial – 12 pt)
xxxxxxxxxxxxxxxxxxxxxxxxxxxxxxxxxxxx
xxxxxxxxxxxxxxx (Text – Times Roman or any serif font 10 pt)
Graphic

Exhibit 3.3 White Space Design Example

Headings or Headers

Headings need to be well written to guide the reader through a document and Web page. It's best to use sans serif fonts for headers and serif fonts for text (see section on Typography for definitions). With current word processing programs, consistent headers are more easily inserted into documents such as lengthy reports. In Microsoft Word, for example, under *View*, the outline view will permit you to see your headers and check for consistent font and size. On the formatting tool bar in Microsoft Word, you can also select the text style to the left of the font and size windows. You can create a header and just apply that style according to the level of information.

In Exhibit 3.3 notice that the point sizes create a hierarchy of information in a similar way that the traditional outline does:

     I.

      A.

      B.

        1.

        2.

With word processing, we can now create the hierarchy visually using point size rather than just tabs and spacing. You can adjust sizes according to content and medium. Here is an example of using point sizes to create a hierarchy of information:

22-point size (centered title)

16-point size (left-justified broad category)

12-point size (left-justified more specific items under broad category)

In addition to being able to place information into a hierarchy or taxonomy for your audience, you help them visualize and remember information in chunks (see *chunking* below). You will also need to pay attention to parallel construction, a grammar term that suggests that if one header begins with a verb, such as *Changing* the Battery in a Holter Monitor, then another header must also start with a verb: *Replacing* the Tape in a Holter Monitor (see Exhibit 3.4). Some technical writers also suggest that using questions as headers is an effective way to reach audiences.

Be careful as you make font choices that they are appropriate for the document. Some elaborate or whimsical fonts are difficult to read. Or if you use fonts such as *Old English Text MT* (𝔒𝔩𝔡 𝔈𝔫𝔤𝔩𝔦𝔰𝔥 𝔗𝔢𝔵𝔱 𝔐𝔗), they signal the reader that the document is a classical one. The emergency equipment instructions headers in that font would be stylistically inappropriate, as well as difficult to read.

     Appropriate: Keep the Defibrillator charged at all times.
     Inappropriate: 𝔎𝔢𝔢𝔭 𝔱𝔥𝔢 𝔇𝔢𝔣𝔦𝔟𝔯𝔦𝔩𝔩𝔞𝔱𝔬𝔯 𝔠𝔥𝔞𝔯𝔤𝔢𝔡 𝔞𝔱 𝔞𝔩𝔩 𝔱𝔦𝔪𝔢𝔰.

Parallel Construction	Unparallel Construction
Holter Monitor Instructions	Holter Monitor Instructions
Hooking up the Patient	*Hooking up* the Patient
Changing the Battery	*How do you change* the Battery
Cleaning the Electrodes	*The Cleaning* of the Electrodes
Replacing the Tape	*To replace* the Tape

Exhibit 3.4 Parallel and Unparallel Construction

Margins and/or White Spaces

White space is also an important consideration in overall document design. In traditional, text-based documents, we are used to planning margins around our documents and know that a consistent margin is attractive and necessary. With word processing programs, standardized forms, and templates, these document design issues are already preset or predetermined. However, as designers of different media, we may have to decide about margins and white spaces. If design elements are mandated for certain hospital and other health-site forms, we need to make sure our computers are set up to accommodate such formatting.

In word processing programs, document designers have choices in formatting text according to justification: left justified, centered, right justified, or full justification. Justification refers to alignment of the text itself along a left, right, or both margins:

Left Justified	Centered	Right Justified	Full Justification
Xxxxxxxxxx Xxxxxxxxxxx xxxxxxxxxx xxxxxxxxx xxxxxx	Xxxxxxxxxx xxxxxxxxxxx xxxxxxxxx xxxxx	xxxxxxxx xxxxxx xxxxxxxxxx xxxxxx	Xxxxxxxxxxx Xxxxxxxxxxx Xxxxxxxxxxx xxxxxxxxxxxx

Text that doesn't line up, or ragged text, is also an appropriate format for some documents. Designers should be aware of the *conventions* of certain documents so that they don't violate them. In patient education or facility brochures, for example, often text is justified on both the right and left. Sometimes though, text is only left justified. With the ease of computers, we can also experiment to determine which format looks visually more appropriate. You'll notice that headings also help with white-space design (see Exhibit 3.3). Also, you can see whether they look better centered or left justified.

In documents that are not simple paragraphing of texts, we also have to consider the space within the document or Web pages, or white space. White space helps readers process information by putting separations between text in ways that also *arrange* or *chunk* text. Even in emails or traditional documents, providing white space as separation between unlike elements can provide easier processing for your audience.

Beginning students in medical writing often write their first formal letters with all the text crowded up into the top half of the page. This lack of attention to layout results in too much white space at the bottom of the page. If you look at Web sites, white space (whether it is actually white or not) is often an issue and points to poor Web design if it's not properly balanced. Scrolling through too much text or through empty space with no visuals or text are both white-space problems.

Typography

Typography refers to font, which includes the style, size, and typeface itself. As we mentioned above regarding headers in inappropriate fonts, the typeface, size, and styles you choose are important parts of your design process. Much of what we know about typography comes from our history of print. Knowledge of these terms is also important to be able to communicate to present-day printers if you outsource your health materials for publication or to use within your own workplace after you have designed them.

To decide on typeface or font, you will want to decide whether to use a *serif* or *sans serif* font. *Serifs* are the little feet that appear on letters: for example f in Times New Roman has small extensions at the top and base, while f in Arial does not. Conventional wisdom says that serifs draw the readers' eyes along and thus are helpful. In many visual documents with small amounts of text, you'll see mostly sans (French for without) serif fonts. Also, because our capacity for font choices has greatly increased with word processing programs, designers are often seduced into using more creative fonts. Again be aware of the appropriateness of the style. You don't want a light-hearted, whimsical, or humorous font choice on a document about living wills, for example. Also, many printers aren't loaded with the less conventional fonts, nor are outsourced, commercial printers always able to duplicate fonts unless you package them up with your other designs. Whatever font or fonts you choose, avoid using many different fonts within one document or changing fonts from one issue of a newsletter, for example, to another unless you plan to inaugurate and continue to use a new font consistently.

Size is another factor that is often determined by conventions. Although we have many choices in word processing programs that make it easy to change font size, we have to use consistency and appropriateness here, too. If the font is too large or too small, it can call negative attention to itself. Size can be used to prioritize important health information, especially in headers and titles (see Exhibit 3.3). Try not to use too many different sizes in a document, or you will hinder rather than enhance usability, unless they are headers that you use to hierarchize information. Also, be aware of the issues mentioned in the earlier chapter on ethics. As we design, we need to be responsive to people with disabilities.

Another consideration in size and style is whether to use lowercase or uppercase letters. Some medical documents use uppercase letters, so again document conventions are important. Often charts that are still handwritten and doctors' scripts are in uppercase. In health education materials and other visual designs, you will want to consider case along with the other myriad variables in the design. In emails, though, the use of capital letters for emphasis can indicate shouting and anger readers, so again, conventions within each genre are important to learn.

Two other terms go back to historical printing processes when each letter was hand placed into a row on a larger printing press. *Leading* means the spaces between lines. You will notice that in document design software (other than word processing programs), you can often set and change the leading if you desire. Depending on what information you need to place into a set amount of space, sometimes this design function helps you balance the white space and make document design more attractive.

Kerning, on the other hand, refers to spaces between letters. You will notice that fonts have different amounts of kerning. If the letters are too far apart, readers can have more difficulty reading the words themselves. At other times, increasing the kerning to spread a title along a page can be helpful in page design and balancing white space.

In typography document designers have additional options available to them that help them emphasize and focus on certain text: **bolding**, *italicizing*, and underlining, not to mention various color choices and highlighting, all found on the formatting tool bars of word processing programs. In Exhibit 3.5, if the titles of the tests and procedures such as "Health Counseling and Cancer Checkup" had been bolded as second-level headers, the guidelines would have been easier to read. In addition, a major guideline is left out, colonoscopy, so the content must be double-checked for accuracy, too. The main problem document designers have is if they overuse these functions (bolding, italicizing, and underlining) in a document or use them inconsistently.

Bullets are another way of organizing items or points in a text. Good usage of bulleting can greatly enhance the processing of information. Again, be consistent with bullet style and don't make bullets too difficult to identify. Statistical information may be more effective in bars and graphs or in a table format.

On some Web pages, bullets are icons that point to or symbolize certain types of information. If the icons are not easily recognizable, they can confuse the navigation for the audience, rather than enhance it. Some Web sites are difficult to read because the icons are not easily recognizable. The use of icons as bullets leads us to our next topic, visual design.

THE IMPORTANCE OF CANCER SCREENINGS

Cancer screenings are vital to detecting cancer early and treating it successfully, says Priscilla Strom, MD, a surgeon with Gainesville Surgical Associates. "The earlier you find cancer, the less chance there is of it having spread to other parts of the body and the better your chance is of treating it successfully."

The following chart is a guideline for screening from the American Cancer Society, but Dr. Strom stresses that these are only recommended guidelines. "Based on your medical history and your family's medical history, you may need to be screened at an earlier age or more frequently," she says. "And don't be afraid to ask your physician to order a test. If you have a feeling something is wrong or unusual, ask for a screening just to be sure."

AMERICAN CANCER SOCIETY SCREENING GUIDELINES

Test or Procedure	Gender	Age	Frequency
Health Counseling and Cancer Checkup	Men and Women	20 - 40	Every 3 years
(to include exams for cancers of the thyroid, testicles, prostate, ovaries, lymph nodes, mouth and skin)	Men and Women	Over 40	Every year
Digital Rectal Exam (manual exam of the rectum)	Men and Women	Over 40	Each year
Stool Guaiac Slide Test (sample of stool examined for the presence of blood)	Men and Women	Over 50	Every year
Sigmoidoscopy (examination of a portion of the large intestine)	Men and Women	50 and over	Every 3 to 5 years
Pap Test (sample of cells from the vagina and cervix)	Women	All women who are, or who have ever been sexually active, or who have reached age 18 should have an annual Pap test and pelvic exam. After a woman has had 3 or more consecutive satisfactory normal annual exams, the Pap test may be performed less frequently at the discretion of her physician.	
Breast Self Exam	Women	20 and over	Every month
Breast Physical Exam	Women	20 to 40 Over 40	Every 3 years Every year
Mammogram (breast x-ray)	Women	35 to 39 40 to 49 50 and over	Baseline Every 1 to 2 years Every year
Endometrial Tissue Sample (sample of tissue from the lining of the uterus)	Women	At menopause for women at high-risk (obesity, failure to ovulate, abnormal uterine bleeding or estrogen therapy)	At menopause for women at high-risk

17

Exhibit 3.5 Communicare Booklet

How Do I Talk About Visual Document Design?

Many of the principles for document design are encapsulated in the above terms, but in addition to these, we need also to consider the visual terms necessary to communicate with visual designers and to be able to design good visuals ourselves. Some of the terminology will be familiar to you, especially if you have had a class in design, advanced technical writing, or visual communication or if you have a background in art or design. Other terms may be unfamiliar to you or used in ways that are not the ways you are accustomed to using them in the past. The following terms are overarching ones that can be applied to most documentation projects: arrangement, chartjunk, chunking, color cuing, color schemes, contrast, conventions, figure-ground separations, focal points, graphics, hues, icons and images, orientation, production, saturation, scale, similarity, value, and visuals. Like the four terms illustrated in the first part of this chapter, these terms are important for you to be able to communicate with other in-house team members or to your outsourced partners. For brief definitions of these visual terms and how they might be applied in health projects, please see Appendix B at the back of this text. You will see these terms used again in Chapter 10 on Presenting Written Materials Visually, but in that chapter, you will be looking more at the design and application of visual communication tools rather than using design vocabulary to communicate to peers and outsourced professionals.

As health communicators, we often use the terms *graphics, visuals,* and *images* interchangeably. Whatever we call them, using such visual communication is very important in technical communication, but even more important in medical writing where patients' well-being often relies upon clear and accurate images. Graphics, like images and visuals, include charts, graphs, photos, line drawings, pictures, icons, and usually anything that is not text. We know that to be able to reach many learning styles, we often need to include graphics to reach visual learners. Graphics do need to match text or illustrate something within the text and need not be used randomly, at least in the Western Euro-American cultures. See the chapter on multicultural design for other perspectives on visual rhetoric.

For example, do include graphics for health education materials and adapt them to the particular audience you are attempting to reach. Few lay audiences within North American culture respond well to text-heavy documents, particularly on the Web; usually only certain professions require such documents. Even professional journal articles are beginning to include more graphics.

When you do include graphics, make sure they illustrate what it is you intend to convey. Sometimes users are confused when graphics don't seem appropriate to the text. Make sure, too, that if a graphic is difficult to decipher (such as the picture of a new piece of equipment), the text carefully explains the parts of the machine. Good labeling and use of diagonal lines drawn between labels and components can be very helpful to your user.

Visuals, like *graphics,* simply refer to the use of visual materials. Technology has greatly enhanced our ability to use visuals, and our audiences tend to rely on them more now than in earlier decades. Many of our health manuals would not make sense without visuals. For example, imagine the instructions for CPR (cardio-pulmonary resuscitation) in text without illustrations or graphics: "Place your other hand on top of the one that is in position. Do not allow your fingers to touch the chest, because that may damage the ribs" (Kemper 1999, 64). Notice how much easier it is to process this instruction with the visual aid (right).

How Do I Manage Large Projects?

There are a number of good models for managing large projects. In this section I suggest two possible models. You'll want to develop a model that works for your project, your workplace, and the team members with whom you work. One of the models is the one that JoAnn Hackos recommends in *Managing Your Documentation Projects;* the other is the proposal model that I use in my advanced technical and medical writing classes. The second model, combining Phase 1 and part of Phase 2 of the Hackos model, has been used in at least 45 projects that developed health materials.

Hackos (28) recommends that project managers divide the project into five phases:

- Information Planning
- Content Specification
- Implementation
- Production
- Evaluation

She also places a percentage of the project development cycle that each phase requires (29).

Exhibit 3.6 CPR Illustration

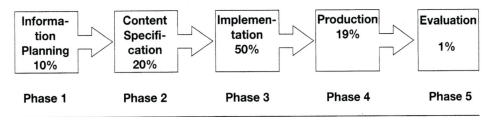

Exhibit 3.7 Hackos' Publications Development Life Cycle

Phase 1: Health Information Planning

Planning health information is a stage some of us leave out when we rush to meet publication deadlines. This omission, however, can lead to projects going badly awry. Although this phase can sometimes be shortened, especially when revising a medical publication that already exists, that prior planning can save you time and resources later.

Hackos divides the information-planning phase into an information plan and a project plan (106). The information plan refers to the needs analysis you perform as well as a description of the project (107). It includes:

- Understanding the goals and objectives of the development project
- Creating the goals and objectives for the publications project
- Analyzing the audience and its environment
- Analyzing the tasks the audience will perform
- Considering the design implications
- Selecting the media

Another way to refer to this part of planning is the research and analysis stage in which you develop a concept for your project. The project plan, according to Hackos, includes the schedule or timeline of the project and the budget. In the second model below, the two parts are combined into one proposal.

For a complete and detailed explanation of the Hackos plan, consult Parts 2 through 6 of her text, *Managing Your Documentation Projects.*

Another Model for Phase 1

It's important to develop a plan or proposal for the project that all team members understand and are willing to work with. Just as nurses develop a plan for patient care, you should develop a plan for your publications. That plan should consider the factors we described in earlier chapters in addition to other components:

- Ethics of the situation
- Audience analysis
- Context analysis
- Timeline
- List of resources both on hand and necessary to procure in order to complete the project

- Persuasive arguments if the publications have not been solicited or if funding is competitive

The model that I use in technical writing classes for health publications is available at www.ncte.org/cccc/servicelearning/clemson. The formatting suggested by this document is not following any particular standard, but rather gives you the opportunity to use graphics and tables and practice other document formatting that will be useful to you in the planning stages. Again, you will want to adapt this document to the particular needs of your project.

The Second Part of Phase 1: Timeline or Schedule

There are many ways to show a timeline or schedule. Often proposal writers choose tables or Gantt charts to represent the time each phase of a project will take.

The more detailed you can make your timeline, the more helpful it will be for you. Also, writing the research agenda can be a good method for determining the various steps you will need to find the information in the early parts of the timeline. Don't assume that once you've done much of the research that there won't be occasions in which you need to reexamine data or do further research, especially in the fast-paced information era in which we live.

Below is another example of a timeline. While it is an effective example of a chart, the information is very general and would be appropriate to any documentation project. Sometimes it is helpful to develop both a generic timeline and a more detailed one for your day-to-day operations.

Small Pox Education Plan

2-May-04	Begin Planning Small Pox Ed Campaign
9-May-04	Determine Experts to Interview
16-May-04	Begin Interviews
23-May-04	Set up Surveys for Audience Analysis
12-Jun-04	Analyze Data and Match with Government Goals
17-Jun-04	Begin Drafting Components of Media
30-Jun-04	Agree on Content of Various Media
15-Jul-04	Beta Drafts of All Sub-projects and Docs
16-Jul-04	Begin User Testing of All Sub-projects
30-Jul-04	Present Alpha Drafts of All Sub-projects
15-Aug-04	Have Media Packages Ready to Deliver to Printers
15-Sep-04	Distribution of Campaign Materials
20-Sep-04	Project Review
30-Dec-04	Surveys of Target Audiences
15-Jan-05	First Analysis and Evaluation

Exhibit 3.8 One Example of a Simple Table Timeline

Now that you have a plan or proposal in place, let's examine the other phases of managing a large communication project within your health profession.

Phase 2: Content Specification

To use the Hackos process model, in Phase 2 you develop a more specific plan of what information you need to put in your publications, whether you are working in print or digital publishing. A detailed plan helps you manage the large amount of information at our fingertips in the Information Age. This part of the project

Project Activities for Diabetes Research Partners Web Site	1	2	3	4	5	6	7	8	9	10	11
Interview Clients re: Health Web Site	■										
Storyboard Health Web Site		■	■								
Develop Templates for Site			■	■							
Usability Test with Patients				■							
Gather Data from Medical Partners					■	■					
Develop Content from Each Partner and Mock Up Pages						■	■				
Review with Client							■				
Make Revisions to Research Data and Partner Sites								■			
Double-check Patient Info, Review Passwords and Security System										■	
Launch Site and Conduct Project and Team Evaluation											■

Exhibit 3.9 Gantt Chart

can only begin after you have gathered enough information and done the analysis of your audience.

In the author's model (see Web site), content specification comes toward the end of the plan as Step VII: "Written Presentation or Final Layout of Project." Either model works as a malleable document, one that changes as your research reveals the important information that will most appeal to your target audience. For example, if you are describing a particular disease such as AIDS, and advocating frequent testing, your target audience won't need or want to know every detail about the history of the HIV virus or exactly what research is being done to find a cure. One of the dangers in large documentation projects is including too much information that is not relevant to your audience. Just because it is important or interesting to you, it may not be an effective way to reach your user. On the other hand, if you are writing a proposal for a grant, the historical and contextual research may provide a valuable background to let the funding agency know you have done your homework and plugged into the current research network.

Some beginning students or communicators assume that once plans are developed, they must be followed to the letter. That's why we emphasize that these phases of the publication process are flexible. Those of us who have completed numerous projects know that you must be able to change the plan as the project develops. In addition, with experience, you will also be better able to predict timelines, know the pitfalls, and understand what information to use for what audience.

Senior designers, who have the experience to forecast well, may plan large projects. However, in smaller workplaces, you may be the leader of a small team or you may work alone by default. Just because you don't have a team of several or many members does not mean you do not need to complete the planning phases of your project. In fact, the planning stages are of maximum importance in such a situation.

If you do have a team of several members, you will want to develop a tracking system that enables you to determine who is responsible for the tasks and content of your project. That should not be just a top-down decision, unless it involves detailed knowledge only particular practitioners would know. Ideally, you will develop this system in a team meeting, so that members have a say in and agree to complete their particular parts of the project. Each member of the team should have the planning document and timeline to refer to throughout the project.

The content of your project will likely involve graphics as well as text. Just as you storyboard for Web sites, even in print publications, drawing rough thumbnail sketches will also help you plan. Using the vocabulary for design that you used above, you can now do a detailed description of the many aspects of document designs. Exhibit 3.10 is an example of the planning of the final layout from a student health education project.

Phase 3: Implementation

Implementation is less easy to document and plan, but project managers and designers should be able to follow the general milestones on the timeline.

**Health Communication
Certificate Program**

- Bold, black, 16-point, Bookman Old Style font

Form that can be filled out to receive applications and other information. Bold, black, Arial, 10-point font

A form to request additional information and an application

Core Courses

Bold, orange, 14-point, Bookman Old Style font

- A bulleted list of the courses required in the program

- Arial, blue, 10-point font, aligned left

Electives

Bold, orange, 14-point, Bookman Old Style font

- A list (not bulleted) of some of the possible electives

- Arial, blue, 10-point font, aligned left

Financial Aid

Bold, orange, 14-point, Bookman Old Style font

- A bulleted list of the financial forms available

Graphic

Mission Statement

Bold, orange, 14-point, Bookman Old Style font

One or two paragraphs describing the mission and purpose of the HCC program, as well as the benefits.

- Arial, blue, 10-point font, aligned left

Graphic

Qualifications

Bold, orange, 14-point, Bookman Old Style font

- A bulleted list of the requirements for getting into the HCC program

- Arial, blue, 10-point font, aligned left

Exhibit 3.10 Plan for Final Layout of Student Project

Depending on the range and complexity of the project, you can set up a system of memos, team meetings, progress reports, and other accountability measures to ensure that each part of the project is being completed on time. In addition, because your work involves human welfare and life-and-death information, you will want to pay careful attention to quality control or accuracy. You may need to schedule team editing and review sessions or hire consultants to ensure that the information is correct.

Just as you want to communicate carefully with your audience, you will also want to keep the lines of communication clear among your health team members. The more skillfully you can communicate with team members, the better your project quality and delivery time will be.

Set up several draft reviewing sessions for the project. Whether you have two drafts (alpha and beta drafts) or more, make sure you have remembered to include reviewers from your target audience so that you communicate the correct messages to your readers. Whether you do this informally, with focus groups, surveys, or formal usability tests, make sure the feedback loop is in place if possible. Because some of our work requires very quick turnaround times, it may be difficult to complete much usability testing. But given our technical capability, you can email attachments to representatives of your audience and ask them for informal and quick reviews. If you have funding available, offering some compensation for the reviews can ensure thorough reviews and compliance with your deadlines.

Make sure all team members receive the client feedback and help them plan revision strategies if adjustments are necessary. For example, if audience members are offended by a particular graphic, make sure the illustrator is aware of this reaction and can develop another graphic and substitute it for the offending one. One group of my students developed a food stamp application brochure that pictured African American faces of all one shade. Feedback from African American clients suggested that representing people this way was offensive. Rather than developing graphics of many different skin shades and possibly offending other groups then left out, the students decided that the brochure could be as compelling and chance less offense by using food graphics instead of people's faces. If we hadn't had the feedback, the agency for which we were designing could have ended up with 50,000 offensive brochures distributed throughout the state.

Phase 4: Production

Depending upon your workplace, you may or may not be involved in the actual production. Even larger hospital systems outsource this function, but some institutions do have in-house production capability, especially with the development of desktop publishing. If you outsource or send your materials to another department within your workplace, you will find the design vocabulary and planning documents especially helpful. Be sure to communicate the project designs clearly to the producers. Make sure you have formatted raw files to fit the needs of production staff members. The project should be planned with these needs in mind. Otherwise you will spend much time reformatting documents that took much time to format in the first place, duplicating your work rather than working efficiently. If you are working in print, you will need to consider time for copy-editing, indexing, reviewing page proofs or galleys, printing and binding, and distribution. If you are working in an electronic media, you will still need to pay careful attention to proofreading. Also, you'll want to view Web sites in both browsers:—Netscape Navigator and Internet Explorer—because hues, saturations, sizes, and spacing can be different from software to software.

If you are doing the production, especially if it is a large health project or involves a number of people, you might want to use a more detailed timeline (see Exhibits 3.8 and 3.9) just to map out the stages of the production itself. Some teams working on large projects prefer using an electronic system or intranet shared among team members to track the stages of the production itself. This system helps keep all participants in the loop, as they can view the progress of the project readily on their own screens. You will want to tailor the communication process of document design according to your own organization's needs, resources, and the size of the project.

Phase Five: Project and Performance Evaluation

This phase, although coming at the end of the process, should be planned for in Phase 1 (see Exhibits 3.8 and 3.9). Part of your plan should include some assessments that determine the effectiveness of your project. The second model of Phase 1 does not require evaluation, because it is used in academic settings, which have short-term, semester-length timelines.

Project Evaluation

There are as many kinds of assessments as there are projects, but often surveys are used to document effectiveness. In some cases, you might be able to set up pre- and post-surveys for your sample audience. (See Hackos 1994, Lauer and Asher 1988, and MacNealy 1999 for good information on survey design.) For Web sites, you might want to institute usability testing, although this method takes time, and if you need to use patients, they may be unwilling or unable to work in lab settings. In this case, field-testing at clinics, doctors' offices, and patient rooms may be more feasible. Ideally, usability is instituted at various, earlier times during the production process, so as to include audience feedback in the revisions of documents. See Chapter 1 on Audience Analysis for more information on field-testing and usability.

Because this type of evaluation and assessment involves the effectiveness of the materials you developed, there are also numerous ways to obtain outside assistance in your evaluation, such as in assessment journals, with consultants, or in the case of public health campaigns, in Chapter 7 on such campaigns in this text.

Here, briefly, are questions that you need to use to evaluate your deliverable (print, electronic, video, or audio) document. These questions can be refined depending on the medium that you use:

- Is the document accurate?
- Is the project well designed and professionally produced?

- Will your readers or users find your deliverable(s) credible?
- Will your audience be able to prioritize the information based on your layout and design?
- Will readers or users easily navigate through the information, whether on a Web site or within a printed document?
- Is the information logically or intuitively accessible?
- Does the document deliver the intended message and does it truly represent the data?

Performance Evaluation

Another part of the evaluation process is a debriefing with your team, or, if you work alone, analyzing your own performances and process of completing the project. Sample questions for this process are:

- Was the timeline accurate?
- What problems were encountered during this project?
- What phases went smoothly, and which ones were problematic?
- Why did problems occur, and how were they solved?
- What were the strengths of the process? In short, what worked well?
- What lessons, changes, or improvements will you make to have the next project go more smoothly?
- Did the project stay within the proposed budget?
- Does the project meet the specifications spelled out in the planning stages?

If you have used a detailed project tracking system, it will be easier to evaluate your performances. You will be readily able to see what problem cropped up, where you fell behind schedule, or who did not deliver on their part of the process. Depending on where you work, you may be required to submit a report on the project. The more data you have collected over the course of the project, the easier the report writing will be. In other work sites, you may do this debriefing and analysis informally.

How Do I Manage Communication Flow?

You may be asked as a professional health writer to solve communication problems within your own or other organizations. If there are glitches in the communication flow of documents within or among various departments of a hospital, HMO, or health insurance company, for example, you need to know where to start. Most professional communicators use a mapping technique to understand how documents and information move through an organization.

Within a work site, think about who communicates with whom. How do the staff members communicate with each other: through memos, email, frequent face-to-face meetings, or conference calls? Trace how and when communication takes place. You can discover the process through on-site observations, interviews, and, if you have staff and time, through shadowing the key players in the various departments or areas you need to study. Exhibit 3.11 is a simple map of a heart catheterization lab that begins this process of mapping out the communication flow.

Once you have a communication process mapped out, you can begin to examine where it breaks down. Through your interviews, you can determine if a certain department needs to change its method of communicating to another department. For example, in the map above, does the catheterization lab clerk or scheduler need to keep a hard copy casebook, inform the catheterization team verbally, and/or post the data to the radiology department and department of nuclear medicine? Perhaps, if this process takes too much time, you could suggest the catheterization lab team also have access to the intranet and all parties involved could view the scheduled cases at the same time.

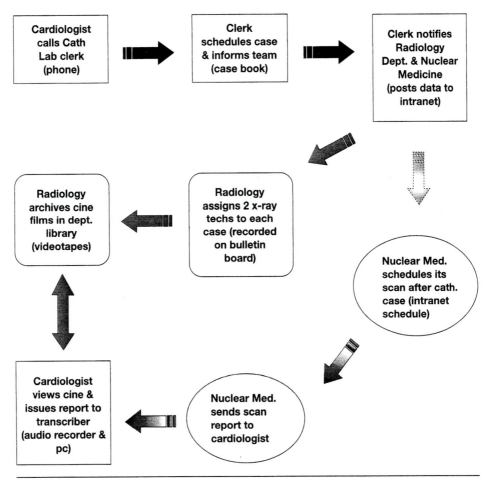

Exhibit 3.11 Communication Flow Map[3]

Summary

At the beginning of this chapter you briefly reviewed the process you go through before you design any health message. You examine the:

- Purpose
- Audience
- Ethics

In addition you learned that in order to make your information architecture successful, it must be credible to your audience, and you must be able to prioritize (or *triage*) the information so that users know what is important. Navigation is another important issue in health message design; help your readers work through information clearly and easily whether they are looking at a poster, scanning a Web site, or reading a research report. Also, remember that data can misrepresent information if it is not well designed.

Document design principles are also key to redesigning good documents. When you do not use predetermined formats such as those used in preprinted forms, word processing programs, and templates, you need to consider overall *layout, headers, margins* and/or *white space*, and *typography*.

- **Layout** helps you visualize the overall structure and ease of navigation of your documents.
- **Headers** are also useful for reader cues to structure and set up a hierarchy of information.
- **Margins and/or white space** make the documents visually appealing and eases user processing of information by chunking text and visual cues.
- **Typography** needs to be carefully considered for processing, too, to aid users in readability and style.

In the section about visual communication and in Appendix B, you learned an important design vocabulary so that you can communicate your projects to both outsourced and in-house printers and publishers. These terms included the following: arrangement, chartjunk, chunking, contrast, color cuing, color schemes, conventions, figure-ground separation, focal points, graphics, hues, icons, orientation, palette, production, saturation, scale, similarity, value, and visuals. Keeping in mind the various examples and explanations of these terms, you will be able to communicate design ideas well and develop projects that turn out the way you envisioned them. Otherwise, if you don't do the detailed and accurate descriptions for your team members, you end up with projects that have to be redone, resulting in cost overruns, wasted time, and late deadlines.

After understanding the terms that apply to a single document or project, you can now apply those to larger projects. In the last sections of this chapter, you learned about setting up and managing a large documentation project or working with a large communication system.

Two planning models gave you different ways to proceed in mapping out your projects and proposals. The Hackos model introduced a five-phase plan. The five phases are: Information Planning, Content Specification, Implementation,

Production, and Evaluation. The second proposal model, which combines Phase 1 and part of Phase 2, gave you an alternative way to write up the planning stages in one document. You can also adapt and combine the various phases and models to your particular project. You learned that the planning stage is the one some designers try to shortcut, only to cause larger delays and expenditures over the life cycle of the project development.

Hackos Phase One includes the following components:

- Ethics of the situation
- Audience analysis
- Context analysis
- Timeline
- List of resources both on hand and necessary to procure to complete the project
- Persuasive arguments if the publications have not been solicited or if funding is competitive

The **second planning model** includes these components:

- Letter of transmittal or cover letter (or memo)
- Description of project
- Audience analysis
- Research agenda
- Timeline
- Budget
- Oral presentation
- Written presentation or final layout of project
- Deliverable (formatting instructions and delivery instructions if necessary)

You learned the importance of developing a timeline to help manage your project; examples included a table and a Gantt chart.

Phase 2 under the Hackos model, Content Specification, is a step usually done after research has been gathered, unless you are already familiar with the content. You have learned that both text and graphics need to be specified for your project. The more work you do up front in these first two phases, the easier the implementation of the project will be.

In the implementation stage, **Phase 3**, you learned it was important to adhere to the timeline and use it to guide your process. In addition, you learned that the feedback loop or audience participation in the document design process is an important component of designing effective documents, whether they are print-based or electronic.

The production phase, **Phase 4**, varies greatly, depending on whether or not you are doing the producing or outsourcing it. You learned that clear communication to publishers is required if you plan to outsource the project using the design vocabulary and clear formatting guidelines.

The last phase, assessment, includes both the assessment and evaluation of the efficacy of the project in reaching its audience as well as the performance of the group members (or your own performance) during the project you just completed. This reflective process is key to developing the most efficient and successful project management and good document design procedures for future endeavors.

In the final section on managing the larger communication flow, you learned that mapping out and developing good research methods are important to successfully understanding a communication process you might be asked to evaluate. Often problems are discovered via a simple mapping procedure. Small changes in the communication project can often solve large communication bottlenecks that seem insoluble to other staff members.

Discussion Questions

1. What are the first questions to ask yourself before you begin a document design project? Give examples of why such questions are necessary before you begin the process.
2. Why are document design considerations important? Why would the access of information be important, and how could issues such as white space, font choices, and graphics make a difference in the world of the health professional?
3. The CDC has asked you to chart the incidence of brain tumors in southwestern Georgia. What kind of graph, chart, or table will you use for this information and why? Show how this discussion can begin a project planning session.
4. You are given a preprinted charting form to describe the location of a patient's unusual rashes. There is a 4" by 6" box for you to complete the description. Will you decide to use words only, draw pictures, or do a combination of both to send to the dermatologist? Why did you make the choices you made?
5. Brainstorm the various ways of assessing your project process and discuss your choices for evaluating a health information billboard, a Web site explaining a new medical procedure, and a preventative health information campaign.

Exercises

1. As a health communicator, you need to write a memo regarding the specifications for a five-page brochure on a new kidney center in your community. Write and include a diagram of the layout (including placement of graphics and text), headers, spacing, and typography for this brochure. Be as specific as possible in your instructions.
2. Research and design a new Web site for a medical innovation. Storyboard the information, instructions, and diagrams for the site. Include a list of font choices and information hierarchy. Your presenter will be a prominent physician and your audience will be the hospital administration. The physician will use the explanation of the innovation to ask for funding.

3. Interview health professionals who have completed successful projects. Ask them the debriefing-evaluation questions. Ask them how they went about evaluating the efficacy of the process. Prepare a brief report (three pages) including quotes from the interview.

4. As a class, have your instructor assign teams or choose your own teams (maximum four people). Choose a health project to work on (either real or hypothetical). Have half of your teams draw up the proposal/plan using the Hackos model and the other half draw up the proposal/plan using the other model. How are the plans different and how are they the same? What type works best for the project? What are the variations within the same model as prepared by different teams?

5. Make a timeline using four different formats; use both a word processing program and a drawing program on your computer. Present these to the class and discuss which ones are easier to read and why.

6. On teams of no more than five people, choose a brochure or poster and develop a list of questions to ask the audience about its design and readability. Then set up a field test in which you access a reasonable number of audience members and ask your document questions. Record or write down answers and prepare a PowerPoint presentation of your findings. Your audience is a document design team making improvements to the present documents.

Works Cited

Berger, John. 1972. *Ways of Seeing*. London: British Broadcasting Corporation and Penguin.

Hackos, JoAnn T. 1994. *Managing Your Documentation Projects*. New York: John Wiley & Sons.

Kemper, Donald W., and the 'Healthwise Staff.' Eds. Jody G. Bower and Steven L. Schneider, M.D. 1999. *Partners Healthwise Handbook*, 13th ed. Boise: Healthwise Incorporated.

Kostelnick, Charles, and David D. Roberts. 1998. *Designing Visual Language: Strategies for Professional Communicators*. Boston: Allyn and Bacon.

Lauer, Janice M., and J. William Asher. 1988. *Composition Research: Empirical Designs*. New York: Oxford University Press.

Lay, Mary M., Billy J. Wahlstrom, Stephen Doheny-Farina, Ann Hill Duin, Sherry Burguss Little, Carolyn D. Rude, Cynthia L. Selfe, and Jack Selzer. 1995. *Technical Communication*. Chicago: Irwin.

MacNealy, Mary. 1999. *Strategies for Empirical Research in Writing*. Needham, MA: Allyn & Bacon.

National Council of Teachers (NCTE) Web Site. Updated September 8, 2002. www.ncte.org/cccc/servicelearning/clemson

Schriver, Karen A. 1997. *Dynamics in Document Design: Creating Texts for Readers*. New York: Wiley.

Tufte, Edward R. 1997. *Visual Explanations: Images and Quantities, Evidence and Narrative*. Cheshire, CT: Graphics.

Endnotes

1. Because many of the documents we develop are forms and/or may include visuals, you will see an overlap here between document design vocabulary and visual communication or graphic communication vocabulary.
2. Some technical writers (such as Mary Lay et al.) refer to layout as grid layout, finding it helpful for students to visualize the layout in a grid pattern.
3. "Cath" is an abbreviation for catheterization, and, in this case, is referring to a cardiac or heart catheterization.

Medical Diagnostic Practices and Charting

 Overview

Chapter 4 covers charting because it serves as the main source of documentation directing the practice of medicine on individual patients and, in fact, helps health professionals form diagnoses. Even if you are not a physician, nurse, nurse practitioner, or physician's assistant, as a medical writer or health communicator, you should be familiar with patient charts. Physical therapists, phlebotomists, radiology technicians, and others may need to consult the chart to complete their own part of the patient care. In the first part of the chapter, several of the most widespread systems of charting are featured. You will need an overview of both print systems as well as the computer-based systems in which computer terminals are set up in each patient room and data are entered on the electronic chart by nurses and others. On-line charting systems are rapidly changing, but I will describe the more generic functions of electronic writing in patient charts. These new on-line capabilities have been one of the most important changes in medical documentation.

The writing of patient histories and physicals (HXP) also falls within this chapter's purview. All clinical visits begin with a taking of histories and physicals to begin the charting process and to enable the formation of diagnoses. Often physicians dictate their HXP information, and transcriptionists then listen to the tapes and log the data into the computers. All of this information leads to the testing, diagnoses, and treatments for each patient.

In this chapter you will find the answers to the following questions:

- What should I know about patient charts?
- What is clinical decision making (CDM)?
- What is the heuristic that makes up a chart?
- What are tentative diagnoses?
- What does a patient history include?

- What does the physical examination include?
- How is the physical examination recorded?
- What is SOAP?
- What are the capabilities of on-line charting?

What Should I Know About Patient Charts?

In this part of the chapter I'll discuss the charts that are not computerized, then move in the last section to on-line patient charts. (Consult Appendixes A and C about medical language conventions.)

In the examples that follow, in many doctor's offices, and even in some hospitals, parts of patient charts are still handwritten, often on preprinted forms. Most physicians and even some other health professionals who interview patients use dictating devices (most often handheld microcassette recorders) to record their histories and physicals (HXPs), as well as their diagnoses and treatments. The data are then transcribed by medical transcriptionists and put into the charts. Many health professionals look forward to the day when the voice recognition software will be more widely used. Perhaps that day is not too far away, but the recognition has to be close to flawless. Because the capability is new, all the results are not yet in (see p. 109).

For those who use some form of handwritten charting, legibility is key, both for avoiding confusion among staff as well as to protect the writer as well as the reader from mistakes and liability. As health professionals, you often record your observations. It's usually necessary to use the jargon and abbreviations common within your particular specialty, those terms that are understandable to those with whom you most frequently communicate. Emotion in charts is unadvisable; if you have particular difficulties with patients, it's best to air those in your team meetings with other staff members (Navarra 1998, 52). Do be aware that patient charts can be subpoenaed for legal cases.

Patient charts function in a number of ways, including as collections of clinical laboratory requests and reports of blood tests, microbiological reports of cultures, pathology reports of tissue analyses, and procedural reports from radiological and other patient testing. All of these data plus the history and physical (HXP), office visits, and daily charting, if the patient is admitted, make up a form of case study. This is one reason physicians regularly refer to their patients as *cases*.

Nurse practitioners and nurses often view the clinical case as the site for clinical decision making (CDM), in which hypotheses about an individual patient's health are formulated and revised (Barrows and Pickell 1991). As Barrows and Pickell also point out, clinical decision making is the medical scientist's or health professional's scientific method. As the site of the written observances and multivalent perspectives (through testing), the chart is key to the process, not only as a reflection or representation of the process, but as a generative entity or invention strategy in itself (Barrows and Pickell 1991).

What Is Clinical Decision Making (CDM)?

In addition to ensuring that the chart remains a site of invention, it must also enable and reflect both the critical thinking and clinical decision making (CDM) in the process of health assessment. Denise L. Robinson maintains in *Clinical Decision Making: A Case Study Approach* that while critical thinking and clinical decision making have some of the same characteristics, such as synthesis, analysis, questioning, and identifying of assumptions, there are differences between the two methods (2002, 2, 3). Those differences are "convoluted," but important to distinguish, because it is those differences that also make a patient chart what it needs to be (2002, 4).

Critical thinking certainly forms the basis of clinical decision making, and both the obvious solutions to problems or diagnoses of health situations need to be considered as well as those that are less obvious (Robinson 2002, 5). Barbara Bates, whose text is cited in Robinson's text, recommends that to get at these less-than-obvious solutions, the health professional should "cluster" the symptoms that don't fit the usual or anticipated data (Robinson 2002, 5). The health professionals, particularly nurse practitioners, physician assistants, emergency medical technicians, and physicians, then go through the process of rejecting and accepting (then revising) diagnoses. "To not examine the data critically is a disservice to the client" (Robinson 2002, 5). A healthy skepticism, or questioning seen as positive not negative, is a trait that the health professional needs to have when going into the diagnostic situation. It's also important not to import one's personal biases and assumptions into clinical settings in such a way as to skew the treatment plans, and when working with patients, a more compassionate, gentle, and subtle approach is the more successful one (Robinson 2002, 5). An example of our personal biases and assumptions interrupting appropriate treatment follows. A nurse practitioner (NP)

> who assumes recurrent head lice in a child is the result of an uncaring or lazy mother is making an assumption about the mother based not necessarily on fact but rather on illogical premises about mothers. This NP is not using all of the nursing knowledge available to critically evaluate why the child has a recurrent condition. When using all of that knowledge, the NP may discover that the mother is unaware of how head lice are transmitted and the importance of thorough bed/house cleaning in addition to the hair treatment. The NP's personal value system may have provoked the attitude that lazy, uncaring mothers do not treat the head lice in the first place. (Robinson 2002, 5)

The situation outlined above calls for a plan based on patient education, and information needs to be supplied to educate the parents of the child, rather than blaming them. The education could be in the form of print materials (brochures, pamphlets, instruction booklets, or charts) or in electronic or video format. If information is not available, or if the patient(s) and parent(s) are from another culture and the existing materials do not communicate cross-culturally, then the medical education or publications departments could design new materials to address the problem. The patient chart would record this decision making and point to an appropriate treatment. By being a site of healthy skepticism and appropriate critical thinking, the chart functions, too, as a method of keeping our personal biases at bay.

In order to see how critical thinking and reasoning work in the clinical setting, we can follow the process by parsing it out. The process also shares many of the characteristics that we see as the basis of technical communication, especially when that basis is predicated on rhetorical theory or a background in rhetoric. Robinson suggests the process includes the following components: "purpose; question or issue; perspective or point of view; data; assumptions; concepts; inference; implication; and consequences" (2002, 6).

- **Purpose**—What is the purpose of conveying the information you want to communicate? Are you trying to decide whether to suggest diagnostic tests? Are you writing to inform the next shift of nurses of significant changes in a patient's psychological attitude toward her disease? The possibilities here are endless.
- **Question or issue**—In addition to routine communication, what question or issue are you working with? This may not be clear from the other data in the charts, such as laboratory results, physical therapy outcomes, etc. If it's not clear what is happening in this situation, communicate the issue to your audience clearly.
- **Perspective or point of view**—Sometimes we record data from a physiological point of view, sometimes from a psychological, economic, or cultural point of view. Don't forget that the patient has a point of view, too. Be sure you consider all these things as you record findings in the chart.
- **Data**—Charts are usually divided into sections. Make sure you record information in the proper sections of the chart. Clarity and accuracy are vital.
- **Assumptions**—Assumptions or conclusions that are based on too little or on biased data can seriously delay or cancel necessary treatment or preventative care. Be careful about drawing conclusions or making assumptions based on inaccurate or culturally biased information.
- **Concepts**—Concepts refer to the background ideas underlying your treatments and reporting within the charts. As you record your clinical decision making, you want to ensure that the conceptual framework within which you work is appropriate to the particular health situation.
- **Inference**—Make sure that what you infer from the data you have recorded from patient observations (and history and physicals) is accurate. In addition, the information should be clear to the next practitioner or health professional who accesses the chart.
- **Implication**—Implication, like diagnosis, regards the clear results coming from your data; unlike inference, which arises from the practitioner based on data, the implication is what you have decided. This decision or implication suggests actions for others to follow.
- **Consequences**—What do you expect as a result of the decision you have made? If the chart acts as a site for synthesis and critical thinking as well as clinical decision making, we ought to see what consequences you predict as happening in the future. Of course most of the fundamental consequences will be restoration to health and a state of wellness. However, sometimes the consequences are short-term or in stages. For example, you might decide to institute a series of intravenous drips in order to change

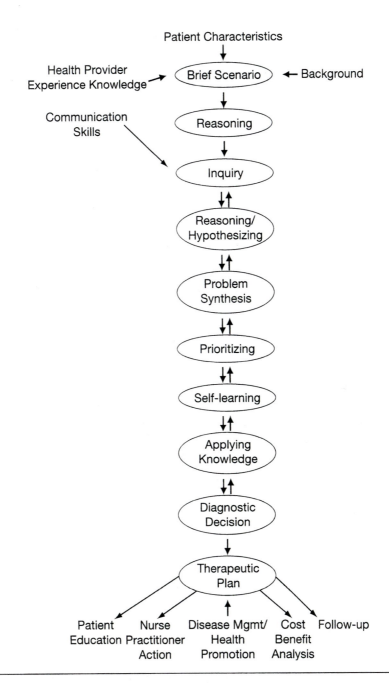

Exhibit 4.1 Schematic of Clinical Decision Making

Source: Used with permission from Lippincott Publishers in Denise L. Robinson's *Clinical Decision Making,* 2nd ed., 2002.

the status of or to stabilize a patient, but it will still be only one of many steps in the process of restoring health.

Exhibit 4.1 illustrates the decision-making process, all of which is charted.

What Is the Heuristic[1] That Makes Up a Chart?

Now that we have an idea of the process that goes into clinical decision making, let's see what kinds of headers make up nursing plans and charts. Because we want to use a heuristic for the information, we will detail the questions underlying the various categories:

- **Tentative Diagnoses:** The first part of the interaction between patient and health professional advances a tentative diagnosis such as "Possible OM." *OM* is a standard abbreviation for otitis media (or middle ear infection). (See Appendix C.)
- **History:** Given the presentation of the patient (how they look and what they are complaining about), what are the important questions to ask? What systems should be covered in the history based on the tentative diagnosis?
- **Physical Examination:** What portions of the physical exam are necessary to complete given the patient's situation?
- **Differential Diagnoses:** What data have you confirmed that support or argue against your tentative diagnosis? Here again, a rhetorical point of view is helpful if it is part of your background, because the argument style will already be embedded in your writing/thinking background.
- **Diagnostic Tests:** Based on the history and physical (HXP), what tests do you need to order (or consult) to verify or refute the diagnosis?
- **Diagnosis:** What diagnosis(es) is appropriate given the information gleaned from the earlier steps? What data support this diagnosis(es)?
- **Therapeutic Plan:** What treatment(s) will you recommend? What patient (or parental) education needs to be instituted? Do specialists need to be brought in or consulted? What about follow-up visit(s)? Will further treatment plans be necessary?

What Are Tentative Diagnoses?

We will follow a patient through the heuristic outlined above to see how the data is charted (based loosely on Robinson 2002, 69–85).

In this scenario, a 35-year-old female presents with diffuse joint pain. The onset was several months ago. She c/o (complains of) stiffness in the a.m. and fatigue at the end of the day. She works part-time as a nurse and is married w one child.

Here are possible tentative diagnoses before the history is taken:

- Bacterial endocarditis
- Depression (can be secondary to chronic illness)

- Fibromyalgia
- Hypothyroidism
- Lyme disease
- Rheumatoid arthritis
- Septic arthritis
- Systemic lupus erythematosus
- Viral infection possibly including "rubella, human parovirus B19, acute hepatitis B, HIV, and some enteroviruses" (69)

What Does a Patient History Include?

Next the health professional takes the patient's history. Many abbreviations are used in both handwritten as well as computerized charts. See Appendix C for the most commonly used abbreviations. In this instance, the full term is in parentheses after the abbreviation.

History

Requested Data	Responses
Allergies	*PCN (Penicillin).*
Current medications	*BCP, MVI qd,* and Advil every 4 hr for joint pain. *(Birth control pill, multivitamin every day)*
Surgeries/transfusions	C-section x 1, no transfusions.
Medical history and hospitalizations/ fractures/injuries/accidents	Chicken pox, mumps, measles as a child. *MVA* 1 yr ago, whiplash injury of neck *(Motor vehicle accident).*
OB/GYN history	*LNMP* 1 wk ago; normal flow and duration *(Last normal monthly period).*
	Menarche, age 13. Last pelvic 6 mos. ago. *WNL (Within normal limits).* Mammogram: none.
	Sexually active; HX 3 partners—used barrier method. Monogamous relationship w husband; no barrier method used. Denies vaginal dryness, discharge, odor, or dyspareunia.
Appetite/Weight change:	No change in appetite. Loss of 10 lbs in 2 mos.
24-hour recall	B: Toast and coffee w milk. L: Turkey sandwich, fruit, coffee with cream. D: Meat, vegetable, starch, pop. S: ice cream.
Family history	Father: Age 60, good health. Mother: Age 58, *RA* since age 45. *HX* osteoporosis *(Rheumatoid arthritis; history).* Sisters 2: Ages 31 and 33. Good health. Child: daughter age 4, in good health. No family history of psoriasis, *IBS (inflammatory bowel disease)* or iritis.

(continued)

Requested Data	Responses
Social history	No tobacco use. Drinks *ETOH* socially (*ethyl alcohol*); wine 1–2 x/wk. No recreational drugs. 2–3 cups coffee/day; 2–3 colas. No regular exercise. Married 5 yrs.; 1 child, 4 yrs old. Immediate family assisting with child care during illness. Works part time as RN on cardiac unit.
Religion	Attends Catholic church regularly.
Income/insurance/home	Husband works as mortgage loan initiator, salary based on commission. Medical and dental insurance through husband's job. Own home.
Stress management	Occasionally walks. Prayer. Enjoys crafts.
Functional status	Has difficulty with fine hand movement because of stiffness and swelling. Decreased strength makes opening jars and picking up child difficult. Dressing self and child becoming more difficult. Painful to walk distances, which has restricted activities, such as leisurely walks and shopping. Unable to do crafts because of pain. Fatigue alters energy level; too tired after work to do activities with child. Tries to make up time on days off. Naps 1–2 hours qd late afternoon. No libido related to fatigue. Family helping with housework and childcare.
History of symptoms	Joint pain started a few months ago. Pain in 2nd, 3rd, and 4th *MCP* and *PIP* (*metacarpal phalanx* and *proximal interphalangeal*) joints at an intensity of "6–8" on a 1–10 pain scale. Has noticed swelling in the hands. Morning stiffness lasts up to 2 hours, relieved by hot shower in AM. Takes Advil q 4 hrs with minimal relief. Also has pain in the balls of both feet aggravated by walking. Some relief with rubbing them. Only wears flats; pain aggravated by heels. Rates pain at "8" on scale with pressure. Symptoms have gotten steadily worse over last several months. Starting to interfere with *ADLs* and lifestyle (*Activities of Daily Life*).
Recent travel	None. No camping, hiking or gardening.
Exposure	Minimal exposure to needles at work.
ENT	Last eye exam: 1 yr ago. No dryness, inflammation or infection of eyes. Wears contacts; denies photosensitivity. Denies dysphagia, dry mouth, or oral lesions (*Ear, nose, throat*).
Skin	Dry, no rashes, lesions, or bites. Denies heat or cold intolerance.
Chest	Denies *SOB*, cough or pain (*Shortness of breath*).
Cardiac	Denies chest pain or recent viral infections.
GU	No history of infections. Denies frequency, dysuria. Color of urine is yellow (*Genitourinary*).
GI	No c/o of constipation or diarrhea. BM pattern once qd, formed, soft stool, and no blood (*Gastrointestinal*).

Requested Data	Responses
Musculoskeletal	Pain and swelling in hands, wrists, elbows, and feet. No deformities. Decreased strength in hands. Decreased fine motor movement. Unable to walk distances because of pain in feet.
Neurological	No *H/A,* loss of memory. Denies tingling or numbness in extremities.
No recent sickness in family	

What Does the Physical Examination Include?

Once the history has been taken, health professionals, whether nurse practitioners, physicians' assistants, nurses, or physicians, initiate the physical exam and record those findings. That format, provided we are not in an emergency situation, as happens with paramedics, EMTs and emergency room doctors, follows this pattern:

- **Vital signs,** which include:
  Height (Chart abbreviation = *Ht.*)
  Weight (Chart abbreviation = *Wt.*)
  Temperature (Chart abbreviation = *T*)
  Respirations—counted per minute (Chart abbreviation = *RR*)
  Pulse/Heart rate—counted per minute and rhythm noted[2] (Chart abbreviation = *HR*)
  Blood pressure—may be taken standing, sitting, and/or lying using an electronic cuff or a manual one (Chart abbreviation = *BP*)
- **General appearance/Skin,** which includes:
  General thinness, obesity, or normal appearance noted
  Color can be remarked upon: pale, wan, flushed, jaundiced, etc.
  Alertness and orientation (Chart abbreviation = *NAD*)[3]
  Skin is noted, too, as to whether it is warm or cool, dry or moist, inflamed, clear of rashes, scratches, lesions, etc.
  Nails should be examined for general nutritional and other clinical information: whether nails are splintered, grooved, clubbed, hemorrhagic, etc.
- **HEENT** (hair, eyes, ears, nose, and throat)
  Hair, especially if markedly thin, distressed looking (e.g., in anorexia, hair thins dramatically and lanugo grows on the body).
  Eyes are examined for dryness, tearing, inflammation, yellowing, discharge, and other signs of infection.
  Ears are checked for infection and inflammation, too.
  Nose—ditto.
  Throats reveal much information, and the mouth and neck are included in this examination as well. Here practitioners are looking at color, signs of infection or inflammation, lesions, swollen lymph nodes in the neck and near the base of the ears.[4]

- **Lungs**
  Lungs are listened to both from the back and the front for many symptoms and problems, too numerous to mention here. (Chart abbreviation is typically *LCTA* = lungs clear to auscultation.)
  Bronchial clarity is noted, too.
  Both chest and thoracic back can be palpated for tenderness, etc.
- **Cardiac**
  Sinus rhythms are noted (Chart abbreviation = *S1* and *S2*).
  Heart sounds are listened for, such as murmurs, rubs, and gallops (Chart abbreviation = *MRG*).
  Pulse is again noted, counted, and commented on. This is the apical pulse.
- **Abdomen**
  Abdomen is both palpated and auscultated.
  Bowel sounds are noted (Chart abbreviation = *BS*)[5]
  Masses and tendernesses are ruled out or in.
  Enlargement of liver or spleen is also noted.
- **Musculoskeletal**
  Range of motion of limbs and spine are noted (Chart abbreviation = *ROM*)
  Tendernesses of joints and other sites are examined.
  Swelling and inflammation are reported on.
  Pulses and reflexes are noted depending on patient complaint.
  Equal strengths and unusual weaknesses of limbs are recorded.

How Is the Physical Examination Recorded?

Next we apply the above criteria for the physical examination to the patient that we have been using as an example in this chapter:

Physical

System	Rationale	Findings
Vital signs	Baseline data. Elevated temp could indicate recent infection or inflammation.	T 98.6° F, HR 60 Regular BP 100/60 rt arm, sitting. Ht 5' 10"; Wt. 145 lbs.
General appearance/Skin	Gives indication of overall status. Also can indicate nutritional status. Skin assessment can indicate specific rheumatic diseases.	Thin woman. Alert and oriented X 3; NAD Skin warm & dry; color natural. No rashes, lesions, bites or scratches. No pallor or jaundice. Nail beds: no clubbing or splinter hemorrhages.

System	Rationale	Findings
HEENT	Dryness of eyes and oral cavity could indicate SICCA[6] syndrome. It is important to check for lymphadenopathy[7] to rule out viral infection or inflammatory disease.	Eyes: No dryness, conjunctivitis, or infection. Hair: thick and evenly distributed. Throat: Positive gag reflex. MM pink & moist. (See endnote 4.) No lesions. No lymphadenopathy. Thyroid nonpalpable, no masses.
Lungs	To rule out pulmonary problems, which can be a systemic manifestation of RA[8] or SLE[9].	Chest symmetrical. No costochondral tenderness upon palpation. LCTA
Cardiac	Quick screen for bacterial endocarditis or systemic manifestation of RA or SLE.	S1, S2 normal. No MRG. Apical pulse 60 & regular.
Abdomen	Quick screen because possible diagnosis is viral infection (hepatitis).	Flat, soft. BS all 4 quadrants. No hepatosplenomegaly.[10] No tenderness or masses.
Musculoskeletal	C/o[11] diffuse joint pain requires a complete musculoskeletal assessment.	Full ROM of neck and spine. No pain upon palpation of tender points (occiput, low cervical, trapezious, supraspinatus, 2nd rib, lateral epicondyle, gluteal, greater trochanter, or knee). Positive synovitis 2nd, 3rd, and 4th MCPs and PIPs bilaterally, warm & tender to palpation. Decreased extension 20° both elbows. Full ROM shoulders, hips, knees, and ankles. No synovitis. Tender MTPs[12] bilaterally. No nodules. Pulses 2+.
Neurological	Important to do for screening. Some diseases may present with neurological deficits, such as hyperthyroidism or entrapment syndrome.	Alert & oriented. Gait, mild limp. Sensation intact. DTRs[13] 2+. No asterexsis, negative Phalen's & Tinel's sign.
Rectal	Since c/o fatigue should check stool for occult blood.	Sphincter tone firm, no masses or hemorrhoids. Light brown stool, guiac[14] negative.

Once the HXP is completed, the data are transcribed or dictated onto the patient's chart. In the hospital setting, once the patient has been "worked up" (slang for the completion of the HXP), the health professional has to make constant assessments of the patient's condition. Usually most of the physicians and others use a traditional method known as SOAP.

What Is SOAP?

A number of different traditions of documentation formats as well as new forms of charting exist. The most common and traditional reporting of follow-up notes is based on SOAP, an easy acronym to remember. SOAP refers to (S) Subjective, (O) Objective, (A) Assessment, and (P) Plan. Dr. Lawrence Reed invented this documentation practice for physicians in the 1960s. "In this problem-oriented format of documentation, each identified problem is charted as a separate note" (Yocum 1999, 54).

Subjective (S)
Record any change in the patient's symptoms and, when necessary, comment on compliance with a particular regimen (e.g., stopping smoking) or tolerance of drug treatment.

Objective (O)
Record any change in physical signs and investigations that may influence diagnosis, monitoring, or treatment.

Assessment (A)
Comment on whether the subjective and objective information has confirmed or altered your assessment and plans.

Plan (P)
After making the assessment, consider whether any modification of the original plan is needed. Structure this section according to the headings listed on page 103 (Dx, Mx, Rx, and Ed).

If there is no subjective or objective change from one visit to the next, simply record "No change in assessment or plans." (Epstein 1997, 6)

Epstein also suggests that in the case of acute or chronic illness a graph or flow sheet is an appropriate way of recording the data so that the information can be reviewed at a glance. However, he cautions that a flowchart with two or more different tests may be more confusing than helpful (1997, 6). Often, the lab sheets are attached to the chart in sequential fashion in the section of the chart marked *lab* or *lab results,* and the data are not transcribed onto another chart or graph. While this method is also not optimum, it does provide an overview.

Nurses have made interesting revisions to SOAP by adding -IE or -IER. I = Intervention, E = Evaluation, and R = Revision (or Review) (Yocum 55). The advantages of the SOAP format are: Documentation remains constant from author to

---

Progress notes

date

11/1/97   S – nauseated, fatigued
             O – less jaundiced
                liver less tender
                taking adequate calories and fluid
                ultrasound liver/biliary tract: normal

             A – seems to be improving
                no obstruction

             P – check liver tests tomorrow
                phone laboratory for hepatitis markers

13/1/97   S – feels considerably better, appetite improving

             O – transaminase levels and bilirubin falling
                IGM antibody to hepatitis A positive
                sigmloidoscopy: bleeding hemorrhoids
                hypercholesterolemia

             A – resolving hepatitis A
                rectal bleeding in young patient likely to be
                hemorrhoids

             P – reassess patient, explain hepatitis A
                consider discharge if next set of liver tests
                show sustained improvement, ask surgeon
                to consider treating hemorrhoids
                recheck cholesterol in 3 months

Exhibit 4.2 An Example of SOAP Notes

author; each problem is directly addressed; it eliminates unnecessary data (chartjunk); it promotes documentation of the healthcare process. Disadvantages include: The format doesn't lend itself to constant care giving situations but works better as summary or shift-end notes; this is not effective for a patient whose condition changes very rapidly; frequent patient turnover creates problems with SOAP notes; routine care is then not documented; problems occur if the notes are not completed or only parts are used (Yocum 1999, 56).

As mentioned above in the *Clinical Examination* section on **P (Plan),** a problem list is helpful to lay out (if the chart itself does not structure it already). Such a list includes: Dx = diagnostic tests, Mx = monitoring tests, Rx = treatments, and Ed = education.

Many of the old systems are now changing as more and more health facilities elect to do their patient documentation on-line. However, even if your first job as a health communicator is with an on-line system, learning the old system is important, because the basic structures of reporting on on-line charts are based on the handwritten, print-based charts. In other words, although you can more quickly record and move from one form to another on-line, you will still have the basic "pages" of the chart, whether print or electronic, such as progress notes, I/O (input/output) records, checklists for admissions and transfers, blocks or pages for narratives, nurse worksheets, and daily and patient assignment sheets. In the next chapter, Chapter 5, you will have a chance to see what traditional forms look like before they're translated to an electronic interface.

What Are the Capabilities of On-line Charting?

The world of on-line charting is an exciting and rapidly changing place. Advances in this area of technology seem to be made even faster than in some other areas of technology. The health and medical sector are now growing at a pace that outstrips the other sectors, even in times of economic difficulty. With the generation of baby boomers aging and needing more health services, this sector will continue to grow for a number of years. Because software packages usually address several areas of documentation in addition to medical records and patient charting, such as order management and laboratory testing documentation, these on-line possibilities are also included in this chapter. Almost all of these documents (or parts and forms of them) appear on the patient chart at some point.

There are a number of ways patient charting has taken advantage of the new technological capabilities:[15] patient registration, order management, clinical documentation—point-of-care reporting, clinical data repositories, laboratory information systems, medical records, nursing language sources for standardization, nursing intervention classifications, office task management systems, patient billing systems, and others.[16]

Patient Registration

Patient registration is of course the first step in any facility-patient encounter. Demographic information is gathered on a patient, and this information is then entered into the hospital or office system in which he or she will receive treatment. This system has long been automated in almost all facilities in the United States, primarily because patients need to pay for the services rendered. With this system, it's been typical practice to generate bills, to contact insurance companies, and to update patient contact information.

Now the patient registration system is also available on-line with browser-based application and a true Web-based architecture (some are not browser-based). As a health writer, you may be able to help your facility choose the best software available (or design a customized program[17]) for its needs. As was true in the earlier automated forms and networks for patient registration, on-line systems provide the same capabilities of integrating patient demographics and financial data. What makes this more helpful than the old systems is the ability to share data across the healthcare facility, while still protecting patient confidentiality. All of the current applications are HIPAA compliant, another important consideration if you have an older system that needs replacing and updating.[18] Another important benefit for using a Web-based registration system is the elimination of duplicate efforts throughout the various departments in the hospital (OpusPtReg 2003, 1). The patient's data are readily available and require no new formatting and entry for those in the other departments that will see the patient during his or her visit. Rather than chancing that important information is not gathered for a particular procedure, for example, the correct and important data will already be on-line for the technician or other health professional to check. Also, the database allows for gathering of other statistics and evidence for physician groups and hospital administrators for billing and assessment purposes.

Order Management

A hospital or health facility runs in some ways similarly to a military establishment. Not only is it heavily dependent on a hierarchy, but it also is run by orders. Physicians and nurses are the primary order generators, especially for patient treatments. Thirty years ago, we used a very primitive order system on triplicate, carbon-sandwiched forms, which we sent from department to department to fill via pneumatic tubes, much like those the drive-through bank tellers use. We had to decipher physician handwriting in order to begin the laboratory tests on the blood (and other bodily fluid) samples that came downstairs to us. Some of us became experts in recognizing certain physician handwriting, but it always seemed risky to me, in case important tests were omitted or wrong tests done—thus delaying vital results impacting patients' lives. We did type the requests, then, into the computer, but think how much more accurate and safe the new systems are that let you order via computer (although of course mistakes can still be made there, too).

Again, when choosing a system or an update, here are the capabilities and benefits to look for:

- A browser-based, true Web interface for ease of use and access
- SQL92 relational data base compatible[19]
- Internet and Intranet integration
- Reports available in different formats (print, on-line, etc.)
- Multiple order submissions
    Functions
        Create orders
        Update orders
        Review orders
        View results
- Automated charge postings

- Medicare compliant
- Graphs and charts viewable and printable
- "Patient and on-line form generation and availability: I/O record,[20] progress notes, in-house transfer checklist, admission checklist, narrative notes, graphic summary sheet, Braden risk assessment scale,[21] nurse worksheets, shift log, patient assignment sheets, daily assignment sheets, restraint worksheet, etc." (OpusOM 2003, 2).
- HIPAA compliant

As you can see from the long list above, an order management system can be quite extensive and useful for the health facility, especially if the facility is large. While software this extensive is often quite expensive, its many functions can be worth the investment. If your facility does not get software that is compatible with its other hardware and software, there can be problems. You'd want to make sure, if you are on a committee making the choice, to determine that. Sometimes it's just a matter of updating browsers and cookies, which is easy to accomplish.

Clinical Documentation—Point-of-Care Reporting

Another frequently used system in larger and especially urban hospitals is the software that enables point-of-care reporting. An on-line system like this, which can also be Web-based, lets nursing staff and others enter data in the patient's room at the time the care takes place. When in a patient's room, you might have noticed the computer stand on rollers into which the nurses type data as they take vitals (blood pressure, temperature, pulse), administer medication, assess the patient, and perform other functions. You may even see a system in place in which staff members scan the items they use for the patient before they open the wrappers on mouthpieces, syringes, vials, and other equipment. All of this reporting via computer in the patients' rooms means that staff is not carrying around charts, clipboards, or slips of paper for recording by them or later by someone else (the ward clerk who is usually on duty at the nursing station).

This system, too, is labor saving for nursing and other staff members already overworked and stretched thin by the nursing shortage. This system makes access easier, record sharing possible, and saves time and space. Some other particular advantages of the clinical documentation capability include:

- SQL92 relational database compatible
- LAN and Intranet integration
- Reports available in different formats (print, on-line, etc.)
- Visual reminders, pop-ups and alarms[22]
- Data can be stored over multiple visits in outpatient clinics or physicians' offices
- Log-on messages to alert caregivers to unfinished charting that can then be completed
- Late and anytime entry of patient data, too

- JCAHO compliant[23]
- Intervention feature that identifies at-risk patients
- Patient task triage and organization
- Improved legibility (OpusClinDoc 2003, 2)

The system above is most useful for hospitals and clinics with in-patient beds. There are other on-line systems for charting in doctors' offices and clinics where patients are treated but not admitted for overnight stays. Such systems are described later in this chapter in the section titled *Office Task Management* systems.

Clinical Data Repositories

Such repositories are on-line databases for clinical results within a facility; these results can then be accessed and consulted via the various search and sorting capabilities of the software. In such a system you can focus your searches and data retrievals without having to spend time entering large amounts of demographic information (OpusCDR 2003, 1–2). Such retrievals can also give you access to trends and statistics necessary for reporting to regulators, media, and funders. You can also then predict future trends for your hospital and better manage and prepare for future demands on your health facility. Many systems do use on-line databases, but they are not necessarily Web-based. This is true of the other systems, too, but it's useful for health writers to be aware of the latest versions and be prepared to work in a Webbed environment. There are a number of systems available, and most are HIPAA compliant by now.

Laboratory Information Systems

Laboratory database systems are developed for hospitals, research labs and multi-physician groups with their own testing facilities. These systems are also specialized for specific laboratories such as the clinical laboratory, microbiology, etc. Like the other systems, these, too, offer easy access to results and help ensure accurate reporting. While laboratory systems have been automated since I worked in them 30 years ago, the user-friendly, on-line databases are more recent. Fewer of the laboratory databases I reviewed are browser-based, but these systems are still very powerful in housing, accessing, searching, and sorting results. Here are some of the features that are most useful for lab databases:

- Server sharing by multiple facilities
- Database region sharing
- Varying test lists depending on site
- Numeric or coded comments
- Abnormal flags, linearity checks, drug interpretations, add-on pathology report
- Collection routes, assignments, labels, barcode specimens (OpusLab 3,4)
- Statistics and epidemiology reports (OpusLab 6)[24]

Nursing Language (Standardization Sources)

Nursing language, like much of the coding systems and standardization of medical language for billing and other purposes, is also becoming standardized. To keep up with and for easy access to these changes, the sources are now on-line in the INFORM NIS database, developed and maintained by Nursing Informatics. The following are available: NANDA (North American Nursing Diagnosis Association), NDEC (Nursing Diagnosis Extension and Classification), NIC (Nursing Interventions Classification), and NOC (Nursing Outcomes Classification). NANDA is "[c]linical judgments about individual, family, or community responses to actual and potential health problems/life processes"; NDEC is the "[r]efinement, extension and validation of NANDA taxonomy of nursing diagnoses"; NIC is "[a]ny treatment, based on clinical judgment and knowledge, that a nurse performs to enhance patient/client outcomes"; and NOC contains "[a] measurable patient or family caregiver state, behavior, or perception that is conceptualized as a variable and is largely influenced by and sensitive to nursing interventions" (Nursing Informatics 2003, 1). All these classificatory systems sound somewhat abstract, so in the next section, there is an example of a nursing intervention classification.

Nursing Intervention Classifications

Basically, as in other coding systems now found throughout health institutions (see the beginning of the next chapter for more information), each diagnosis, intervention, procedure—almost any action in medical care—now has a numerical code. The same is true in nursing, too. For example, "Fluid Monitoring" is number: 4130. If you are charting according to this code as a nurse or as a health writer, this number defines the term as the "collection and analysis of patient data to regulate fluid balance" (Nursing Interventions 2003, 1).

> Determine history of amount and type of fluid intake and elimination habits; Determine possible risk factors for fluid imbalance (e.g. hyperthermia, diuretic therapy, renal pathologies, cardiac failure, diaphoresis, liver dysfunction, strenuous exercise, heat exposure, infection, postoperative state, polyuria, vomiting, and diarrhea); Monitor weight; Monitor intake and output; Monitor serum and urine electrolyte values, as appropriate; Monitor serum albumin and total protein levels; Monitor serum and urine osomolality levels; Monitor BP, heart rate, and respiratory status; Monitor orthostatic blood pressure and change in cardiac rhythm, as appropriate; Monitor invasive hemodynamic parameters, as appropriate; Keep an accurate record of intake and output; Monitor mucous membranes, skin turgor, and thirst; Monitor color, quantity, and specific gravity of urine; Monitor for distended neck veins, crackles in the lungs, peripheral edema, and weight gain; Monitor venous access device as appropriate; Monitor the signs and symptoms of ascites; Note presence or absence of vertigo on rising; Administer fluids as appropriate; Restrict and allocate fluid intake, as appropriate; Maintain prescribed intravenous flow rate; Administer pharmacological agents to increase urine output, as appropriate; Administer dialysis as appropriate noting patient response. (Reed and Sheppard; Titler; McCoskey and Bulechek; Nursing Interventions 1–2).

Office Task Management Systems

Such a system needs many of the more common browser-based, multiple format, and HIPAA compliance capabilities above as well as other capabilities for these work sites, mostly physician offices and clinics: Patient folder with entries for allergies, past medical history, surgical history, immunizations, diagnoses, treatments, alerts for the patient, remarks about the patient, patient overview, and patient summary. Specialized capabilities include those for pediatrics: (growth charts and immunizations) and for obstetrics (OB, menses, and genetics histories; estimated due dates; fetal movement, position, heart rate; prenatal lab values; mother's weight, b.p., fundal height; contraceptive history; labor and delivery duration, complications, gender, birth weight, and delivery method) (Medinotes 2003, 2, 5).

Some systems include excellent features that are especially user-friendly for some of the nondegreed personnel in the office or clinic: color coding of various entries, pop-up boilerplate for letters, and WYSIWYG[25] (Medinotes 2003, 1). Other features that are useful to office tasking can include: physician referral letters, patient instructions, patient letters, physician orders, prescriptions, billing, graphic annotators for physician drawing, E & M coding, duplication and additions of HXP and SOAP notes, office text messaging, and daily patient lists. All of these are now available as is the capability to interface with voice recognition software for dictation (Medinotes 2003, 2–4).

Patient Billing Systems

These systems have been available the longest and are found in the most locations. Software is available for private and state insurance billing, patient billing, HIPAA compliancy, and Medicare and Medicaid billing.

At some point print charts will be a thing of the past, but, surprisingly perhaps, both systems now exist side-by-side with smaller practices and facilities unable to afford the initial costs of the charting and documentation software.

Disadvantages

The disadvantages do not outweigh the advantages of on-line-charting, registration, etc., but there are some significant ones. The most obvious, of course, is that computers and networks crash, leaving a lag before the backup system can be generated. In addition, other problems ensue; nurses often have to be almost as adept at computers as they are at nursing, and often health professionals are in healthcare because they want contact with patients, not with machines. Also, there are infectious disease problems—keyboards are not wiped down often enough. With computers that are based in patient rooms, every technician, nurse, and caregiver who enters the room usually has to log information into the patient's chart. This means more frequent hand washing should take place between touching the patient and touching the keyboard. Unfortunately, this doesn't happen as much as it should. This computer usage means increased risk of hospital-borne infections for patients.

Because we are suffering, and have been for awhile, from a nursing shortage in many states, we are increasingly recruiting and hiring nursing staff from other countries, where less computerized systems exist. These nurses not only have language barriers to overcome, but they may also have technical literacy problems, too.

Employers and hospital administrators believe computers have decreased nursing loads because of the ease and speed of reporting. Although there aren't many studies on the subject at this time, even a couple of weeks of observation at a large teaching hospital in the Midwest convinced me that the nursing load is as difficult as it has always been and even heavier in some cases. Nurses have often had to keep track of equipment used by patients for billing purposes, although ward clerks handled many of those functions before we were so computerized. Now, in addition to the many other duties, many nurses are busy scanning UPC codes into the computer for every 4 × 4 sponge they unwrap and every new length of IV tubing. The bookkeeping functions once located at nursing stations have moved directly into individual patient rooms and are more practically accessed by nursing and other staff directly involved with patient care. As I've mentioned before, history has shown that nursing shortages are cyclic. Hopefully, once we have more adequate staff numbers and computer programs become even less demanding, these extra burdens will be more equitably shared.

Summary

In this chapter you have gained an overview of the main systems used in the gathering of information for a chart. Charts are the central "data bank" of the patient's medical activities while being treated by health workers, and each patient is referred to as a case with the requisite case number.

As a health professional using the diagnostic and charting systems—even if peripherally—it's useful to know how you need to think in order to process the information in the manner appropriate to charting. To help you understand the thought processes, this section featured clinical decision making (CDM) and framed it by comparing it to critical thinking. The diagnostic system used by proponents of CDM is especially useful in the less routine cases.

To illustrate this method of diagnosis and thought processes, you followed the same patient through the history and physical (HXP). Then you saw how the data that was gathered in the HXP was charted in much detail (in this case and with this method more time-consuming than some facilities allow). But CDM does give you the parameters for the most complete patient case analysis. If you have a form set up for this data and the practitioner has the time for a thorough HXP, this data will be more easily entered on the chart.

Another of the most well-known and traditional sources of data generation is the system used by most physicians known as SOAP (or in the case of nurses,

SOAPIE and SOAPIER). These acronyms simply make the organizational structure or heuristic of the ongoing encounters with patients easy to remember and common to almost all doctors and nurses. These are used in assessing patients for triaging, and for progress notes and pre- and post-treatments. They form a type of shorthand easily accessible to health workers.

Ideally, health professionals could use flow charts for this data, but, unfortunately, so much data are generated, it is difficult to see the information in an inclusive graph. Instead, such data can be more easily seen in tables with few categories, such as the problem-related plan, which organizes the information according to Problem, Dx (Diagnostic Tests), Mx (Monitoring Tests), Rx (Treatments), and Ed (Education). This system gives a more global overview to the medical processing of one case.

Print documents are limiting in that they are not easily shared, and cannot be easily sorted for statistical purposes. Many hospitals, clinics, and doctors' practices are opting for on-line charting and documentation. It's possible now to find an on-line system for almost every medical documentation function. Many automated systems have been in medical facilities for many years, but the Web-based systems are relatively new. In this chapter examples were given for patient registration, order management, clinical documentation—point-of-care reporting, clinical data repositories, laboratory information systems, medical records, nursing language sources for standardization, nursing intervention classifications, office task management systems, and patient billing systems.

In the next chapter, we take up the forms used for charting various patient self-reports, encounter forms, and procedures.

Discussion Questions

1. As a student or health professional, you have experienced the emphasis on critical thinking in your education. What does that mean to you? What are your strengths and weaknesses? Give examples in the discussion. How will (or have) critical thinking skills help your clinical decision-making practices? Give examples.
2. What abbreviations and terminology do you have problems with? Discuss the terms in class. What can you do to help yourselves remember these terms? How would charting on-line help the understanding of terminology? Discuss the pros and cons of using these medical abbreviations so widely.
3. Do you or others in the class have examples of difficult cases in which the more thorough parameters of CDM would have helped or were those cases solved using this method or others? Would CDM help in cross-cultural health care? Why or why not?
4. What do most charts have in common? Where would you find differences given the patients have the same diseases? How about when the patients have different diseases? How would you divide a chart to make it the most logical and accessible? What information would you want up-front?

5. Why is SOAP a good strategy for progress notes? Can you think of more effective ways to do progress notes? Why do you think nurses have changed SOAP over time? Is this format appropriate for on-line charting? Why or why not?
6. The advantages of on-line charting and documentation are obvious. What are the drawbacks to on-line systems? Look up several different forms of on-line charting and documentation on the Internet. Discuss how systems compare and the advantages and disadvantages of each.
7. Which of the systems in this chapter would be the first choices of healthcare providers if they had limited budgets? What could be done to help smaller providers be able to afford on-line systems?

Exercises

1. In a team of several people, draw up an informal plan of how to make the patient history portion of the CDM chart more accessible as a print document.
2. In a team of several people, draw up an informal plan of how to make the patient history portion of the CDM chart more accessible as a digital or on-line interface.
3. In a team of several people, draw up an informal plan of how to make the patient physical portion of the CDM chart more accessible as a print document.
4. In a team of several people, draw up an informal plan of how to make the patient physical portion of the CDM chart more accessible as a digital or on-line interface.
5. Look up SOAP on the Internet including the databases mentioned in Chapter 1. Do the same with clinical decision making. Graph your findings (citation frequencies) and present them to your class. Read two on-line articles and draft a brief report on any new information in the articles.
6. Choose one of the commercial on-line charting and/or medical documentation systems that you find on the Internet. Prepare a PowerPoint presentation in which you make an argument for why the commercial system you have chosen would be a good choice or a poor choice for a small health facility.
7. Explore what AMWA (American Medical Writer's Association) says about writing for electronic interfaces. Present the information in a memo to your class. Find examples for one of the suggestions you find in AMWA literature.

Works Cited

Barrows, Howard S., and Garfield Pickell. 1991. *Developing Clinical Problem Solving Skills: A Guide to More Effective Diagnosis and Treatment.* New York: W.W. Norton.

Bates, M.D., Barbara. 1995. *A Guide to Clinical Thinking.* Philadelphia: J.B. Lippincott.

Braden, Barbara, and Nancy Bergstrom. Braden Risk Assessment Scale. In *Ageing and Aged Care Division of the Commonwealth Department of Health and Ageing* Web site. © Commonwealth of Australia. Draft National Framework for Documenting Care in Residential Aged Care Services. 8/12/2003. August 15, 2003. http://www.health.gov.au/acc/reports/download/assesstls3c.pdf

Epstein, MBBCh, FRCP, Owen. 1997. *Clinical Examination* 2nd ed. London: Mosly.

McCoskey, J. and G. Bulechek, eds. 2000. Iowa Intervention Project: *Nursing Interventions Classification (NIC)*, 3rd ed. St. Louis: Mosby-Year Book, Inc.

McLellan, Tim. "What Is an Oracle Relational Database?" © 1994. August 15, 2003. http://www.islandnet.com/~tmc/html/articles/orareln.htm

Medinotes. © 2003. 8/13/2003. http://www.microwize.com/medinotes

Navarra, Tova. 1998. *Toward Painless Writing: A Guide for Health Professionals.* ThoroFare, NJ: Slack, Inc.

Nursing Informatics. © 2003, University of Iowa. August 13, 2003. http://www.uihealthcare.com/depts/nursing/informatics/index.html

Nursing Interventions Classifications. © 2003, University of Iowa. August 13, 2003. http://www.nursing.uiowa.edu/centers/cncce/nic/nicintervention.htm

Opus Healthcare Solutions. 8/7/2003. August 13, 2003. http://www.opushealthcare.com

Reed, G. M., and V. F. Sheppard. 1971. *Regulation of Fluid and Electrolyte Balance.* Philadelphia: W. B. Saunders.

Robinson, Denise L. 2002. *Clinical Decision Making: A Case Study Approach.* Philadelphia: J. B. Lippincott.

Titler, M. G. 1992. "Interventions Related to Surveillance." *Symposium on Nursing Interventions. Nursing Clinics of North America* 27, no. 2: 495–516.

Yocum, Fay. 1999. *Documentation Skills for Quality Patient Care,* 2nd ed. Dayton, OH: Awareness Productions.

Endnotes

1. A heuristic is simply an outline or a list of questions that enables one to deduce certain information from various inputs.
2. During this technique of checking the radial pulse, it is possible to pick up marked arrhythmias such as ventricular tachycardia (fast heart rate), PVCs (preventricular contractions), bradycardia (slow heart rate), and others. A normal rhythm is noted as *regular.*
3. *NAD* stands for no acute distress. Of course, if patient is neither alert nor oriented, or is in acute distress (anxious, withdrawn, comatose, etc.), such a state is duly noted.
4. Mucous membranes (*MM*) are also revelatory of much information. Certain diseases, such as strep infections, leave recognizable clues in the mouth, on the tongue, and in the throat. The gag reflex is also noted, although that's not why the practitioner uses a tongue depressor. In fact, the tongue is held down to better view the throat and especially the tonsils and other throat glands.
5. *BS* can stand for both *bowel sounds* and *breath sounds.* Context is the determinant here.
6. *SICCA* = Keratitis Sicca/Dry Eye Syndrome, a form of ocular disease.
7. *Lymphadenopathy* = swollen lymph glands typical of viral and other infections.
8. *RA = rheumatoid arthritis.*
9. *SLE = systemic lupus erythematosus.*
10. *Hepatosplenomegaly = liver and/or spleen enlargement.*
11. *C/o = Complains of, complaint of, complaining of* and refers to what the patient articulates as the problem.
12. *MTPs = metatarsal phalanx.*
13. *DTRs = deep tendon reflexes.*
14. *Guiac = blood in stool microbiological test.*

15. Although I'm quoting a number of sources here, in no way am I advocating the use of these particular products. These examples were chosen because they are representative of what is currently available.
16. This list grows virtually every time I log on.
17. This does not mean that health writers have to know or be able to write programming language. At this point in our technology, this would be an unusual expectation. However, you may be able to work with an in-house or outsourced computer engineer to design a customized software program for your work site.
18. If you are unfamiliar with or foggy on HIPAA 1996, review that section in Chapter 2 on Ethics in Medical Writing.
19. SQL92 is a relational database, meaning that you can store data in various tables and define relationships between and among tables based on one or more field values common to the tables you are relating (McLellan 2003, 1).
20. Intake/Output to measure patient fluid production, elimination, and balance.
21. Braden risk assessment scale assesses a patient's risk of developing bedsores or pressure ulcers on the skin. The score is based on skin sensitivity, moisture, patient activity level, mobility, nutrition, friction, and shear (Braden et al, 1995, 7–9).
22. As a former health professional who did patient care, I can assure you this function is particularly important. Because we get so busy with the many duties and because there are often surprises, especially with critically ill patients, having reminder systems and alarms will help ensure that important tasks get done. When you are tired, too, as you often are doing double or emergency shifts, it is riskier to be working without some kind of alert system.
23. JCAHO stands for Joint Commission on Accreditation for Healthcare Organizations. Systematic data collection can actually help facilities be more JCAHO compliant.
24. Laboratory databases tend to be among the most extensive and have the most capabilities. For more detailed information, view the Web site at: http://www.opushealthcare.com/solutions/opuslab.htm
25. What You See Is What You Get.

Medical Forms and Reports

 ## Overview

You've learned how the process of putting information on a medical chart works, what the thinking behind the process is, and why certain traditional styles of recording data have evolved. Next that information is shaped to be put into somewhat constrictive forms that don't always exactly mirror the process health professionals used to gain information. Because much of medical care and healthcare is fast-paced, health providers often are limited in the amount of time they can spend getting this information down, whether it is on printed forms, in computer programs, or dictated via audiotapes. As mentioned earlier, there are attempts now to use voice recognition software to cut out the middle step of transcription. As a medical writer, you can perhaps work with your organization and design forms more in line with its particular needs. Linda Crew, Executive Director of the Joseph F. Sullivan Center for Nursing and Wellness, often has my students customize forms for specific demographic and treatment purposes.

The primary purpose of this chapter is to analyze the document styles, genres, language, and functionality of those forms in professional health settings. We will be examining different medical forms as to their use, how styles are adapted to that usage, the style of language used in the reports, the intended audiences, and possible future changes in conventions for reporting.[1]

This chapter is by no means comprehensive, nor could it be, given the glut of medical forms in the health profession, but it does include a small, representative sampling of what might be seen in the medical field, documenting different procedures and areas of medicine. You will begin with patient encounter forms, then consider procedural forms and progress notes. You'll learn the advantages and disadvantages of all the forms. There are portions of reporting forms embedded in the text for close reading and analysis.

The following questions are addressed here:

- What are encounter forms?
- How are encounter forms customized?
- What are procedural reports?
- What are progress notes?
- What about forms that patients fill out?

What Are Encounter Forms?

As the name implies, encounter forms are generated when a patient encounters a health professional or vice versa. This is the form used for taking patient histories and physicals (HXP).[2] The encounter form genre usually has entries for some patient demographics, many bodily organs and systems checks, a place for comments, and lists of areas that can be circled or check-marked to save time. Some encounter forms still are in triplicate if the system is not automated or computer-ready. Most are printed with billing codes so that the document is at least dual purposed. Often it has other applications, too, if lab requests, diagnostic tests, and procedures are circled.

How Are Encounter Forms Customized?

The first form we'll discuss is from the Joseph F. Sullivan Center at Clemson University in Clemson, South Carolina. Shown in Exhibits 5.1 to 5.4, it gathers fairly extensive information, even though it is only one page in length. The health practitioner conducting the examination completes the encounter form. Thus, there is no patient fill-in section, except at the very end on the back of the sheet, where the patient would authorize the completion of further medical testing without Medicare supplemental payment and/or also sign a consent for further treatments or procedures.

The bulk of this form is designed for individual diagnostic tests and their results. It features sections for providing information on the purpose of the visit, the medications the patient is taking, and further subjective analyses of the patient performed by the doctor or nurse practitioner. There is also space for the name of the interpreter at the top of this form. The line for an interpreter is unique to this form in the rural upstate area of South Carolina and demonstrates the Sullivan Center's work with local Latino populations. In their rural health outreach, an interpreter is often needed to translate Spanish to English and vice versa.

The first part of this form has limited space for purpose (or "complaint" [C/O]),[3] medications, blood pressure, weight, height, temperature, pulse, respirations, LMP—last monthly period, and the subjective analysis (S of the SOAP work-up).[4]

This section of the form is well laid out with room for the practitioner to record her or his own impressions in line with the SOAP assessment. Unlike many other forms, the vital signs are on the left rather than across the front. The

ENCOUNTER RECORD   Interpreter _____

_____FP   _____PROB   _____PE   _____REPAP   _____BCN   _____COLPO   _____LAB ONLY   _____OTHER

Purpose for Visit		BP		
Meds / ACHES				
Subjective		Wt.	Ht.	W/H
		Temp	Pulse	Resp.
		LMP		

Exhibit 5.1 Sullivan Center Encounter Form Part 1

divisions on this part of the form apply the principle of "chunking" well, with the vital signs grouped together on the right rather than spread across the top of the chart.

The rest of the front side of the Sullivan Center form (not shown) has sections for the individual tests performed by the doctor. There is a section for a hearing examination, where the decibel and pitch readings for each ear are recorded, and then the hearing system is classified as "Normal" or "Abnormal" depending on the results. There is a section for vision in which each eye is examined for farsightedness, nearsightedness, color vision, and phoria.[5] The results are then classified as "Corrected" or not.[6] There is another section for laboratory tests and results: urinalysis, pregnancy and strep throat screening, a wet prep Pap smear for gynecological examinations, and several other categories. Although the Sullivan Center is mostly geared toward women's health, parts of the form can be used with patients of either gender.

The next section of the Sullivan Center form, shown below, examines the different body systems, such as skin, eyes, neurological system, gastrointestinal

NL	AB	Review of Systems		NL	AB
		General:			
		Skin:			
		Eyes:			
		ENT			
		Endo./Metab..			
		Neuro:			
		MS			
		Cardio.			
		Resp.			
		GI			
		Urol			
		Psych			
		Genitalia:			
		Breast:			
		GYN:			
		Other:			

Exhibit 5.2 Review of Systems

system, genitalia, and respiratory system, to name a few. This section is well laid out and labeled for speed and thoroughness for the examiner. If you look at the whole form, you'll see that this represents only the left side of the form, and the second list of "normal" and "abnormal" are for the results of the physical. For the extensive work-up possible on the form, it is clear the developer of the form is not guilty of "chartjunk." Every aspect is necessary and usable.

Further to the right on this form is a section for the physical exam (see Exhibit 5.3). It includes (far right) tests and their results. These tests examine such things as the skin, eyes, ears, nose and sinus, spine, and joints and strength. A small section to the side of the report further details different tests such as those for cholesterol, high density lipoproteins (HDL), electrocardiagrams (EKGs), X-rays, HIV, and various others. The only problem with usability here would be the grouping of physical exam features on the same line with a test that has no relation to the part being examined. Since this is counterintuitive for users, there may be a more appropriate grouping (a separate box after this—perhaps a horizontal, narrow box, below this part, labeled tests?).

The back of the page (Exhibit 5.4) has a very large section for a list of "Activities" with the date that they were performed, with a substantial amount of space for writing. This space is for detailing any follow-up information or appointments that could not be detailed on the rest of the form. There is also a small section in the lower left corner of this section that allows the practitioner to assess and circle descriptors of suspicious skin lesions, as well as biopsies performed. This is a very important section of the exam; even though it receives only a small amount of the page, it focuses attention quickly.

This encounter form contains many useful features, primarily stemming from the heavily detailed but streamlined format. This report provides substantially more information than many documents, because not only is it quite comprehensive and useful as the first sheet on the patient's chart, but it also contains the Medicare authorization and a consent form. This form is often represented in

Physical Exam	Initial	Test	Result
Appearance:		T. Chol	
Skin:		HDL	
Eyes:		BS	
Ears:		A1C	
Nose/Sinus		Hgb	
Oral:		Pap	
Nodes/ Thyroid		Mamm	
Chest:		GC/Chl	
Heart:		STS/HIV	
Abdomen:		Veni	
Extremities / Pulses:		Colpo	
Neuro / Reflexes		Biopsy	
Psych:		Cryo	
Spine/Joints:		Histofz	
Strength:		EKG	
Other:		PFS	
		XR/US	

Exhibit 5.3 Sullivan Center Physical Exam Section

Date	Description of Activities

C	**Code**		**Reid**		
O	--- SCJ		Margins	0-1-2	
L	=== Suspicious		Color	0-1-2	
P	// Metaplasia		Vessels	0-1-2	
O	/VV Columnar		Iodine	0-1-2	
S	xxx Biopsy		Score		

MEDICARE ADVANCE BENEFICIARY NOTICE (ABN)

Medicare does not pay for laboratory tests and other services associated with routine screening and/or annual physicals. I have been notified by the clinician that I have requested a non-covered service. I choose to proceed with the requested service and I agree to be personally and fully responsible for payment.

_____   _____
Signature of Medicare beneficiary   Date

_____
Signature of service coordinator

CONSENT FOR SPECIAL PROCEDURE

The risks, benefits, and potential complications for a _____ have been explained to me. My alternative options have also been explained. I have been given the opportunity to ask questions and have them answered to my satisfaction. I hereby consent to the procedure.

_____   _____
Signature of client   Date

_____   _____
Signature of client   Date

11/2000
Revised 12/02   Name_____ Chart #_____

Exhibit 5.4 Encounter Form—Description of Activities

other health facilities by at least three different forms: one for the history and physical, one for test results, a Medicare authorization, and a separate consent form. Otherwise, this report is characteristic of many encounter forms in the health professions.

The "Description of Activities" section will be filled in both in narrative form and with abbreviations, and even with the space for comments, this form could be easily entered into computerized databases; however, there is no patient identification information other than Name and Chart number provided for this purpose. Thus, in actuality, this form will be added to the patient's printed chart, rather than entered into a database. As was explained by some doctors, most HXP information is still not entered into databases. The medical information is generally filed away in print-based charts, while the identification numbers, visit history, address, phone numbers, and other pertinent information is put into the facility database for quick reference. Compiling all of the medical information into a database will eventually be feasible, but hard copy charts are still most common. The document, in reality, represents only one vital part of the communication that needs to take place in the hospital setting.

As we know, it is the responsibility of the doctor, nurse, or other health practitioner to explain the results of the tests to the patient, so that he or she understands the medical concerns, the treatment options, and the expected outcome. This way, the patient will know the problems well enough to cooperate with the practitioner to complete the treatment. Without a clear explanation, patients are often resistant to further medical tests and treatments. Clearly understanding the problems is a key issue for doctor-patient interactions, since trust must be developed to ensure patient compliance. The physician or the nursing staff would normally follow up with an explanation of the diagnoses to the patient and would also educate the patient with a simpler handout or verbal description.

The next form, shown in Exhibits 5.5 to 5.7, is another one from the Joseph F. Sullivan Center. It follows the initial encounter called the "Health Encounter Summary." It is a very general form that is used in an initial health screening to provide a basic understanding of the patient's personal and medical history for the healthcare provider.

It is very interesting to note that the first items requested on the form are "Personal Strengths/Social History" and "Primary Language." The "Personal Strengths/Social History" request offers a large blank on which the medical professional is to give a brief description of the patient's social abilities, past and present. The "Primary Language" request offers the following selections: "English," "Spanish," "Some English," and "Other _____." This reflects the region in which the form is utilized. It can be assumed that if this form was being used in another area of the country, a longer list of language options might be offered. Requesting

Joseph F. Sullivan Center
Health Encounter Summary

Personal Strengths/Social History: _____

_____

Primary Language: English   Spanish   Some English   Other _____

Exhibit 5.5 Social History and Language

this language/cultural information on the top of the form is key, because this basic information about a patient is critical to providing healthcare to all patients.

Again, from a document design standpoint, this form could benefit from some attention to establishing focal points via good headers, increased point size on titles, and bolding to provide contrast between the category (e.g., Primary Language) and the choices (e.g., Spanish).

The next set of requests for information that the form asks of the medical professional is found in a grid set (see Exhibit 5.6). The grid initially asks for the following: "Problem List," "Current Medications," and "Allergies." Given the previously requested personal information, it is good, intuitive navigation for the form to progress to the basics of the patient's medical history and needs. The "Problem List" column is asking the medical professional to cover the basics of the patient's needs, be they personal, mental, or physical. The "Current Medications" area is typical, as is the "Allergies" column, enabling the healthcare provider to properly and safely treat the patient without unintentionally bringing about any allergic reactions due to prescribed treatment. Often this information is put on a bright

PROBLEM LIST	CURRENT MEDICATIONS	ALLERGIES
SCREENINGS:		
Pap		
PSA		**SUPPLEMENTS:**
Mammogram	**IDENTIFIED RISKS / CHANGE:**	
Glucose / Hgb A1C	Nutrition / Obesity	
CBE / CTE	Stress	
Immunizations	Physical Activity	
Lipids	Mental Health	
PMDs:	Substance Abuse	

Exhibit 5.6 Sullivan Center—Problem List

orange or red tag on the front of a hard copy chart and also on the patient's wrist-band if these are worn in the health facility.

Within the same grid the report requests the following information: "Screen-ings," "Identified Risks/Change," "Supplements" and "PMDs."[7]

While the form appears to be laid out well, especially for people with many problems and medications, many other patients will not have so many problems and medications (the area allows for up to 19). This form, though, is used to record numerous encounters, not just one. On the lower part of the form, the "Risks/Change" section gets only a tiny space to record a positive or negative change. It would seem this page could be redesigned for better usability. Because over time the risks and changes can and will be different, a timeline or graph may be useful here.

The bottom section on the front page of the health encounter form, shown be-low, is designed well for maximum usability. It is a lone grid in which the exam-ining clinicians are to sign off on their assessment of the patient. It provides 15 different spaces for the clinician to date and sign the form, thus creating a long us-ability for this particular form.

The fact that the "Name," "Date of Birth," and "Chart #" are on the bottom of the chart lets the health professionals use the standard "flip book," a clipboard chart in which the forms are clipped in at the top of the document. This usability factor, while appropriate for print charts, is not usable in an electronic chart, where the name is at the top to keep from having to scroll down to the bottom of a screen to establish patient identity.

On the reverse side of this encounter form (not shown) there is room for the "Health History." The information on this page covers the basics of past, present, and expected conditions the patient has experienced. It features both full words and abbreviations, such as "Cardiovascular (MI/HTN/Stroke),"[8] "Lung Prob-lems (Emphysema/TB/SOB),"[9] and "Vascular/Blood Clots/PVD."[10] The only unusual feature about this section is that, in addition to the family history cate-gories for patient and family, there are also spaces for commenting. I'm not sure this is often used, as the narrative report following such an encounter would typi-cally read: "Pt. w family hx of MI." I doubt the practitioner would comment on this form, "Schedule EKG or Stress Test" after the family hx box is checked, but

Review of History

Date	Clinician's Signature	Date	Clinician's Signature	Date	Clinician's Signature

NAME_____

DATE OF BIRTH_____CHART #_____

Exhibit 5.7 Sullivan Center—Review of History

instead would pull out an order sheet, circle it, and then dictate the comment as part of the report.

The next section of this same page is a series of questions concerning the present health condition of the patient. The "Allergy History" names environmental allergies and allows the clinician to quickly circle the appropriate items, such as "Pollen, Grass, Weeds, Trees, Yeast, Molds, or Latex." Space for allergies to medications was on the earlier form; however, it's good practice to include it again under the allergy category. The rest of the queries quickly give the healthcare provider the necessary information to assess the patient's condition and prognosis by accounting for the patient's dietary and lifestyle habits.

Finally, the last section on the report covers "Gynecological History" for female patients, including "DES[11] Exposure," something you don't see on every chart. It also asks for the patient's information regarding "Tubal, Menopause, and HRT.[12]" These physical characteristics are more typical of gynecological forms; usually, however, "tubal" is referred to as "ectopic" pregnancy, since, although rare, not all ectopic pregnancies are in fact in the Fallopian tubes. The final part of the "Gynecological History" is the grid asking for information about the female patient's history of pregnancy. It leaves five blanks to answer the questions of "Date of Pregnancies," "How Far Along?" "Birth Weight," "Complications/Miscarriage/Abortion," and "Now Living," usually enough blanks for most women.

The encounter forms featured above represent most of the features typical of this genre. The content is fairly standard, although there are a number of formatting variations among different facilities' forms. Of course medical specialties, which we have many of now, feature questions and diagnostic tests that are tailored to their specific HXPs. As a health communicator, you will want to make sure all forms are clearly and professionally reproduced. Approach any changes you might think appropriate to the forms cautiously, as clinicians can be resistant to suggestions of change in documentation.

What Are Procedural Reports?

In addition to encounter forms, another frequent form you will find on the chart is the procedural form. This form is crucial to the nursing staff for proper aftercare on the floor after a procedure, and it also documents what happened during the procedure. A good procedural form will include all the screens necessary to ensure a safe procedure (medications, blood-type check, etc.). In addition to providing guidelines or an outline for filling in quick notes to document the procedure, the form serves as a good data set from which doctors can later dictate their narrative report. Often procedural forms are subpoenaed for evidence in malpractice suits, as they are the official record of the medical facility. Also, today many surgical suites are equipped with video cameras that record the procedure, again to ensure that a documented version is available to cover the institution.

Our first procedural form, shown in Exhibits 5.8 to 5.12, is a labor and delivery record form from a women's hospital. Procedural forms, this one included, are very detailed; the nursing staff and physician fill this form out during the delivery of the infant. This form also has two carbon copy sheets attached to the back of it, so that one copy can be placed on the mother's chart and one on the infant's. The form is divided into five parts: "Antepartum," or pre-delivery, "Labor," "Delivery," "Infant," and "Signatures." In examining the "Antepartum" section below, you can see that there is space for giving information regarding characteristics of the "Present Pregnancy," along with possible medical complications. There is also a section detailing past medical history, abnormal past pregnancies, and abnormal infants, with space for writing under each of these headings. There is also a section entitled "Obstetrical History on Admission," and a section covering the blood type, immunizations, antibody screens, and other information vital to preventing problems with Rh negative mothers and blood transfusions.

The predelivery information in Exhibit 5.8 enables healthcare professionals to prepare for any problems during this delivery based on past ones, so it is filled out prior to the actual delivery itself. In the formatting of this form, unlike the previous forms, the headers on the various subsections are alphabetized and set in a larger point size to make it easier to separate them from the rest of the text. Although it again has the section labels running vertically along the edge of sections, the lettering is in white on a black background; thus the contrast helps improve the usability of this form compared to the earlier forms.

The next section of this form covers the actual labor (see Exhibit 5.9). The first section asks whether the labor was spontaneous or induced, and the next two sections detail how the labor was induced or accelerated, such as using a "Cervical Ripening Agent," Rupturing the Membranes," or "Oxytocic Agent."[13] This section of the form also asks whether any analgesics were administered six hours or so before delivery and, if so, which ones. This section further requests information about how labor was monitored and features a larger section for writing on the "Complications of Labor/Delivery." Both uterine contractions and the infant's heartbeat are monitored, and each has a number of different monitoring capabilities: internal and external, manual, or none. Again appropriate point size on headers, bolding of headers, and contrast on section titles enhance the usability of this form.

ANTEPARTUM	**PRESENT PREGNANCY**		**COMPLICATIONS**		**OB Hx ON ADM.**		**BLOOD**	
	LNMP		❑ None ❑ Bleeding ❑ Preeclampsia ❑ Diabetes Mellitus ❑ Other:		Grav.		ABO & Rh	ANTIBODIES ❑ Yes ❑ No
	EDC				Para.		RUBELLA IMMUNE ❑ Yes ❑ No	HEP SCREEN ❑ None ❑ PND ❑ POS ❑ NEG
	Gest. Age at Del.	wks.			Abort.		GBBS ❑ N/A ❑ Neg. ❑ Pos.	SEROLOGY ❑ Neg. ❑ Pos.
					Living Infants			
	PAST MEDICAL HISTORY ❑ Normal ❑ Abnormal, specify							
	ABNORMAL PAST PREGNANCY ❑ No ❑ Yes, specify							
	ABNORMAL INFANTS ❑ No ❑ Yes, specify							

Exhibit 5.8 Labor and Delivery Procedural Form—OB Hx (History)

LABOR

	*INDICATION FOR INDUCTION	LABOR ACCELERATED
❑ Spontaneous Onset	❑ Cervical Ripening Agent _____	❑ Artificial Rupture of Membranes
❑ Labor Induced	❑ Artificial Rupture of Membranes _____	❑ Active Labor Management
	❑ Oxytocic Agent	❑ Oxytocic Agent

ANALGESIA within 6 hours of delivery	MONITORING		COMPLICATIONS OF LABOR / DELIVERY
	Uterine	Fetal Heart	
❑ None	❑ None	❑ None	
❑ Yes, specify	❑ Palpation	❑ Fetoscope	
	❑ Ext	❑ Ext	
	❑ IUPC	❑ Int.	

Exhibit 5.9 Labor and Delivery Procedural Form—Labor

The third section of this form, shown below, documents the actual delivery and is the most complex segment of the form. It first asks for any blood loss and the quantity, along with the presence of meconium.[14] The number of spaces for times not only documents the delivery for legal purposes, but also helps nursing staff and physicians determine how long to wait before setting up for a cesarean section. Of course, many other physical parameters are measured throughout. This form also supplies space for details of episiotomies, tearing of the birth canal, and their repairs. The largest portion is devoted to the "Type of Delivery," which gives room for specifying whether the delivery was spontaneous or forced, using "forceps," "vacuum extraction," or "rotation," and the degree to which each of these was used. It also asks for the position of the baby when delivered, such as "breech," or whether a C-section was performed, and whether any additional procedures, such as tubal ligations, were performed.

What is quite remarkable about this form is its overall economy. Staff can record a large amount of information here on one single sheet. In addition to being compact, the information on this document is well grouped, using proximity as an organizing feature, along with a chronological structure that is intuitive for the user.

This next part of the form, shown in Exhibit 5.11, allows space for the characteristics of the infant after the delivery: gender, weight, and whether the infant

DELIVERY

BLOOD LOSS ___ cc	MECONIUM ❑ Yes ❑ No		ANALGESIA/ ANESTHESIA	IN LABOR	FOR DEL.	AFTER DEL.	TIME SUMMARY	DATE	TIME
			General				Onset Labor		❑ AM ❑ PM
OXYTOCIC AGENT		EPISIOTOMY	Spinal				Rupture Membranes		❑ AM ❑ PM
	Intra Uterine	❑ None	Pudendal				Onset 2nd Stage		❑ AM ❑ PM
IM IV		❑ Med-Lat.	Epidural						❑ AM ❑ PM
❑ Pitocin ❑ ❑ ❑		❑ Lt. ❑ Rt.	Local				Delivered		❑ AM ❑ PM
❑ Methergine ❑ ❑ ❑		❑ Med.	None				Placenta		❑ AM ❑ PM
❑ Prostin ❑ ❑ ❑									

TRAUMA	REPAIRED		REPAIRED
❑ Cervical Laceration	Y / N	❑ Labial Laceration	Y / N
❑ Periurethral Laceration	Y / N	❑ Hematoma	Y / N
❑ Vaginal Laceration	Y / N	❑ Sulcus Tears	Y / N
Degree:		❑ R	Y / N
		❑ L	Y / N

PLACENTA			UTERUS EXPLORED
❑ Spontaneous	❑ Manual ❑ Abnormalities, specify: _____		❑ Yes ❑ No

TYPE OF DELIVERY	❑ Precipitate ❑ VBAC	❑ Breech (vaginal) ❑ Spont ❑ Assisted
❑ Spontaneous		❑ Extraction ❑ Forceps Forceps Type _____
❑ Forceps Indication: _____		❑ C-Section
❑ Outlet ❑ Low ❑ Mid Forcep Type _____		❑ Primary ❑ Repeat Indication: _____
❑ Vacuum Extraction Indication: _____		❑ W / TOL ❑ W / OUT TOL
❑ Outlet ❑ Low ❑ Mid		❑ Low Tranverse	OTHER PROCEDURES ❑ Tubal Ligation ❑ Hysterectomy
❑ Rotation		❑ Classical ❑ Low Vertical	❑ Other ❑ Curettage
❑ Manual ❑ Forceps Forcep Type _____			

Exhibit 5.10 Labor and Delivery Procedural Form—Delivery

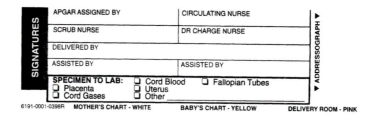

INFANT	❏ Male ❏ Female	**WEIGHT** gms.	lbs.	oz.	❏ Liveborn ❏ Stillborn	**PEDIATRICIAN IN ATTENDANCE** x		**O₂** ❏ None ❏ Laerdal

APGAR

HEART	COLOR	RESP.	TONE	REFLEXES	TOTAL SCORE	**INFANT EXAMS**	**DISPOSITION OF INFANT**	
1 min. 5 min.	1 min. 5 min.	1 min. 5 min.	1 min. 5 min.	1 min. 5 min.	1 min. 5 min.	❏ Appears Normal ❏ Pediatrician Consult ❏ Other _____	❏ 1. OBRR ❏ 2. NICU ❏ 3. Other _____	VAG DEL. ❏ C/S ❏ TUBAL ❏ RETURN TO DR ❏ OTHER ❏

Exhibit 5.11 Labor and Delivery Procedural Form—Delivery Record, Infant

SIGNATURES	APGAR ASSIGNED BY	CIRCULATING NURSE	ADDRESSOGRAPH
	SCRUB NURSE	DR CHARGE NURSE	
	DELIVERED BY		
	ASSISTED BY	ASSISTED BY	
	SPECIMEN TO LAB: ❏ Placenta ❏ Cord Gases	❏ Cord Blood ❏ Uterus ❏ Other _____ ❏ Fallopian Tubes	

6191-0001-0398R   MOTHER'S CHART - WHITE   BABY'S CHART - YELLOW   DELIVERY ROOM - PINK

Exhibit 5.12 Labor and Delivery Procedural Form—Signatures

was "Liveborn" or "Stillborn." There is also space for the attending pediatrician to sign. Further spaces are for the use of oxygen on the infant, the apparent condition of the infant, the disposition of the infant,[15] and the "APGAR," which details different reflexive, hearing, vision, and physiological tests, and how the infant responded to them after one minute and five minutes, respectively.

The final segment of this form, shown in Exhibit 5.12, is for numerous signatures documenting labor and delivery. This has become necessary because of legal problems resulting from deliveries such as birth injury to the child or the mother, mix-up of newborns if not properly tagged and footprinted at birth, etc.

To further elaborate on this form, this is one of the more complex forms because of the variations possible in labor and delivery, and it has evolved from numerous encounters with infant delivery. The check-box format permits rapid documentation with some spaces for writing in the report. Patient information, because this form is put on a top-bound chart, is stamped in the space on the lower right, permitting readability and easy access. This form is very high in utility and functionality as a specific report for labor and delivery.

What Are Progress Notes?

The fourth form featured here, shown in Exhibits 5.13 to 5.15, is an "Infant Clinical Progress Record" from a women's hospital. This form differs from the others in that it is more similar to a data sheet, one that allows for frequent and repetitive monitoring. Overall, these progress notes record newborn vital signs, which must be recorded and authorized every hour of the day to satisfy the requirement for close monitoring of infants following their births. Like all progress notes, they follow a patient over time after a health crisis, birth, or procedure.

TIME	12 MID	1 AM	2 AM	3 AM	4 AM	5 AM	6 AM	7 AM	8 AM	9 AM	10 AM	11 AM	12 NOON	1 PM	2 PM	3 PM	4 PM	5 PM	6 PM	7 PM	8 PM	9 PM	10 PM	11 PM
INITIALS																								
TEMP — INFANT TEMP.																								
TEMP — BED																								
HEART RATE																								
RESPIRATORY RATE																								
ACTIVITY																								
PHYS/BEH PAIN MGMT — FACIAL EXP.																								
PHYS/BEH PAIN MGMT — CRY/STATE																								
PHYS/BEH PAIN MGMT — TONE																								
PHYS/BEH PAIN MGMT — HEART																								
PHYS/BEH PAIN MGMT — TOTAL																								
PHYS/BEH PAIN MGMT — INTERVENTION																								
PHYS/BEH PAIN MGMT — RESPONSE																								
COMMENT PROCEDURE																								

Exhibit 5.13 Infant Clinical Progress Record—Vital Signs

The format of this form allows for easy observation of a continuum of results, which will enable quick evaluations of those results. The front of this form (not shown) consists of a large data table. The most basic information, such as weight, weight change, and the time of day, along with the date, age, head circumference, and length of the baby, is recorded carefully.

The next section (see above) records infant temperature, heart rate, respiratory rate, and different activities (Phys/BEH—physical activities/behavior), such as facial expression, cry state, tone, heart activity, total activity, and any intervention and/or response. There is also space for comments. Because newborns are very vulnerable and, in many ways, unknown, nurses record their vital signs (including their facial expressions and behaviors) for assessment purposes.

While this form is adequate for its purposes (basically to provide a grid to fill in), it's not the easiest to read, especially with the categories in smaller font than the individual components. In addition, abbreviations are used in headers/labels (especially the vertical ones), which could be difficult to decipher at a glance.

The next section of this form (see Exhibit 5.14) applies to more specialized recordings, such as circumcision on male babies, with categories for "Circ. Done," "Circ. Healing," and "1st Void After Circ." For the umbilical cord stump, there are recordings for whether the cord is dry and intact, if it is moist, if there is a clamp on it, and if the clamp has been removed. There is also a section for photo treatment or for an eye patch, used when infants are monitored under bilirubin lights to aid in liver function, which is slow to begin in newborns. If lights (or natural sunlight) aren't used, the infant may be jaundiced. Sections for "Bonding time with the mother" and "I.D. Band Verification" are among the more recent additions. In the past, "bonding" of the human infant with its mother was not a well-known subject; now psychologists have weighed in on its importance, and nurses are expected to observe and record the mother bonding. I.D. band checks are crucial to legally ensure the right baby goes to the right mother, especially if the infant is not rooming in with the mother at the health facility. The final part of this section records information about bathing the infant,

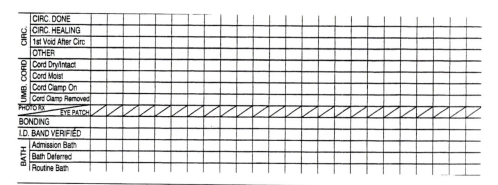

	CIRC. DONE
CIRC.	CIRC. HEALING
	1st Void After Circ
	OTHER
UMB. CORD	Cord Dry/Intact
	Cord Moist
	Cord Clamp On
	Cord Clamp Removed
PHOTO RX	EYE PATCH
	BONDING
	I.D. BAND VERIFIED
BATH	Admission Bath
	Bath Deferred
	Routine Bath

Exhibit 5.14 Infant Clinical Progress Record—Post-Birth Care

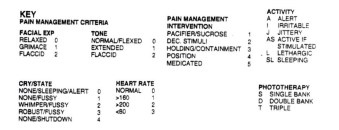

KEY

PAIN MANAGEMENT CRITERIA

FACIAL EXP		TONE	
RELAXED	0	NORMAL/FLEXED	0
GRIMACE	1	EXTENDED	1
FLACCID	2	FLACCID	2

PAIN MANAGEMENT INTERVENTION

PACIFIER/SUCROSE	1
DEC. STIMULI	2
HOLDING/CONTAINMENT	3
POSITION	4
MEDICATED	5

ACTIVITY

A	ALERT
I	IRRITABLE
J	JITTERY
AS	ACTIVE IF STIMULATED
L	LETHARGIC
SL	SLEEPING

CRY/STATE		HEART RATE	
NONE/SLEEPING/ALERT	0	NORMAL	0
NONE/FUSSY	1	>160	1
WHIMPER/FUSSY	2	>200	2
ROBUST/FUSSY	3	<80	3
NONE/SHUTDOWN	4		

PHOTOTHERAPY

S	SINGLE BANK
D	DOUBLE BANK
T	TRIPLE

Exhibit 5.15 Key to Observations and Interventions

with spaces for recording the admission of the bath, whether the bath was deferred, and if it was a routine bath. All of these recordings are basic functions and tasks carried out daily with newborns, and this form makes carrying out those procedures and keeping track of them much more efficient.

This next section of the medical progress report, Exhibit 5.15, is unique among forms because babies are unique. They can't tell us what's wrong, so we have to depend on their body language. Numbers are assigned so frequent checks of the infant can record their activity over hours. For example, if an infant was constantly crying or had extended limbs, he or she could be expressing some degree of pain and discomfort. If the baby was constantly sleeping, never alert, or had flaccid instead of normally flexed muscles, serious problems could be expected and the infant would need quick intervention. Because infants sleep so much, without some monitoring system like this, it would be hard to catch serious and not-so-serious problems. On the phototherapy section, "single bank" and "double bank" refer to banks of bilirubin lights. These qualitative values of observation are transformed into quantitative values for recording onto the form. This greatly streamlines and standardizes these forms for comparison and ease of use. Pain management interventions include "Pacifier/Sucrose," "Dec. Stimuli," "Holding/Containment," "Position," and "Medicated." "Dec." means decrease stimuli; "Holding/Containment" means holding and wrapping the infants securely in blankets, which, resembling their tightness in the womb, often comforts them.

On the reverse side of this form (not shown) there is a chart for "Intake and Output," where oral intake is recorded, specifying whether feedings are from the breast, formula, or sucrose. These feedings are recorded throughout the day, with the time being recorded as well. For breastfeeding, there are many different feeding patterns, and these are noted on the form. For the output section, there are headings for spitting, voiding, and stools. All of these are important functions for infants and must be monitored regularly to ensure that the infant is behaving and functioning properly.

Along with detailed information about feedings, there is a key that specifies how the information should be recorded, just as on the previous page. It also gives a very detailed charting system for recording breastfeeding. Breastfeeding is a very important function for infants and is monitored as accurately as possible via this report since amounts cannot be measured. Like APGAR scores, LATCH is a mnemonic device for remembering to check the various important parameters of the process; L = Latch, A = Audible swallowing, T = Type of nipple, C = Comfort (of breast nipple), and H = Hold positioning. Latching otherwise refers to the infant taking hold of the nipple to begin nursing. There's even a chart that assigns numbers to breastfeeding characteristiscs.

Overall, this report form is highly functional and encourages thoroughness, speed, and precision with the recording of information. This form is vital to recording the normal (or abnormal) development of the infant. It is also interesting to note that very little of the language within the report is cryptic (except for a few cryptic-looking abbreviations mentioned above), and it can be understood easily by those who are labor and delivery or pediatric staff.

As a document designer or health communicator, you will want to become adept at designing and customizing tables for both procedural and progress forms at your facility. Although these forms are often preprinted by a printing house, you will still need to design and revise certain forms as changes occur in procedures and care giving.

What About Forms That Patients Fill Out?

The next form represents the genre of patient assessment in which the patients fill in their own health information on all or part of the form. You are all familiar with going to the doctor's office for an office visit or a health facility for a procedure, and, if you are a new patient, you must go early to fill in an extensive questionnaire regarding your health history. This questionnaire is placed at the front of your chart for easy reference. Such assessment or history forms come in preprinted formats as well as those designed in-house. You may be asked at some point to better customize a form for your workplace.

The primary function of this document is to give an overview and history of a patient who is being referred for surgery by another doctor. The form is often well structured visually and has three distinct sections that are easy to distinguish with a quick glance. Such a form is used to assess everything from whether or not the patient has leg cramps to the date of the patient's last Pap smear. This assessment is completed before the first contact with the surgeon. Another interesting thing about some forms is that a barcode is located in the bottom left corner. This allows for the forms to be scanned into a computer. On this form, the instructions to the patient need to be bolded for better figure/ground separation. Otherwise, the instructions would fade into the background and would be indistinguishable from the surrounding text.

There should be a portion that lists different conditions as well as test dates, providing a detailed HXP. The conditions can be listed in two columns with boxes to simply check yes or no next to each. The conditions should be extensive, and the list should be well organized with the conditions grouped into subjects and areas of the body (although they are not specifically labeled this way, there are subtle divisions). For example, have the first few conditions pertain to cardiology and the circulatory system, while the next group pertains to conditions of the head and brain. This makes filling out the form easier because the patient can concentrate on one area of the body at one time. The patient can fill the form out quickly, and the physician can read it quickly, gaining important information before going into surgery. Filling in their own forms has become routine for patients in this culture, but it wasn't always this way. With the expense of and demands on physicians' time, this time-saving self-documentation is a valuable part of the communication system.

A final section should contain a place for the patient to comment on his or her family history as well as other health concerns. The placement of this section at the end of the list also saves time because only things that were not covered in the previous list need to be written down. Such a form can be condensed but easy for patients to fill out.

Summary

The forms shown in this chapter represent but a few of the many possible forms used in medical and health settings, but the four genres featured here are among the most common in health professions: encounter forms, procedural reports, progress notes, and forms filled out by patients to provide information to health providers. By studying some of these forms, you can gain an awareness of what the range is and what considerations are important when designing and filling out forms. Most of the forms featured here are not set up for scanning or inputting into a computer database.

You learned about encounter forms, which document the first official "encounter" between a patient and a health professional or vice versa. This is the site for history taking and physical exam (HXP). The encounter form genre usually has entries for some patient demographics, HXP, systems checks, a place for comments, and lists of areas that can be circled or checkmarked to save time. Some encounter forms still are in triplicate if the system is not automated or computer-ready, and they include billing codes for a dual-purposed document. Often they provide a section for the health professional to circle lab tests, diagnostic tests, and procedures.

The procedural form in this chapter documented labor and delivery in a hospital setting. This example demonstrated a document design to enable maximum information in a minimum of space. Although the physician later dictates a narrative report that also goes into the patient's chart, this form helps the professionals by providing all the possibilities attending a birth in an easy-to-use format. As writers we can see that procedural forms need to be customized to reflect the particular details typical of each special procedure.

This chapter included examples of progress notes, which differ from the others in being similar to data sheets allowing for frequent and repetitive monitoring. Any progress records will include vital signs, and, in this example, they allow for close monitoring of infants following their births. Progress notes provide a continuum of results for quick evaluation purposes.

The last genre discussed in this chapter is the form that the patient, rather than the health professional, fills out. This form, too, provides a type of assessment, but also includes the patient's own history and screening for particular maladies that affect procedures such as surgery and the anesthesia required for surgical procedures. These forms are typically called *Patient Assessment Forms,* but they may be called by other names, too, depending on the health facility. This assessment genre is similar to the HXP encounter forms, except that the "encounter" is more specific, i.e., for a surgical procedure and/or anesthesia.

While forms such as these remain a key part of print-based patient charts, in the future, you may be working with or even designing software formats that allow for electronic documentation. Also, of course, as medical and surgical techniques and capabilities change, the forms must change to reflect the new technologies.

Discussion Questions

1. What issues are important ethically in working with forms in a medical setting? What actions could cause violations of the HIPAA regulations? What options do you have as a health writer to design forms and ensure HIPAA compliance?
2. What is the history of chiropractors in this country? Why do you think there have been different branches of medicine that would oppose each other when medicine is geared toward curing people and keeping them healthy?
3. Discuss the encounter forms presented here. Do you agree with the analyses in the chapter? What other advantages and disadvantages do you see in the forms? What changes would you make in the forms if you were asked to make them more user friendly?

4. How would you prepare a form for a group of patients from another country besides changing the language? Discuss, too, how you would research the other culture to design your form in ways appropriate to a very different society.

5. Think about a medical procedure with which you are familiar. Describe the parts of the procedural form that would be necessary to have for this particular procedure. What does "form follows function" mean and how does that saying apply to medical procedural forms?

6. Take any of the forms in this chapter and discuss how the form would work as an electronic document. What interactivity could you imagine being possible for that particular form?

7. How do you think medical forms will change in the future besides changing to a computer-based format? Be creative here and use your imagination.

Exercises

1. In your team, search the Internet and your library to find sources for the historical development of chiropractics, Western biomedicine (the one with which you may be most familiar), and osteopathy. Prepare a PowerPoint or other presentation, perhaps a timeline, that illustrates the development of the different branches of healthcare.

2. Create a chart or graph that shows statistics of the popularity of the different areas of medicine, chiropractics, and osteopathy with the American public. Present a compare-and-contrast analysis of your findings.

3. Research the history of the American Medical Association. Do an in-depth analysis of some of its strategies over the years. How long has it been an organization and what kinds of events made such an association possible?

4. In a team, prepare a charting form, a synthesis that combines the best aspects of the forms featured in this chapter.

5. Choose one of the procedural or assessment documents in the chapter. Storyboard a possible interactive Web design that would work for moving this document to an electronic interface.

6. As a team, choose one of the forms here, redesign it, and take notes on why you made the choices you made to change or keep parts of the present document. Give a PowerPoint presentation featuring your new design, the old design, and explanations of your design choices.

7. If possible, find other medical/dental/veterinary forms in your community and bring them into class. Compare the forms you have with the ones in the chapter. Can you imagine there would be regional differences in forms or do you think that they would be the same throughout the country?

Works Cited

Crew, Linda, Executive Director of the Joseph F. Sullivan Center for Nursing and Wellness. Personal Interview. May 13, 2003.

Joseph F. Sullivan Center Encounter Form. Joseph F. Sullivan Center for Nursing and Wellness at Clemson University. Clemson, SC. February 2003.

Joseph F. Sullivan Center Health Encounter Summary. Joseph F. Sullivan Center for Nursing and Wellness at Clemson University. Clemson, SC. February 2003.

Endnotes

1. Many thanks to Chris Welch, Andrew Mathias, and James Galloway, two premed and one political science major, in my Spring 2003 Honors Technical Writing class. They encountered many obstacles in obtaining medical report forms for this chapter, much of which they wrote. (I have taken the liberty of revising their prose to suit the textbook, but the research is largely theirs.) I remain deeply grateful for their diligence and persistence.
2. If you haven't done so, read Chapter 4 or review it.
3. C/O is often seen on encounter forms and means: "Complains of."
4. SOAP explanation can be found in Chapter 4.
5. Phoria is the relative directions assumed by the eyes during binocular fixation of a given object in the absence of a neural stimulus.
6. "Corrected," of course, here means with glasses or hearing aids or other assistive devices.
7. PMD = Primary medical doctor or primary care physician.
8. MI stands for myocardial infarction or heart attack. HTN is hypertension.
9. TB is tuberculosis and SOB is Short of Breath.
10. PVD is peripheral vascular disease.
11. DES refers to a drug given to pregnant women in the fifties to ensure against miscarriage. Unfortunately, daughters of these mothers have had an unusually high rate of ovarian and cervical cancers.
12. Tubal refers to a tubal or ectopic pregnancy; the gamete or embryo begins to grow in a Fallopian tube rather than in the uterus. HRT = Hormone Replacement Therapy for menopausal women.
13. Oxytocin causes the uterus to contract and speed up labor, as does rupturing of the membranes of the amniotic sac that can be accessed once the cervix is dilated.
14. Meconium is the fetal waste product and is a sign of fetal distress if it is present during labor and delivery.
15. Disposition of the infant means where the baby goes after its birth: with the mother to the OBRR (Obstetrical Recovery Room) or to the NICU (Neonatal Intensive Care Unit).

CHAPTER 6

Health Education Materials

 ## Overview

Reports and manuals represent much of the writing in hospitals, clinics, and HMOs, but frequently document designers in the health professions are responsible for creating health education materials. The preventative health industry and its attendant writing tasks have grown exponentially, especially in the last decade. In the past, medical practitioners communicated most health information verbally to patients. With the shortage of nursing staff, time limits enforced by HMOs, and other factors, health professionals frequently do not have time to communicate verbally to patients except in brief sound bites. Often when doctors do communicate more extensively with patients, the information is overwhelming, either because of technical jargon or because of the large amount of information, not to mention the emotional valence of negative news. In addition, any patient who reads the newspaper, watches television news broadcasts, or surfs the Web hears about every new study regarding lifestyle issues in healthcare, new pharmaceuticals, genetic breakthroughs, and medical discoveries. Often, too, patients are confused as studies are reported that contradict each other. Many of them complain that they alter their diet, exercise, medication, and other aspects of life to adjust to one set of study results, only to have that "new" knowledge contradicted by another study several months or years later. While patients have more and more access to health information, their education and understanding, because of the overwhelming nature and fast pace of the Information Age, is limited and spotty.

Now when patients prepare for surgery or leave the hospital after a procedure, they are often given instruction sheets that the health professional goes over with them, as opposed to educating them verbally. In some healthcare settings, patients view videotapes before surgeries or to understand diagnoses, such as diabetes, that require large adjustments of their lifestyles for proper management of the disease. Before cataract surgery, for example, patients are sometimes asked to view a videotape (the irony inherent in this situation does not escape us). In the following chapter, you will learn about the types of health education materials, both traditional and on-line, which medical writers develop for targeting groups of health information consumers. I'll outline methodologies here that emphasize education of and training for health education professionals-to-be and patients, e.g., the types of materials mentioned above. Thus in this chapter you will look at examples of genres and their health content. In Chapter 3 you already read about how to design and format documents in general.

In this chapter you will learn the answers to the following questions:

- What are well designed traditional materials?
- What are well-designed healthcare handbooks?
- How are booklets and magazines designed?
- What do good newsletters look like?
- What do well-designed brochures look like?
- What about instruction sheets?
- What other health materials are available?
- What about on-line materials?
- How do I find reliable health Web sites?[1]

What Are Well-Designed Traditional Materials?

Traditional materials include handbooks, booklets, newsletters, brochures, pamphlets, posters, instruction sheets, and update letters that are used to educate and inform patients. Writers based in pharmaceutical companies develop various charts, diagrams, anatomical drawings, and other posters to hang in offices where patients are examined. In addition to the anatomical and physiological posters, the drug companies also send out plastic models of body parts, so that doctors, nurses, and other health professionals can use them to explain diagnoses and treatments to their patients. Because such companies have healthy budgets, they can afford the best document designers, materials, and printing for their health education products. Sometimes materials are developed in-house if the company is large enough to afford the printing equipment. Often, however, once the design and content are developed, the actual printing and its preparation are outsourced. Either way, professional writers are responsible for checking and double-checking content and writing the information in as clear, accessible, and effective form as possible.

What Are Well-Designed Healthcare Handbooks?

In the chapter on designing for your audience and regional context (Chapter 1), you learned about a relatively small but well-funded hospital that decided to target the population of their geographic area (a small-but-growing town in the

Southeast and the towns in its adjoining counties) in an effort to reduce the number of inappropriate emergency room visits. Hospital administrators had heard of a campaign in Boise, Idaho, that had successfully addressed this issue. A major part of the campaign included handing out a 371-page book, *Healthwise Handbook*, which listed healthcare problems, detailed prevention strategies and care that could be given at home, and included a list of symptoms that would signal the need for trained professionals. The purpose was to keep patients from wasting valuable time and resources, both their own and those of the health professionals, by going to the emergency room when it was not necessary. The book was well done and very successful among Idahoans in Boise. In the southeastern region, the book was printed and distributed by the thousands in the entire multiple-county area, one delivered to every resident. Unfortunately, unlike the target population in Boise, Idaho, which spent a significant amount of time reading, in the small towns of the rural Southeast, the television set was on constantly in the homes when people were not at work. In short, a book was not the first choice for consultation in any type of health situation (or perceived emergency). Fortunately, the hospital system added other features when they discovered the book was not being consulted, such as a hotline, which included a nurse-on-call 24 hours per day to answer questions. The hotline has been much more successful than the book. However, *for the right audience*, such books are valuable sources of patient education.

As you can see in the illustration on the following page, this book cover, although somewhat text-heavy and "busy," is making very clear its intent on the top center of the cover as well as on the lower right: to educate people about when they don't need to come to the emergency room and to educate them about what symptoms do demand professional or emergency care. The top center reads: "Eight out of 10 health problems are cared for at home. This book will help you do a better job." On the lower right we read in an even larger font: "Over 180 health care problems. Prevention. Home care. When to call a doctor." In the full-color cover, these two text boxes are red, drawing the readers' eyes to these purposeful labels.

The text is divided into five parts according to anatomical complaints:

Part I—Self-Care Basics

- Making Wise Health Decisions
- Prevention and Early Detection
- First Aid and Emergencies

Part II—Health Problems

- Abdominal Problems
- Back and Neck Pain
- Bone, Muscle, and Joint Problems
- Chest and Respiratory Problems
- Eye and Ear Problems
- Headaches
- Mouth and Dental Problems
- Skin Problems

Eight out of 10 health problems are cared for at home.
This book will help you do a better job.

PARTNERS

HEALTHWISE®
HANDBOOK

An initiative of PARTNERS FOR A HEALTHY COMMUNITY

More Info Plus more information from health help lines, Internet Web pages, information stations and more.

Over 180 health care problems.
Prevention. Home care.
When to call a doctor.

healthwise®
PUBLICATIONS

Exhibit 6.1 Cover of *Healthwise Handbook*

- Infant and Child Health
- Chronic Conditions

Part III—Men's and Women's Health
- Women's Health
- Men's Health
- Sexual Health

Part IV—Staying Healthy
- Fitness and Relaxation
- Nutrition
- Mental Self-Care and Mental Wellness

Part V—Self-Care Resources
- Your Home Health Center
- Index

The above list gives you a good indication as to the areas of prevention and patient education that you will most likely address in your work with patients. The emphasis on the home as the center of healthcare is another recent development, although up until the end of the eighteenth century and the beginning of the nineteenth century (and later in many countries), most of the healthcare was centered in the home. Home health service is now the fastest growing sector of healthcare in the present economy. Once acute situations are stabilized and procedures are completed, because of the high cost of hospitalization and treatment, patients are hurried out of the hospital and into the care of family and/or friends. This situation provides those of us who are professional health writers with even more opportunity to work on materials for patient and home caregiver education, as the site of much of the follow-up care has changed back to the home.

In Exhibits 6.2 and 6.3 you see the way each individual health problem is laid out:

- A description of the problem
- Prevention
- Home treatment
- When to call a health professional

This consistency gives patients and caregivers an easy cognitive organizer to follow as they use the book over time. In addition, the footer for each page includes the following information: "For more information, see inside the back cover." Each footer message includes a small graphic or icon that draws the readers' eyes. The graphic is a small starburst that is labeled "more info." When the reader turns to the inside back cover, she discovers that graphic again, now bigger and at the top of the page. It says, "More questions? Here's (icon is inserted here with the "More Info" label)." As you see in Exhibit 6.4, this page is nicely designed to be a quick reference via various media (telephone, the Web, at home or at the library) and allows access to more print copies. The text is clearly designed with the

Healthwise Handbook

- If you experience vertigo (sensation that the room is spinning around you) that is severe or persists for more than three days, has not been diagnosed, or is significantly different from previous episodes.

- If you have repeated spells of light-headedness over a few days.

- If your pulse is less than 50 or more than 130 beats per minute when you are feeling lightheaded.

Ear Infections

Ear infections can occur in the middle ear or the ear canal (see Swimmer's Ear, page 175).

A middle ear infection (otitis media) usually starts when a cold causes the eustachian tube between the ear and throat to swell and close. When the tube closes, fluid seeps into the ear and bacteria start to grow. As the body fights the infection, pressure builds up, causing pain. Antibiotic treatment stops bacterial growth, relieving pressure and pain.

Left untreated, the pressure can cause the eardrum to rupture. A single eardrum rupture (perforation) is not serious and does not cause hearing loss. Repeated ruptures may cause hearing loss.

Young children get more ear infections because they get more colds and their eustachian tubes are more easily blocked.

Symptoms of a bacterial ear infection include earache, dizziness, ringing or fullness in the ears, hearing loss, fever, headache, and runny nose. In children who can't yet talk, tugging on the ear may be a sign of pain, especially if they are sick.

Otitis media with effusion (**serous otitis**) is a collection of fluid in the ear that often remains after an ear infection. There are often no symptoms, or there may be a feeling of fullness in the ear and some minor hearing loss. Effusion is not a cause for concern and may not require treatment unless it lasts longer than three months or causes significant hearing loss.

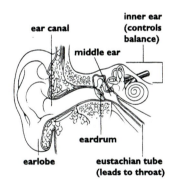

Ear infections can occur in the ear canal or middle ear. Dizziness can be caused by an inflammation of the inner ear.

170   **For more information, see the inside back cover.**

Exhibit 6.2 Ear Infections

Source: Healthwise Handbook, page 170.

Eye and Ear Problems

Prevention

- Teach your children to blow their noses gently. This is a good idea for adults, too.

- Breast-feed your baby. Breast-fed babies have fewer ear infections.

- Feed infants in a relatively upright position to prevent milk from getting into the eustachian tubes. Do not allow infants to fall asleep with a bottle. (Nursing infants may fall asleep at the breast.)

- Avoid exposing children to cigarette smoke, which is associated with more frequent ear infections.

- If possible, limit your child's contact with other children who have colds.

Home Treatment

- Apply heat to the ear to ease the pain. Use a warm washcloth or a heating pad set on low. Don't leave a child alone with a heating pad.

- Rest. Let your energy go to fighting the infection.

- Increase clear liquids.

- Acetaminophen, aspirin, or ibuprofen will help relieve earache. See dosage on page 341. Do not give aspirin to a child or a teen under 20.

- Oral nasal decongestants may help relieve earache pain. Avoid products that contain antihistamines.

- If dizziness occurs, see page 168.

When to Call a Health Professional

- Anytime an ear infection is suspected. If the exam confirms an infection, antibiotics will be needed.

- If a severe earache lasts longer than one hour or any earache lasts longer than 12 to 24 hours. If the pain is severe at night, call the next morning even if the pain has stopped. The infection may still be present.

- If an infant repeatedly rubs or pulls on the ear and is not just playing with the ear.

- If a child has an earache and fever and appears ill.

- If a headache, fever, and stiff neck are also present, which may be signs of meningitis (see page 150).

- If home otoscope exam shows redness in the ear of a small child who cannot describe ear pain. See page 333.

- If there is a white or yellow discharge that is not wax, or a bloody discharge from the ear. This may indicate a ruptured eardrum.

- If there is no improvement after three to four days of antibiotics.

- If stuffy ears or hearing loss persist without other symptoms for more than 10 days after a cold has cleared up.

 For more information, see the inside back cover.   171

Exhibit 6.3 Ear Infections continued
Source: Healthwise Handbook, page 171.

More questions? Here's More Info

It's Confidential - and It's Free

Call Partners Nursewise.
Talk with a nurse 24 hours a day.
Call **261-2001** in Anderson County
or **1-888-525-1333** in Pickens,
Oconee and Hart Counties

Go to the Web:
www.healthy-community.org
Click on the Healthwise
Knowledgebase icon shown at
right and reach the most comprehensive consumer health
database in the world.

Don't have Web access of your own?
Visit your public library and use an on-line
computer to reach www.healthy-community.org
and the Healthwise Knowledgebase.

Need extra books? You can buy a copy of the Healthwise®
Handbook for $10.00 at the following places:

AnMed Health Resource Center at Anderson Mall, Anderson
Oconee Memorial Wellness Center at Oconee Memorial Hospital, Seneca
Palmetto Baptist Education Center at Town and Country Plaza, Easley
Hartwell Pharmacy, 18 Athens Street, Hartwell
Cannon Memorial Hospital, Hospital Coffee Shop, Pickens

Exhibit 6.4 Inside Back Cover
Source: Healthwise Handbook, 1999.

reader in mind, so that she has as much access to reliable health information as is possible. Because these texts were passed out to every household in a three-county region and a hotline and Web site were set up, anyone seeking information in this area should be able to do so.

Notice the language in all of the exhibits featured so far in this chapter is "unjargonated." People who are not health professionals can clearly understand the language without being versed in medical discourse. In Exhibit 6.3, the Prevention, Home Treatment, and When to Call a Health Professional sections are bulleted to again provide the simplest access for the reader. In this way the format gives patients and caregivers a checklist that can be quickly perused for the information that fits the particular situation. This information architecture is well designed for maximum accessibility.

At this point we are waiting for the assessments of how well the recent introduction of the *Healthwise Handbook* has done to decrease emergency room admissions. We suspect, given this community, the 24-hour call line has been more effective, although the statistics show that the book itself was effective in Boise, Idaho, where families still read books together. Also, given the economic changes—huge job layoffs with the attendant loss of healthcare benefits, most likely emergency room visits have increased. We do know handbooks will be effective among some communities, but definitely not all.

How Are Booklets and Magazines Designed?

Some of the patient education materials serve two purposes. Besides the obvious purpose of educating health consumers, the materials can also serve to advertise hospitals, HMOs, health centers, and other medical facilities and products. The pharmaceutical companies produce educational materials for patients, too, although they more frequently target the physicians who will prescribe their medications.

One local hospital in the rural upstate of South Carolina publishes a professional-looking, full-color, glossy quarterly magazine that both advertises its facility as well as educates the public. These are delivered by snail mail to a wide area of residents. The publication is *OMH TODAY: A Quarterly Publication of Oconee Memorial Hospital*. Each issue centers on a particular health problem and also serves as a guide and update for programs and available health services. The Winter 2003 issue, for example, was centered on cardiac health. The front cover shows a man's chest (he is dressed in a white shirt and tie) with his hand clutching his left upper quadrant. The caption that is centered on this color photo says in large letters: "You're Having Chest Pain: What to Do Until the Ambulance Arrives." Given the distances between rural citizens and the facility, such an issue is especially useful, because in this area the ambulance could be many miles and minutes away. The booklet is 10 pages long not counting the back cover, which also contains helpful information.

The standard format for this publication includes a central article that features the lead article advertised on the front cover. This covers four of the 10 pages in the publication, pages five through eight. On the first two pages, every quarter the reader gets an update of the new physicians on staff at OMH and the profiles of other doctors on staff as well. *OMH TODAY* features a short paragraph for each of the new physicians without photos, while the on-staff physicians have their pictures along with their education, experience, and Board certifications. Below the new physicians, the publication advertises its passing the accreditation procedure and the opening of its new endoscopy suite. Along side the fuller descriptions of staff physicians and their photos, the magazine aligns its table of contents with two features:

- "You're Having Chest Pain: What to Do Until the Ambulance Arrives."
- "OMH Cath Lab: Providing a High Standard of Care."

The regular quarterly features, or "In Every Issue," include

- Hospital News
- Health Tips/Recipes
- Wellness Center Services

The format and layout stay the same from quarter to quarter, although the color choices change.

The feature article on chest pain conveys very practical information on what to do after you call the ambulance. The article begins with the scenario of a husband with chest pain, who is referred to in the third person and a wife who calls the ambulance, who is referred to in the second person. The audience seems to be women, who are most often the caregivers. The scenario principle is used effectively here (Flowers et al., 1983).[2] The following paragraphs illustrate the well-written and practical advice given throughout the article:

> According to Wayne Garland, OMH EMS Director, you've made the right decision [to call an ambulance]. Now, here's what to do next: Before you call 911, make your husband comfortable and find out exactly how he's feeling so that you can relay this information to the emergency dispatcher. Try to have him lie down and remain quiet, as exertion of any kind can extend the potential damage of the cardiac event.
>
> After you talk to the emergency dispatcher, make sure your husband remains calm and quiet. If people nearby are upsetting him, ask them to leave the room so he won't be disturbed. (Sometimes having others "hover" nearby makes the situation more stressful for the victim, which may worsen the situation.) Remain calm and ensure he's as comfortable as possible.
>
> If your husband is clammy, lethargic, sweaty, pale, hot, or not responding appropriately, have him lie down and elevate his legs. Loosen restrictive clothing. Keep him still. Don't give him fluids as he may become nauseous. . . .
>
> Once he's settled and before EMS arrives, you might want to gather his medications; the paramedics will need to know what he's taking. Make sure you know the rest of his medical history as well, including other medical problems that may affect his "normal" pulse, blood pressure, and respiration rate. (The paramedics will want to compare what he's presently experiencing with his "normal" readings.) (*OMH TODAY,* 7)

The text is clearly written, easily comprehendible, and targeted correctly for its audience. In the rural upstate of South Carolina, it is appropriate to target married women, since the demographics support this choice. If you were to write a similar article in New York City, it would most likely be appropriate to target males and females with the victim a stranger or someone perhaps less well known (for instance someone on the subway or on a New York street).

The last two pages of the publication feature the following five main topics each in their own column with a light-colored background:[3]

- "OMH Cath Lab: Providing a High Standard of Care"
  Diagnostic Procedures
  Interventional Procedures
  Device Placement
- "Cardiac Health Terms"
  Angina
  Arteriosclerosis
  Coronary Arteries
  Coronary Artery Disease [CAD]
- "Risk Factors for Heart Disease"
- "Tips for Decreasing Your Risk of Heart Disease"
- "A Heart Healthy Dinner"
  Pineapple Marinated Grilled Salmon
  Rotini Primavera
  Sorbet with Fresh Fruit

One advantage of this booklet or small magazine is that it contains helpful information, but it is also not overwhelming for readers.

The back cover also contains useful information and is a regular feature of the publication. The layout here is three-columns, two of which are the Wellness Center Programs with descriptions of the programs and the instructors:

- Weight Management Programs
  Healthy Solutions 5 + 5
  The Ultimate Diet
- Wellness and Prevention
  Family Health: Helping Our Children Live Well
  Pedometer Program: Moving and Losing
  Individual Health Plan and Coaching with Health Risk Appraisal
- Diabetes Self-Management Program
- Medical Nutrition Therapy
  FREE! Healthy Eating: Nutrition Basics
  Grocery Store Tour
- Monthly Wellness Screenings
  Cholesterol and Blood Sugar
- Parents and Infants
  Preparation for Childbirth and Parenting Classes
  FREE! Boot Camp for Dads
  Free Information Night on Pregnancy, Childbirth, and Newborn Parenting

In addition to the Wellness Center Programs, the back cover also lists phone numbers for all the Community Health Services Continuum and, lastly, OMH events over the last quarter, such as Immunizations, Auxiliary News, and Awareness Events.

As a health professional or as professional health writer, you will want to design or contribute to such a booklet or magazine with a solid awareness of your particular audience and context (see Chapter 1 of this book). You can also ask for a previous issue, if one exists, or a model publication that the health administrators find appropriate, so that you can write or contribute in a way that fits well into the existing or desired genre. Lee Clark Johns emphasizes the usefulness of models in his article on using templates from the file cabinet to reproduce the documents necessary for your job. Be sure you get feedback on your designs and content from those commissioning such work before you go to press. This step will prevent surprises; if you get time, you might want to conduct focus groups or do some formal or informal usability testing on the project depending on your budget and timeline.

What Do Good Newsletters Look Like?

Newsletters are less costly and more quickly written documents that accomplish some of the same purposes as the larger publications mentioned above. Health writers and health professionals can do simple newsletters via desktop publishing with a range of software beginning with the low-end and low-learning curve of Microsoft Word to Adobe PageMaker to Quark Express (most typical software for magazines and other such print publications with a medium-to-high learning curve) to the high-end Web designer programs such as Macromedia FireWorks. Newsletters are typically columnar with two to three columns the usual layout choice. Graphics and photographs (including free clipart graphics or ones you purchase via the Web or on a CD) can be easily imported with most programs. A scanner for photos and other artwork not yet digitized is helpful for building a file of useful visuals for your newsletters. Also, you'll want to use a consistent logo to make your newsletters visually and instantly recognizable.

The following example is a simple newsletter designed in Microsoft Word that is cheaply produced on $8\frac{1}{2}$-by-11-inch paper whose weight is heavier than the usual copy paper used for letters and other daily writings. A heavier weight paper is important especially if you plan to mail the newsletters.

The background on the Greenville Gynecology Group newsletter is buff; light colors are best if you plan to do the publication using black/white reproduction (the cheapest route). Many physician groups now send their own newsletters to patients, and it's especially useful when the group moves its office or takes on new partners, as is the case in this first newsletter from this physicians' group. As simple and easy to produce as this newsletter is, it's well designed and professional looking. The logo is quite a sophisticated and detailed design. A photo of the physicians is not of the highest quality resolution, but it is still passable in the copy being mailed out to patients. On the back page of the newsletter, doctors have included their email addresses, which is especially useful for patients who are outside the city (many of their patients are included in this group). More and more physicians are communicating with their patients via email (administrative assistants often screen the emails, to save time for busy doctors); however, one should not expect all doctors to communicate in this way.

The front page of this newsletter is the two-column layout with narrow margins to maximize the space for text. A photograph is inserted in the center of the newsletter with text wrapping around all sides of it. In most programs, when you embed the graphic or photo, you can resize the text wrap if you want it to come closer to or stay further from the edges of the visual. You want to keep the amount of space or margin on each side equal and the same for all photos and graphics. This particular newsletter uses two fonts, which is a good stylistic practice. Some newsletters' authors who have little design skills will use too many fonts and not have a consistent look to the document. The header font here can be ornate because it is targeted for a female audience. Again audience and context are the main factors here. A text box on the front page contains contact information, which is more typically found on the backside of a newsletter, but since this particular issue is a relocation one, it makes sense to highlight it by putting it on the front page. There are problems with the alignment on the bottom of the front page, i.e., the two columns do not match or do not end along the same gridline. This practice is not consistent with good newsletter or page design. Problems of this kind can be solved if the writer/designer simply puts the two headers at the top of the page along the same gridline (as is also appropriate). That change would create appropriate space between the article and the text box below.

As a health professional or as a health writer, good document design can be learned, and experience will give you more skills as you practice. Newsletters are an easy and often used choice, so it's important to learn and practice the basic design skills that make this genre successful and professional looking. Often, too, it is useful to develop a file or know where to get brief blurbs of health information (women's health in this case) to fill in newsletters in which you have more space than required information.

What Do Well-Designed Brochures Look Like?

Another frequently used method of patient education, advertisement, or outreach is the brochure. Brochures are effective tools because the writer, designer, or contributor can use colors and graphics with brief but important amounts of information that are quickly read and comprehended by your target audience. One large series of brochures for health advice can be found on pamphlet racks in many discount and national-chain pharmacies. One such series is produced by On Target Media, Inc., which produces extensive, three-color brochures or pamphlets of many pages and with extensive health information. For example, the one on cholesterol (see Exhibit 6.5) comes under the On Target Media series on Health Advice for Women. The cholesterol pamphlet includes 10 pages, featuring the following serif headers:

- Is Cholesterol Increasing Your Risk Of A Heart Attack?
  Get it checked
  Comparing the risks
  What is cholesterol?
  If I feel well, I have no reason to worry, right?
  Atherosclerosis

- What Are The Risk Factors For Heart Disease?
  The ones you can't control are:
  The ones you can control are:
- How Is Cholesterol Measured?
  What is "good" cholesterol?
  What is "bad" cholesterol?
  What are triglycerides?
  How often should I get my cholesterol levels checked?
  How much fat and cholesterol should I eat each day?
- What Can I Substitute For High Saturated Fat And High Cholesterol Foods?
- Work With Your Doctor To Reduce Your Risk of Heart Disease

You can see as a health communicator that the subject of cholesterol is covered quite thoroughly in this small booklet/brochure. A woman picking up this text gains information covering all aspects of this particular health issue. She's also encouraged to talk to her doctor about cholesterol. That message is featured prominently on the front page of the document, right under the 18-pt. title "Cholesterol." Within the text box is a size 14-pt. font, "A helpful guide when talking to your doctor." In addition, the brochure features four quotes from women aged from 45 to 61 years old. These quotes are somewhat emotional as opposed to the rest of the information in the brochure. For example, Terri, age 45 says, "I didn't realize I was at risk for a heart attack until it happened—suddenly and without warning. I never thought anything like that would happen to me."

The booklet also includes a checklist with a series of nine questions that women can mark, blank spaces for an appointment time, cholesterol levels, and questions for the physician. In addition the booklet features five graphics (three with statistical comparisons) and two photographs. Color reproduction is excellent and clearly this document is professionally printed. Data sources are well cited. As a health professional and/or health writer, if you are responsible for developing such informational pamphlets, you may want to accumulate your own file of other health education materials on the market so that when you design and develop your own line of materials, you have some models. Showing a team of coworkers or your supervisors these models is also useful in order to elicit a clearer idea of what their expectations are. If you are expected to develop a number of such products, you may also want to create a logo (or outsource this task to graphic artists if it is not your forte) and a style sheet so that your publication line is distinguishable and consistent. If you are to supply only the health information content and not the designs, you should ask that you be included in the development loop so that you can verify the accuracy and clarity of the information. If the area in which you are to supply content is relatively new, do thorough research to background yourself and verify your information with several physicians and nurses whose area of specialty is the same as the one you are researching. You'll want to review materials several times during the process, including the penultimate draft, so that information is not accidentally skewed as a result of design choices. For example, if you are developing work on warning signs of heart attack, you want to make sure your designers do not bury such information as a result of wrong hierarchical arrangement of data via design choices. (See Chapter 3 for more discussion on visual arrangement.)

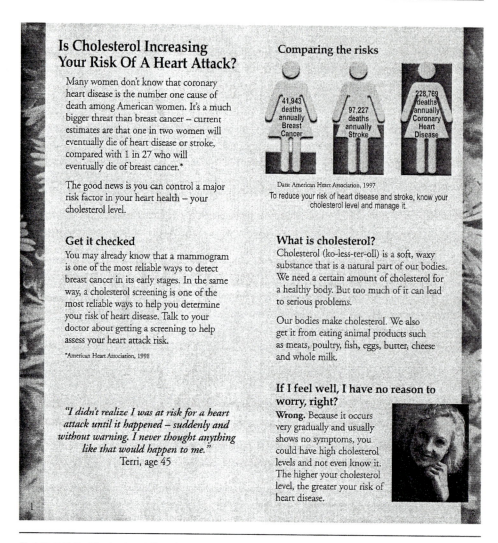

Exhibit 6.5 Cholesterol Brochure pages 1 and 2
Source: On Target Media, Inc. 2001

What About Instruction Sheets?

Frequently both before and after procedures, patients are given instruction sheets to educate them on how to prepare for a procedure, what to expect from it, and how to manage care after a procedure. These are usually at most one- or two-page sheets that are brief and to the point. This information is often extremely critical. For example, if a patient is told to fast before a procedure and does not follow the instructions, if he or she is put under general anesthesia, the possibility of vomiting and aspirating stomach contents during a procedure can endanger the patient's life. On some exams, for example a colonoscopy, if the patient does not follow the preparation directions to the letter, he or she may have to repeat what is at best an uncomfortable procedure because the doctor did not have the full viewing capacity necessary for accurate diagnoses. Usually doctors or nurses give the patient the instruction sheet in the exam room prior to the procedure. Proper medical practice mandates that a nurse, physician, technologist, or other health

---

COLONOSCOPY INSTRUCTION SHEET
PHYSICIAN'S NAME

PICK UP FLEETS PHOSPHOSODA® PREP AT LEAST 2 DAYS PRIOR TO
YOUR PROCEDURE

AVOID ASPIRIN, MOTRIN®, OR ARTHRITIS MEDICATIONS TEN DAYS
PRIOR TO PROCEDURE IF POSSIBLE (TYLENOL® MAY BE TAKEN)

Please read this information sheet very carefully. You will be asked to sign a **consent form** stating that you have been informed of **all the items on this sheet and that you give your consent for the procedure.**

The preparation of the colon for this procedure is extremely important in order to assure an adequate examination and to make the procedure as safe as possible. It is **absolutely essential** that you comply with the following instructions: One (1) day before that procedure, drink only a clear liquid diet. DO NOT EAT SOLID FOODS. **You may have** broth, bouillon, strained fruit **juices** without pulp (white grape, apple), lemonade, black coffee – No cream or creamer, Jell-O®, Sprite®, 7UP®, ginger ale, Gatorade®. **Avoid any red or purple liquids or red or purple Jell-O®. Ginger ale or apple juice may be taken with prep.**

Fleets® Preparation for Morning Procedure (times are suggested and may be modified for patient convenience.)
1. The night before your procedure, mix 1 tablespoon of Fleets PhosophoSoda® with 8oz of ginger ale, Sprite® or apple juice and drink at 4pm or earlier. Repeat this twice more within 20 minutes (total of 3 glasses). One hour after taking Fleet PhosphoSoda®, drink at least three more 8oz glasses of clear liquid.

2. Mix 1 tablespoon of Fleets PhosophoSoda® with 8oz of ginger ale, Sprite®, or apple juice at 8pm and drink. Repeat this 2 additional times in the next 20 minutes (total of three glasses). Drink 3 more 8oz glasses of clear liquid 1 hour later. **You may drink approved liquids to within 4 hours of your scheduled arrival time to_____ Hospital.**

3. **If your stools still have particles or are brown in color, you should take an over the counter Fleets® enema prior to arrival.**

(continued)

Exhibit 6.6 Colonoscopy Instruction Sheet

Fleets® Preparation for Afternoon Procedure
1. The night before your procedure, mix 1 tablespoon of Fleets PhosophoSoda® with 8oz of ginger ale, Sprite® or apple juice and drink at 7pm. Repeat this twice more within 20 minutes (total of 3 glasses).

2. One hour after taking Fleet PhosphoSoda®, drink at least three more 8oz glasses of clear liquid.

3. On the morning of the day of your procedure, mix 1 tablespoon of Fleets PhosophoSoda® with 8oz of ginger ale, Sprite®, or apple juice and drink at 6am. Repeat this twice more within 20 minutes (total of three glasses). One hour later take 3 more 8oz glasses of clear liquid. **You are encouraged to drink approved liquids to within 4 hours of your arrival time at _____ Hospital.**

4. **If your stools still have particles or are brown in color, you should take an over the counter Fleets® enema prior to arrival.**

Day of Procedure
1. **Do not eat breakfast. Only approved clear liquids if allowed.**

2. **For morning procedure drink only sufficient water to take essential medications.**

3. **Report to Ambulatory Surgery (2nd floor) at _____ Hospital.**

4. **A companion must accompany you to the examination because you will be given medication to help you relax. It may make you drowsy, and unable to drive yourself home. Even though you may not feel tired, your judgment and reflexes may not be normal. Failure to make these arrangements will result in the cancellation of your examination. You will not be allowed to leave the recovery area unescorted.**

5. **Please tell us of drug allergies or peculiar reactions that you may have.**

6. **If you have any questions or need to cancel, please call us as soon as possible.**

Ambulatory Surgery _____ Hospital: Phone #.   Physician's Name: Phone #

Exhibit 6.6 (continued)

professional review the instructions with the patient to make sure of the person's complete understanding. Following a procedure, if the patient has had anesthesia, often the instruction sheet is reviewed with the caregiver or with whomever accompanied the patient. Again, this practice ensures compliance, because after anesthesia, it's possible a patient will not remember talking to the doctor or receiving instruction.

The instructions in Exhibit 6.6 are reader/patient-centered as patients must follow them carefully, and the instructions are given in steps. The preparation includes two different timelines. One timeline is geared toward a morning procedure; the other timeline is geared toward an afternoon procedure. When the nurse goes over the instruction sheet in the particular surgeon's office, he or she checks the appropriate material for patients to read depending on whether they are facing a morning or afternoon procedure. These instructions are vital, as mentioned above, so that patients and the physician do not need to repeat the exam because of poor patient compliance with instructions.

As you can see in Exhibit 6.6, this instruction sheet, while mostly providing clear steps for patients, does have a few problems in document design. In the second paragraph under the title, this wording is confusing: "**You may have** broth, bouillon, strained fruit **juices** without pulp (white grape, apple), lemonade, black coffee—No cream or creamer, Jell-O, Sprite, 7UP, ginger ale, Gatorade." The way *no* is situated before cream or creamer makes the reader wonder if the *no* also modifies Jell-O and the clear drinks. It does not, but the structure of the sentence could confuse patients. One of the good document design functions is the use of bolding and underlining for emphasis. For the most part, the document is well designed. To be accurate from a technical writing standpoint (and legally), product names should include a trademark ™ or registration symbol ®. (I added them here.) The morning and afternoon preparations are almost identical except for times. In the morning procedural steps, there are only three of them; in the afternoon preparation, there are four steps. The last sentence of step number one is split into a new step in the afternoon preparation. It would be best to keep the preparations, since they are identical, in the same format and/or number of steps.

As a health communicator, you will want to notice if patients are confused by certain phrases. If you don't answer the phones yourself, you can certainly query receptionists to ascertain if patients have called to ask questions about certain forms and patient education materials. The physicians will also be thankful for your vigilance, because patient compliance ensures fewer complications and fewer malpractice suits. Patients are less likely to sue health and medical facilities with which they have good communication. Such clarity keeps patients safer and healthier. Also make sure you update text if there are changes. We often develop boilerplate text that we can easily cut and paste. It's important, though, to keep that information current as procedures and medical technology change so rapidly. Many facilities also tend to reuse forms until the copies are barely legible. This practice is neither cost effective nor safe in the long run.

What Other Health Materials Are Available?

Many genres are available to educate patients and to remind them of important health information. Whatever materials keep patients advised of health practices should be used. Imagination and creative invention in document design can help save lives and improve the quality of those lives. Many doctors' offices and other health facilities feature flyers and posters reminding patients of important health information. Because of the prevalence and ease of desktop publishing, health professionals and writers can utilize many new forms easily and quickly.

Featured here are just a few more examples of the possibilities. Sometimes short phrases and clever wording can grab readers' and viewers' attention in ways longer tracts cannot. Especially in our fast-paced, mobile, sound-bite-based culture, often the writer and designer get only a few minutes of the audience's time to deliver a message. As health message designers, we need to take more advantage, too, of the media employed by advertisers to sell products. In fact, we are selling health, a product or process that is the most valuable. Hospitals and other public health facilities need to direct more campaigns via billboards, public service announcements, and ad spaces on buses and subways. Television ads can be used for more than selling pharmaceuticals.

One small print medium that's been helpful in raising breast cancer awareness, especially among some strata of females, is the shower card, for example, the one published by Wyeth-Ayers Laboratories. This format is a not a new idea, but has been published since the middle 1980s. The breast exam is illustrated both via graphics and text on a waterproof, plastic hanger card that hangs off the shower nozzle inches in front of one's face. This has been a good reminder for women to conduct the all important monthly breast exam. Because this exam can also be conducted in the shower, this unusual site for health education materials is ideal.

The Department of Health and Human Services, along with a number of other agencies, publishes a bookmark (see Exhibit 6.7) that features an older woman reading a book in brightly colored graphics (reproduced here in black and white).

Exhibit 6.7 Mammogram Bookmark
Source: U.S. Department of Health and Human Services. Pub. 2/4/98.

The title sticks out of the top of the book when the bookmark is used and says "MAMMOGRAMS, Not just once, but for a lifetime." The flipside has clearly spaced text that repeats the front title and gives short phrases of additional information without being text-heavy:

- You are never too old to get regularly scheduled mammograms.
- Cancer can show up at any time—so one mammogram is not enough.
- **Medicare helps pay for yearly mammograms**. (in bold print for emphasis)
- For Medicare information, call **1-800-638-6833**.
- For information on breast cancer and mammograms, call NCI's Cancer Information Service at 1-800-4-CANCER (1-800-422-6237). (Dept. of Health and Human Services)

What About On-line Materials?

As we moved from oral discourse about health education (the wisdom passed down in villages and families) and into printed materials, health education services understandably relied heavily on print. Now, however, we have another major shift in our technology thanks to the advent of digital publishing. As computers became more and more prevalent and accessible, patients have turned to the Internet and to email for health data. As noted in the newsletter example, some medical practices are listing their email addresses for patients to communicate their concerns and questions. In addition to email access, many facilities have their own Web sites. The larger public sites like WebMD and others feature support groups via discussion boards and synchronous chat rooms for patients suffering from similar ailments to gain comfort, advice, and contact with others who understand the specific difficulties of their conditions. These sites have been invaluable, particularly for patients with more unusual diseases who may be separated by large distances geographically. Although most sites are positive, unfortunately negative support groups exist as well. One particularly disturbing example is a Web site that supports anorexia as a positive lifestyle, as opposed to the debilitating and deadly disease that it really is. Fortunately, such sites are few. Sometimes, too, because it is still difficult to check the veracity of sources on the Internet, false or misleading information is posted. Often it is worthless at best, but some misinformation can cause very serious consequences.

Health writers can assist patients' education by letting them know which Web sites are accurate and reliable and by steering them away from sites that are problematic. As Web designers or architects of other electronic health education materials, we can make sure our own information is accurate and ethical before we post it electronically. In addition to these considerations, you might want to reread the section in the chapter on medical ethics about copyrighting and verifying sources of on-line materials. At this point in our Information Age, documenting sources on the Internet is still difficult, unlike citing print-based work. This situation, too, adds to confusion for patients about which on-line sources are reliable. Often there are no easy ways to check. One strategy is to go to well-known and highly respected Web sites to follow links from those sanctioned sites.

One thing we do know for sure, patients are doing Internet searches when they have health problems, and they're arriving at health facilities with vocabulary, questions, and issues that most physicians and health professionals were not hearing a decade ago. If information is unclear or false, a professional must spend valuable time straightening out the confusion and errors. To give a perspective on Internet usage by patients, a May 2002 research report conducted by Pew Internet & American Life Project (www.pewinternet.org) found that "About 6 million Americans go on-line for medical advice on a typical day." That means more people go on-line for medical advice on any given day than actually visit health professionals (on a given day)," (Pew "Vital Decisions" Report). Both the Pew survey organization and Harris Interactive have coined names for people who frequently (and sometimes too frequently) consult the Internet for health information. Pew calls them "health seekers," while Harris has named them "cyberchondriacs" (greenvilleonline.com 6D). One 44-year-old Colorado woman reported that when she searched the Web for information on anemia, "[the] process created as much fear . . . as the symptoms did" (greenvilleonline. com). She elaborated:

> "All you have to do is key in your pet fear or symptom, and up pop little abstracts with scary information in them. . . . Even if you are trying to find comforting statistics, you have to wade through all this bad. Of course we pick up on the bad, only."

> After several hours of picking up "on the bad," she had enough. [She] asked a friend to visit some health information sites and do the pointing and clicking for her. (greenvilleonline.com)

Even though many patients have used health reference books in the past to check on their symptoms or condition, this woman characterized the Internet as "much worse." The frightening terms she had seen would not disappear from her mind once they had flashed on her screen.

Some patients act on faulty recommendations based on a less-than-clear understanding of what their real diagnoses are. In addition, added fear and stress, such as that experience by the aforementioned woman, retard rather than expedite healing. Although she coped by having a friend do this research for her, such a decision would depend on the friend's ability to weed through information and form a good understanding.

The research data and health experts disagree somewhat on what the negative aspects are of using the Internet for health information. The Pew report indicated that only two percent of "health-seekers" know someone who has been harmed as a result of Internet health research. But Howard Wolinsky, author of *Healthcare Online for Dummies,* suggests that the main problem is that patients who are desperate become vulnerable to Web sites promoting questionable or fraudulent treatments (greenvilleonline.com).

Dr. J. Donald Capra, M.D., an immunologist and president of the Oklahoma Medical Research Foundation, noted that patients now come in with stacks of printouts from Web sites and expect their physicians to read a large sheaf of information during the visit. If the physician doesn't spend an hour perusing the documents, some patients feel short-changed. As Capra says, though, "If patients use the information to ask questions and [are] reasonable, that is fine" (greenvilleonline.com). Dr. Richard Glass, M.D., deputy editor of the *Journal of American Medical Association*, is seeing the same phenomenon in his practice. He asks patients to bring in a few questions only and that they be based on reliable Web sites.

How Do I Find Reliable Health Web Sites?

There is an organization, the Utilization Review Accreditation (URAC), which screens health-related Web sites for best practices in health information. It doesn't, however, certify the accuracy of the information presented so much as it screens sites for the following criteria:

- Sites do not allow potential "sponsorship-spawned conflicts of interest."
- Sites provide security for any personal information the patient may share on the site.
- Sites are structured so users have easy access to information (provide good navigation for patients) (www.urac.org)

The senior vice-president of URAC, Guy D'Andrea, said that a medical site's credibility stems largely from:

- "[W]hat the perspective of the site is,
- [W]ho is behind it,
- [W]hat is their motive,
- and [W]ho creates the [site's] content." (*Coping with Cyberchondria* 7D)

Although the research data and the experts' opinions do not always agree on the pros and cons of on-line health information, on one policy they do agree: if patients use the correct balance of exercising good judgment and self-control, they can be helped by on-line health information, especially if they use the information to formulate a few questions for their medical practitioners. The Pew survey data reveals that 51 percent of people reported they themselves or people they knew had been "significantly helped by following advice they found on the Internet" (Pew 2002, 7D).

The following Web sites have received accreditation from the URAC Health Web site Accreditation Program:

- www.adam.com
- www.healthyroads.com
- www.ashplans.com
- www.ashnetworks.com
- www.ghi.com
- www.hayesonhealth.com
- www.hiaa.org
- www.health-intl.com
- www.hhni.com
- www.healthwise.org

- www.intelihealth.com
- www.medlineplus.gov
- www.veritasmedicine.com
- www.laurushealth.com
- www.webmd.com
- www.wellmed.com

As a health practitioner and/or medical communicator, you will want to become aware of these sites and aware of patient practices. Then you will be able to steer patients toward the accredited sites, and you can interpret information, correct misinformation, and comfort those patients whose searching has generated more fear than knowledge.

Because a later chapter in this text covers less well known electronic forms of writing, we will not go over those again here. Except for the use of the Internet, online communication is as yet relatively untapped, and the possibilities for health education games, on-line videos, tutorials for postsurgical care, and other imaginative approaches are endless. We will discuss these possibilities in Chapter 11.

Summary

In this chapter you have learned that there are many genres of traditional or print-based health education materials, including handbooks, booklets, newsletters, brochures, pamphlets, posters, instruction sheets, update letters, shower cards, and bookmarks—all used to educate and inform patients. In addition, writers based in pharmaceutical companies may develop various charts, diagrams, anatomical drawings, and posters.

As a health communicator, both format and content are important to get your messages across. In the previous chapter, you learned about designing and formatting information. In this chapter you examined health content in several genres of traditional, print-based health education materials to determine if the information was accurate and clear, and if the language was well targeted to its audience. You gained knowledge of the areas of prevention and patient education that you will most likely address in your work with patients. You know that the emphasis on the home as the center of healthcare is a recent development and that such a move is the result of economic changes. Home health services are now the fastest growing sector of healthcare. Once acute situations are stabilized and procedures are completed, patients are encouraged to take advantage of home healthcare. This situation puts even more importance on our jobs of designing and communicating patient and home caregiver education.

You read critiques of a general health handbook, a hospital quarterly publication, a booklet for women on cholesterol, instruction sheets for colonoscopies, shower cards for breast self-exams, and a bookmark reminding women of the importance of mammography. In the handbook, you learned how to organize information and what information was important to include. In the hospital quarterly, you were reminded how effective the "scenario principle" is for effective communication that reaches your target audience (Flower 1983 and Hayes 2003). In the instruction sheets on colonoscopies, some information was confusing, and you learned that in such detailed instructions, every word makes a difference. In addition to the content, you also know now that you can employ many creative genres to reach an audience.

In addition, you looked briefly at the pros and cons of one of the current, most consulted health educational medium: health sites on the Internet. Several surveys (from two well known polling houses: Harris and Pew) showed both problematic and useful applications of patient searches. Problems develop when patients overdo their searches, consult unaccredited health sites, and frighten themselves. As a health instructor and/or practitioner, you learned that you need to be aware of the role, both positive and negative, medical Web sites can play in patients' physical and psychological well being. You also have an accredited list of health sites plus the URL of the accreditation organization, so that you can consult the best Web-based data for your work. In addition, you can guide patients who want to use the Internet to find out more information about their particular health problems.

Discussion Questions

1. Discuss and develop an outline of factors that will influence how you design three brochures describing open-heart surgery for patients who are on a pediatric cardiac ward (children under the age of 16), patients who are African American males between the ages of 25 and 40 years of age, and geriatric patients over 60 years old.
2. What are some alternative ways of representing the information on the chart on the top of page two in the brochure in Exhibit 6.5?
3. Which genre should you choose for the following tasks and why did you make this choice?
   a. Daily reminders for dietary control
   b. Information on an upcoming gall bladder surgery
   c. Blood donation recruitment for a local agency
   d. Strategies for persuading teens to use safe sex
   e. Newborn care issues
4. Why is it important to develop both print and electronic sources of health information?
5. What ethical issues can occur when you are working with health education materials? Research the Tuskegee syphilis study[4] and suggest what ethical health education materials should have been developed.
6. Discuss what rhetorical strategies you would use if you were leading a support group of men recovering from prostate cancer who had all surfed a Web site that contained false and frightening information.

Exercises

1. In teams create storyboards of a six-page brochure that addresses one of the three audiences described in the first Discussion Question above. If access to computers is available in the classroom, conduct demographic research, research about open-heart surgeries and find clip art, graphics, and photo options for your brochures.
2. Develop a service-learning project for a local clinic or hospital that screens for or performs such cardiac surgeries or has other needs for patient education materials.
3. Use Microsoft Office Excel to find other ways of presenting the statistics on page two of Exhibit 6.5. Compare the effectiveness of horizontal bar graphs, pie charts, vertical bar graphs, and other graphic representations of the data.
4. Interview senior citizens and show them several health materials developed for a senior patient audience. Ask them about which materials are most effective in reaching them and why. Take extensive notes and/or ask permission to video- or audiotape them.
5. Prepare a group report based on feedback from various students or student teams gathered in Exercise 4.
6. Gather together some examples (or create some) of health education materials developed for low-literacy or non-English-speaking patients in this country.
7. Go to one of the accredited Web sites and choose a topic of interest for you. Then use that information to design a booklet, pamphlet, newsletter, poster, shower card, bookmark, or web page. Then write a reflection on the process and the product as a whole. Be sure to articulate why you used the genre you used to convey your health information.

Works Cited

A.D.A.M., Inc. © 2003. August 26, 2003. www.adam.com

Aetna IntelliHealth. © 1996–2003. August 26, 2003. www.intelihealth.com

American Specialty Health Networks (ASH Networks). © 2001. May 10, 2003. www.ashnetworks.com

American Specialty Health Plans of California, Inc. (ASH Plans). © 2001. May 10, 2003. www.ashplans.com

Capra, Donald J., M.D. *Coping with Cyberchondria.* August 26, 2003. www.greenvilleonline.com.

Flower, Linda, et al. 1983. "Revising Functional Documents: The Scenario Principle." *New Essays in Technical and Scientific Communication,* 41–58. Eds. Paul Anderson, R. John Brockman, and Carolyn Miller. Farmingdale, NY: Baywood.

Glass, Richard, M.D. Quoted in *Coping with Cyberchondria.* August 26, 2003. www.greenvilleonline.com.

Group Health Incorporated (GHI). 7/1/2003. August 26, 2003. www.ghi.com

HAYES OnHealth. Limited Access: Members Only. August 26, 2003. www.hayesonhealth.com

HealthHelp. © 2003. August 26, 2003. www.hhni.com

Health Insurance Association of America. 8/26/2003. August 26, 2003. www.hiaa.org

Health International. © 2003. August 26, 2003. www.health-intl.com

Healthwise n.d. August 26, 2003. www.healthwise.com

Healthyroads. © 2001. August 26, 2003. www.healthyroads.com

Johns, Lee Clark. 1989. "The File Cabinet Has a Sex Life: Insights of a Professional Writing Consultant." *Worlds of Writing: Teaching and Learning in the Discourse Communities of Work*, 153–87. Ed. Carolyn Matalene. New York: Random House.

LaurusHealth: VHA's Health Information Resource. © 2003. August 27, 2003. www.laurushealth.com

MEDLINEplus Health Information: A Service of the U.S. National Library of Medicine and the National Institutes of Health. 8/26/2003. August 27, 2003. www.medlineplus.gov

OMH TODAY: *A Quarterly Publication of Oconee Memorial Hospital* 7, no. 1 (Winter 2003). Seneca, SC: OMH.

On Target Media, Inc. 2001. "Cholesterol: A Helpful Guide When Talking to Your Doctor." *Healthy Advice for Women.*

Partners for a Health Community. 1997. *Healthwise Handbook*. Boise, ID: Healthwise, Inc.

Pew Internet & American Life Project. (www.pewinernet.org). Greenvilleonline.com: 6–7D. June 25, 2002.

U. S. Department of Health and Human Services. Public Health Service. National Institutes of Health. National Cancer Institute. Health Care Financing Administration. "Mammograms: Not Just Once, But for a Lifetime." Bookmark. Government publication number Z498.

URAC. Formerly, the Utilization Review Accreditation. Now URAC Health Site Accreditation, http://websiteaccreditation.urac.org. August 26, 2003.

Veritas Medicine. © 2003. August 27, 2003. www.veritasmedicine.com

WebMD Health. © 1996-2003. August 27, 2003. www.webmd.com

WellMed Health Hub. © 2003. August 27, 2003. www.wellmed.com

Wolinsky, Howard. *Healthcare Online for Dummies*. August 26 2002.

Wyeth-Ayers Laboratories. 1987. "Do It Yourself: Monthly Breast Self-Exam." Shower card. Philadelphia: Albert Einstein Healthcare Foundation.

Endnotes

1. Please be aware that Web sites are subject to change, links break, and whole sites are taken down. Some sites do not receive proper updates. Check with www.URAC.org to keep current on approved sites. It is also always appropriate to check with your doctor before you initiate changes in your own treatment. Do not use poor judgment: do not diagnose yourself and begin treatment without professional medical advice.
2. The scenario principle refers to putting a human agent in a scene, who is taking some action, rather than speaking in abstract generalities with no human agent present.
3. *OMH TODAY*, 10–11.
4. More on the Tuskegee study can be found in endnote 2 in Chapter 2.

Public Health Campaigns

 ## Overview

This chapter also covers health education, but unlike Chapter 6, Chapter 7 discusses campaigns targeting large audiences as opposed to educational efforts targeting individuals and smaller groups. Methodologies are featured here for the campaign developers/writers who are usually part of agencies, large hospital systems, nonprofit organizations, and the government. Because these campaigns attempt to reach such large audiences with multiple genres, each campaign must begin with a plan. Because you have read Chapter 3 about handling larger documentation projects, we will use that as a base to teach you how to plan a health campaign.

As in the previous chapter on health educational documents, the purpose of public health campaigns is usually to change health behaviors and/or to persuade citizens to change preset habits of thinking about health issues. Some familiar and recent campaigns are those that target cigarette smoking. Other such campaigns include educating parents about the importance of infant and child car seat restraints and raising public awareness about breast, prostate, testicular, and colon cancer. All such campaigns ask us to change behaviors and institute healthier and/or safer practices. For example, campaigns may encourage us to:

- Stop smoking
- Place babies in properly installed infant car seats
- Buckle children up in car seats, child restraints, or seat belts
- Schedule regular mammograms to screen for breast cancer
- Practice a monthly breast exam for suspicious lumps[1]
- Take the test for prostate cancer
- Perform self exams for testicular cancer
- Have a colonoscopy after 50 years of age

- Eat more fiber
- Get checked for colon cancer via laboratory tests

After 9/11 and the anthrax letters of 2001, we now realize that our public health is at risk from bioterrorism, chemical warfare, and nuclear exposure. This current threat of a terrorist attack makes large public health efforts even more important, because knowledge on the part of the citizens who are the victims of such attacks could prevent massive exposures, injuries, and deaths. For example, at the time this text was being written, the Centers for Disease Control (CDC) was planning a public health campaign to educate Americans about smallpox. Tom Skinner of the CDC said, ". . . the plan here is to embark on a mass-media education campaign that could include posters, advertisements, and public service announcements" (Skinner 2002). As health communicators, we need to plan campaigns ahead of possible disasters, so if one does strike, citizens know how to initiate the right behaviors in response in order to avoid the behaviors that could make the situation worse. In the case of smallpox, we need to educate both about the vaccination itself as well as about the disease of smallpox (Heifferon 2004, 25).

What does it take to move people to action? As health communicators who have grasped some classical concepts of rhetoric and rhetorical strategies, you know that persuasion is key. In this chapter you will learn the answers to these questions:

- How does a team identify and analyze a problem?
- What ethical considerations are important?
- How does a team develop good analyses and proposals?
- How can I implement projects effectively?
- What media choices are best for campaigns?
- What are some possible strengths and weaknesses?
- How can I troubleshoot a campaign?
- What assessment techniques should I use?
- How do I debrief and learn from a campaign?
- What grants and foundational support are available?

Campaigns are not the province of single individuals but require teamwork, resources, timelines, and good internal communication. In addition, they require excellent demographic research that thoroughly analyzes the target audience(s).

How Does a Team Identify and Analyze a Problem?

William J. McGuire suggests that to choose an appropriate problem for a public health campaign, designers should start with the following three criteria:

1. Seriousness of the problem
2. Effectiveness of the solution
3. Suitability of mass persuasion as a way to solve the problem (McGuire 1999, 62).

An example of applying the first criterion is in the problem of vitamin deficiencies among American children. Although it is a major issue in some other countries, the problem is relatively rare in this country. The problem of vitamin deficiency would not pass the first criterion if one were addressing the issue here because it is not as serious a problem in America (McGuire, 62). For a solution to be effective, it must be one that human behavior can alter. We could not at this point in time, for example, do a public health campaign to stamp out Alzheimer's or toxemia for pregnant women because these are both problems for which preventive measures are not yet known (McGuire, 1999, 62). Therefore, these problems do not pass the second criterion. For the third criterion, a one-on-one form of treatment, such as that found in successfully breaking a heroin addiction, would not be possible for a public health campaign. What would be useful, though, is a public health campaign telling how to find treatment centers and what kinds of help are available (McGuire, 62). In other words, although some problems do not lend themselves to public health campaigns, sometimes shifting the focus onto a more appropriate goal within the problem/solution set could work. A public health campaign might be effective for the families of Alzheimer's patients, if the campaign educated families about coping skills and issues necessary to solve when dealing with an Alzheimer's or senile dementia sufferer (McGuire, 1999, 62). It is important to analyze the health problem at the very beginning of the campaign; otherwise much effort and many resources could be wasted on a problem that is not effectively addressed in a mass persuasion campaign.

As a health writer, by the time you have addressed these criteria for your campaign, you should record your analysis and the arguments you made to prove that the issue fit each criterion. These writings will form the basis for your other organizational tasks in the campaign and especially in fund-raising efforts as you solicit support for your large project. Even if the solution to a health problem seems obvious, you will need this analytical documentation to sell your ideas to those who can supply the material support you will need.

What Ethical Considerations Are Important?

Equally important to a successful campaign as the analysis of the problem is the analysis of the ethics. Mass persuasion is designed to manipulate large groups of people; therefore whatever you plan has to be for the public good with relatively miniscule chances of the behavior changes you recommend resulting in harm to any individual or group. As you choose an appropriate problem and solution for a campaign during your planning stage, it's equally important to consider the ethics of such a plan very early in the process. You do not want to discover halfway through the project, or, even worse, after its launch, that you and/or other team members have some moral uneasiness regarding the campaign or that it has ethical drawbacks (McGuire, 1999, 63). Sometimes you will need to modify the goal or restate the problem in a different way so that the team takes the ethical high road. As we discussed in the ethics chapter, the stakes are much higher for health communicators than for most other technical writers or document designers. Do not ignore a team member who has a problem with the ethics of any aspect of a campaign, because in this situation, hypersensitivity to both negative and positive outcomes is a virtue, not a drawback. The second discussion question in this chapter should become part of the routine for the planning or preplanning process of a public health campaign.

Because we devoted a full chapter to ethics earlier in the text, here let's consider an issue more specific to large health campaigns that involves questions of health policy or health legislation.[2] Roxanne Parrott, Mary Kahl, and Edward Maibach are especially eloquent on this topic. They suggest we ask certain important questions before we choose our audience or audiences and begin designing our health message. They've developed these questions based on what they saw as failed or poorly targeted campaigns in the past that addressed relatively powerless audiences of consumers when lawmakers and policy writers were a more appropriate audience. Ethically, too, we have the responsibility to use our resources where they will do the most good.

1. Does the absence of domestic laws to support a national health standard reduce the likelihood that a message will enable health? If yes, an appropriate starting place and audience for health messages is the lawmakers rather than the general public.
2. Is there a conflict among present domestic laws that reduces the likelihood that a health message will enable health? If yes, once more, the appropriate starting place and audience for health messages is the lawmakers rather than the general public.
3. Do federal administrative practices inhibit the design of appropriate messages? If yes, proceed if the health of most audience members could still potentially be improved within the constraints of message design; otherwise, do not proceed.
4. Do institutions exist to provide the services recommended to enable health? If not, the appropriate starting place and audience will be the group(s) responsible for allocating support for the institutions.
5. Do state and local rules and actions support the use of institutions designed to provide services to enable health? If not, an appropriate starting place and audience will be the group(s) responsible for enforcing federal policy mandating services.
6. Do cultural norms support the actions advocated by state and local health promotion efforts? If not, revise the practices and aim to acknowledge cultural traditions. (Parrott, Kahl, and Maibach 1995, 281–82)

The authors write very strongly about these issues and admit to a type of media advocacy to alter administrative policy and health practices. They emphasize, "Recognition is given to the fact that when administrative conditions do not warrant an emphasis on individual enablement, health message designers are ethically obligated to cease such promotions" (282). This focus places health campaign planners under an ethical obligation to research the health legislation, health policy, and cultural norms for the site or sites of their campaigns.

How Does a Team Develop Good Analyses and Proposals?

Good planning for a public health campaign involves a number of steps, some of which we have already discussed in previous chapters. (See Chapter 1 on audience analysis and Chapter 3 on large documentation project management; also, see Exhibit 3.7, Hackos' Publications Development Life Cycle, for a good outline of the process.) Because in all message design and again in mass message planning, audience analyses are key, it would be appropriate to briefly review what you learned in the first chapter and see how that knowledge fits in this rhetorical situation of planning a large public health campaign. With that knowledge firmly in your head, we can discuss audience analysis as viewed through a marketing or communications lens called *segmentation.*

What Is Audience Segmentation?

If you are working with professionals trained in marketing or mass communication, you might find them talking about audience analyses in a different way from those trained in technical communication or rhetoric. The marketing or mass communication term is typically called *audience segmentation.* Segmentation refers to the notion that audiences are in fact not really an undifferentiated mass, but rather differentiated groups within groups. This idea is widely accepted by media scholars (Windahl and Signitzer 1992, 181). Although this specific term is not used widely in rhetorical or technical writing circles, the approaches we have discussed throughout this text are somewhat similar in focus. Although the two methodologies in Chapter 1 center more on the individuals telling their perspectives, segmentation at least recognizes that individuals within groups vary enough to be categorized together as subgroups. As Windahl and Signitzer maintain: "A key concept in communication planning is segmentation. The concept relies on differentiated audiences with their subaudiences; an audience is divided into subgroupings that are internally homogenous but differ from each other (Frank, Massey, and Wind 1972; Haley 1985; Kotler 1986). The subgroups receive different messages and/or are reached through different channels" (Windahl and Signitzer 1992, 181).

One drawback with segmentation is that it does take time, resources, and effort. It's generally easier to prepare one uniform message and send it out rather than to tailor several messages and design different media to reach several different audiences. But if the resources and time are available, segmentation is usually well worth the effort. There are many ways to segment, similar to the approach toward audience analysis described in Chapter 1:

- Demographics
  Age
  Gender
  Occupation

  Family size
  Position in the family life cycle
  Income
  Education
  Geographical location
  Religion
  Race
  Nationality
- Beliefs
- Attitudes
- Behavior
- Principle of access (organized according to how many resources it takes to reach a particular group)
- Public's resources (organized according to a group's purchasing power or ability to use resources to act on the message)
- Media use (sometimes there are media limits, such as if only newspaper and radio are available) (Windahl and Signitzer 1992, 182–87)

The drawback to simply segmenting audiences is that not all individuals are identical within a certain age range or within certain income levels. However, given pragmatic concerns, you cannot always spend time interviewing individuals to get their "read" on a situation. A compromise between segmentation and the individual or small-group interviews would be setting up focus groups based on your audience segments.

How Are Segments Applied?

Researchers need to move from theoretical and abstract concepts of what an audience looks like to a practical, operational, or working definition. The campaign planner asks: Who are these people? What do they look like? How can they be reached? Often our first step is to make assumptions. If a campaign is targeting shoplifting as a mental or addictive behavior, the planners may assume that most shoplifters are young males between the ages of eight and 15 years of age. But is that true? They would need to find statistics, call some police departments, talk to store managers in malls, etc., to verify such an assumption. And, the planners may be surprised at what they discover.

  The audience segments themselves depend on identifying groups based on the criteria above. So within the demographics category you might target a campaign to get more preteens and retired people to use the age-appropriate exercise programs at recreation centers to reduce stress and health problems caused by lack of physical activity. The focus of another campaign might be to convince college students to help lobby governments for more clinics for diagnosing and treating STDs.[3] When you segment an audience according to beliefs, it doesn't usually mean religious beliefs, but beliefs such as worldviews, cultural assumptions, etc. For example, urban dwellers often have somewhat different belief systems from those of rural or small-town dwellers. Urban worldviews tend to be widely focused, while small-town dwellers can be more focused on the local area.

People's attitudes lead them to approach their own health and safety behaviors in different ways. For example, you may think in terms of risk-takers and the cautious. Risk-takers will need a very different approach regarding health and safety concerns compared to those who are already cautious. Attitudes are closely linked to behaviors, and these different behavioral categories need to be defined for the particular purposes of your campaign proposal and plan. Behaviors can be divided according to inner-directed (introvert) versus outer-directed (extrovert), and need-driven (people who think about basic survival and sustaining their present status) versus achievers and even over-achievers.

How Do We Implement Our Research Plan?

Once you have decided whom to target, you will want to figure out how to do the research necessary for the plan and proposal. In our discussion of audience and contextual analyses, we described the various analytical and research tools available to you:

- Library research
- Internet research
- Field research
- Participant observation
- Surveys
- Interviews
- Focus groups
- Usability strategies

These tools are essential in the development of public health campaigns as well. Remember that often geography, environment, and history are important, too. For massive campaigns, in which you target a whole country, the demographics will differ from region to region. You will need to decide whether the similarities and differences will make it possible to develop a "universal" message or if the local differences (they could be language differences, too) make it necessary for you to create different versions of your informational campaign to reach various groups. In other words, do you segment your audience or not? Such decisions often depend on whether or not the health issues you focus on affect certain age groups, one gender, etc. For example, a disease that strikes only geriatric patients means that you can usually count on a number of similarities in the aging process in planning your campaign.

Sometimes communication theorists recommend that the target be not so much the audience itself as the particular behavior you are targeting in the campaign. Whether you are targeting a negative health behavior that you want people to stop or asking people to initiate and maintain a beneficial health practice, it's important to narrow or segment your audience and contextual focus to concentrate on that behavior and its modification. Rhetorical theorists talk more about *rhetorical situa-*

tion and *context,* whereas communication theorists use other terms: *personal and situational variables.* Whatever terms or theoretical foundation you use, the methodology and importance of good preplanning research and analysis do not change.

How Do I Include Multiple Perspectives?

Although this text is more practical than theoretical, it's necessary to complicate the picture by mentioning current debates in both theoretical camps regarding public health campaigns. As health communicators, you may or may not know much about the theories of postmodernism. Most likely you have heard the term; it's even in the common parlance of stand-up comedic routines and comic strips in newspapers. In our case, we only need to know that postmodern thought challenges the idea of Truth, the Truth that Plato believed existed in an ideal form, the Truth that most religions espouse they alone possess.[4] Instead of one Truth, most postmodernists believe that truth is a shifting idea or set of ideas that is socially constructed by a group of like-thinking individuals. As Brenda Dervin writes regarding public health campaigns:

> This means that whatever one individual group calls "information" or "knowledge" at any given point in time is applicable only to that time and space and the self-interests and observing capacities of the "observers." Untangling which factors account for the "observation" is impossible, and even when observation is prescribed by a set of rules (as in scientific observation), most scholars agree that even . . . the so-called empirical facts . . . are subject to the same limitations. (Dervin 1989, 70)

So, what does this standpoint mean for those of us wanting to design health messages and particularly large public health campaigns? It means that as professionals, it is appropriate for us to consider multiple truths. That is, we need to be able to recognize different perspectives that exist among the audience members we want to address. We need to consider these points in order to understand the need for change in how we conceptualize both audiences and the campaigns themselves (Dervin 1989, 71).

One example Dervin uses is that of a breast cancer information campaign that targets all women, but particularly older women with family history of the disease. In this campaign in the United States, one overriding message has been: *Women should have breast cancer checkups* (Dervin 1989, 71). This health message is based on a number of assumptions:

- Women get a disease called breast cancer.
- The rate of the disease goes up as women get older.
- Early detection means early treatment.
- Early treatment means a longer life. (Dervin 1989, 71)

Many of us would plan a campaign based on such assumptions, not realizing there are other perspectives out there in addition to the ones we find so obvious. One other perspective is from that of alternative or even integrative medicine.[5] Another perspective is from feminists or environmentalists who think that perhaps we are focusing the problem in the wrong area. Instead, we should be looking at corporations who pollute the environment and use problematic chemicals on our food sources (Dervin 1989, 71). Such discourse communities believe there are links between these pollutants and chemicals and breast cancer in women.

How Do Multiple Perspectives Change the Agenda?

Using the breast cancer example, we need to consider how knowing that there are multiple perspectives changes the ways we set up our research agenda. That agenda needs to include as much background on the issues as is possible to gather, even if you are constrained by a short amount of time. (See Chapter 1 for research methodologies.) Knowing the history of earlier campaigns on the topic as well as the various changes in treatment in recent times is invaluable. Sometimes your audience, especially if it involves senior citizens, has a longer memory of certain issues than you do. Also, most issues involving disease are not as black and white as they may seem.

For example, women's magazines and other journals carry stories in which women report feeling offended by the way medicine has objectified and defined them or their problem. For many years surgeons carried out unnecessary radical mastectomies, and what seemed necessary in one decade, the 1970s, was no longer true a decade or two later (Dervin 1989, 71). Even recently, radiation therapy given to women after they have had lumpectomies has been shortened from the usual six-week routine as a result of changes and upgrades in technology and in the results of follow-up, longitudinal research. This history of unnecessary suffering on the part of many women in our culture means the audience may not be sympathetic to your latest efforts. If you take into account this history and sensitivity in your campaign, you are more apt to be successful than if you ignore your audience's understandable misgivings on this topic.

My technical writing students and I undertook a service-learning project to increase awareness of the risk of breast cancer in African American women in our community; in doing so, we quickly discovered that the women were very suspicious of the medical establishment.[6] We worked with sociologists, historians, and the women from the community so that part of our research included an understanding of the impact of the Tuskegee experiment and forced sterilization on the African American community.[7]

What Are Sense-Making Interviews?

One way to take attitudes into account, Dervin suggests, is to conduct "sense-making interviews" (Dervin 1989, 76). In this research method, like the one we also featured earlier, she calls for centering the audience concerns as an effective way to develop a plan for public health campaigns. By interviewing the people you are trying to reach, rather than assuming and predicting how people think, you can let them determine the focal points and even the genres and messages that will be most effective in your campaign. Like the Healthwise campaign in our area, large campaigns require millions of dollars for implementation. Without doing the in-depth research, these large fiscal resources, not to mention staff time and effort, can be wasted.

For example, one blood donor center felt it was losing donors because they were leaving with some negative feelings about the donation experience. Several problems came to light in the sense-making interviews. Donors were confused about when and where they were to answer questions at the various stations in the donor process (Dervin, Nilan, and Jacobson, 81–82). Also, access was a major problem. Instead of planning a campaign that pushed people to donate again who had dropped out of service, health professionals changed the access to the center and posted the questions on the stations so that donors knew what happened when and where (Dervin, Nilan, and Jacobson, 81–82). The campaign then advertised greater access and ease of donation, thus drawing back its previously dissatisfied donors.

How Are Campaign Proposals Different?

Like the answers to many questions: it depends. Sometimes campaign proposals are not that different from the proposal model we developed in the chapter on project management. Because campaigns tend to be large and involve multiple media, the proposals can be more lengthy and detailed. Many proposals also request funding or attempt to justify the use of resources from the parent company. Since proposals differ whether they are requesting corporate, government, nonprofit organizational funding, and/or are large grants applications, no hard and fast rules can be made. Often the funding entity has forms or requirements that you must follow. We will discuss grants in depth in the next chapter. The National Institutes of Health, the National Institute of Mental Health, and the National Science Foundation have numerous grants and funding possibilities for health campaigns. The larger funders often have on-line forms, and NSF now requires on-line applications exclusively. If you are proposing the campaign in-house, you should request or check the file cabinet for previous models.

Since large campaigns involve numerous partners, often the proposal takes extensive time to get the multiple partners, funding issues, tasks, etc., sorted out. One proposal and grant I worked on was a three-year United States Department of Agriculture (USDA) grant that involved federal, state, university, and nonprofit agency funds. Its purpose was to reach the upstate counties in South Carolina and increase the applications for food stamps. The initial exigence[8] was a statistic showing increased malnutrition among elementary school children. Food stamp monies were not being spent, and many myths and historical problems seemed to contribute to the lack of interest in applying. With all these partners, the technical writing/health communication piece of this project involved my engaging undergraduate and graduate students over numerous semesters to design and help produce the campaign materials. We designed and developed 100,000 brochures of numerous kinds—public service announcements, or PSAs (both videotapes for television and audiotapes for radio), posters, Web sites, PowerPoints, and newspaper ads. We targeted different segments of the audience: senior citizens, single mothers with small children, African Americans, and others. The students did research, including field interviews in soup kitchens, Wal-mart's, senior day cares, and community centers; design work; and usability testing. Often they were involved in production and distribution as well.[9]

The proposal for this good-sized campaign was made up of 41 single-spaced pages (not counting the extensive budget spreadsheets) and was divided into the following components:

I. Face Sheet—1 page

    I. Criteria for Approval—$^3/_4$ page

    II. Scope of Work—1 $^1/_4$ page

    III. Plan of Action—38 pages

        A. Statement of Need—11 pages

            1. Need for Food Stamp Program Information and Access Activities

            2. Target Audience: Characteristics and Location

            3. Projected Number of Individual Adults and/or Children to Be Impacted by the Project

            4. Describe How Arrived At Projected Impact and Provide Data to Support Statement

        B. Description of Outreach Activities—23 pages

            1. Organizational Background

            2. Organizational Diagram

            3. Project Operation, Training, and Communications

            4. Activities, Materials, and Reaching Target Audience

            5. Maintaining Records and Reporting to DSS

            6. Expected Outcomes and Measurements

            7. Program Timeline

            8. Job Descriptions

        C. Budget Narrative—4 pages

            1. Salary

            2. Fringe Benefits

            3. Travel

            4. Other Direct Costs[10]

            5. Indirect Costs: Facilities and Administration

The food stamp proposal also included numerous tables and charts.

- Table 2: Estimated Number and Percent People of all Ages in Poverty by State and the Upstate Counties (see Exhibit 7.1). (U.S. Census Bureau, March 1998 Current Population Survey)
- Table 3: Estimated Percent of Children Below Poverty for South Carolina and Upstate Counties. (U.S. Census Bureau, 1997 Model-based Estimate)
- Table 4: County Race Profile. (U.S. Census Bureau, *Census 2000*)
- Table 5: County Age Profile for Persons 65 Years and Over. (U.S. Census Bureau, *Census 2000*)
- Table 6: Third-party Payer for Services Provided at Greenville Hospital. (System FY2000. Source: GHS System Records)
- Table 7: Percentages of GHS System Discharges by County of Residence. (FY1999. Source: GHS System Records)
- Food Stamp Outreach Community/University Partnership (FSOCUP) Process Map—Greenville Hospital System.

All the tables compared the same seven counties in the Upstate. Most of the tables in this lengthy proposal were used to document the need for food assistance and social services and to explain emphases within the target audience section of the document.

How Can I Implement Projects Effectively?

Again using our example of the food stamp application campaign, the best way to set up the entire project for effective implementation is to develop a clear plan of activities and timelines for the partners. In the FSOCUP proposal, such a plan is clearly laid out. Because of the numerous agencies, faculty members, and institutions that

Estimated Number and Percent of People of All Ages in Poverty by State and the Upstate Counties

State and County	Estimate	Percent
South Carolina	569,945	14.9
Anderson	17,611	10.8
Greenville	36,925	10.5
Laurens	8,869	14.3
Oconee	7,246	11.1
Pickens	11,312	11.1
Spartanburg	28,666	11.6

Exhibit 7.1

Source: U.S. Census Bureau, March 1998 Current Population Survey

were partners, each of the partners had a plan within the larger plan. For example, here are articulated goals and their concomitant activities scheduled to achieve those goals. One of the agencies was a countywide volunteer information service, and this agency's plan included the following goal, outcome, and activities:

Goal 3: To increase Food Stamp Program participation among the English and Spanish-speaking population in XXXXX county.

Outcome 1: Provide a "satellite" office space to the XXXXX County DSS[11] Food Stamp Office at least eight hours per week to provide information and increase access to the Food Stamp Program.

Activities:

1. Work with DSS to set up office space and procedures.
2. Work with DSS to publicize the site.
3. Accept phone calls about Food Stamps.
4. Explain the program and prescreen people for potential eligibility.
5. Assist people in understanding the information and documentation they will need to give the DSS worker.
6. Set up appointments for people to meet the DSS worker.
7. Conduct follow-up with people after their appointment with the DSS worker.

Another example from that same campaign plan included the following plan for a small community center serving low-income African Americans.

Goal 4: Increase awareness of Food Stamp Program and Food Stamp applications among the XXXXX Community Center (XCC) members.

Activities:

In the next funding cycle, 10/01/01–09/30/02, XCC plans to expand its Food Stamp work by developing the following activities and programs:

1. Informational Presentations to:
   a. Parents of the Head Start children at XCC
   b. Adult Education students attending classes at XCC
   c. English as a Second Language (ESL) students and work with their teacher to explain the program to people struggling with the English language
   d. Homework Center parents and older children
   e. Special community presentations just for Food Stamp Program outreach x 2
   f. Opportunities made available at seven area churches to distribute brochures and present brief announcements during church services

2. Work with XXXX County DSS to increase access to the Food Stamp Program by:
   a. Publicizing XCC as a place to learn more about Food Stamps and begin the application process.
   b. Accepting phone calls about Food Stamps.
   c. Explaining the program and prescreening people for potential eligibility.
   d. Assisting people in understanding the information and documentation they will need to give the DSS worker.
   e. Setting up appointments for people to meet the DSS worker, either at XXXX County DSS or at XCC.

f. Conducting follow-up with people after their appointment with the DSS worker.

3. Explore the possibility of the ZZZ Baptist Hospital and XXXX County Health Department, including Food Stamp Program prescreening in their admissions assessment.

Each partner submitted its own plan and then negotiated and revised it with the rest of the partners and the principal investigators. In addition to the detailed activities list for each partner, there was a somewhat detailed timeline. The first two months of the campaign timeline, which began October 1st and ended September 30th in each of the three years, included the following:

October–November

- Principal Investigator submits the FSOCUP project to Clemson University Institutional Review Board (IRB) for review and approval.[12]
- Initiate development of Food Stamp educational video to be completed by March 30, 2002. (This video was developed by my honors technical writing students as a service-learning project.)
- Identify existing SCDSS[13] Food Stamp materials to distribute until newly developed materials are available. (tech writing students)
- Public hospital staff, partners, project directors and coordinator, and agency-based outreach workers receive training in the Food Stamp Program from the SCDSS.
- University faculty trains students in the Food Stamp Program.
- Faculty meets with community agencies (and their outreach staff person) to: identify staff and volunteers to train; discuss programs, services, and clients of the agency; and decide on methods faculty and students will use to provide information and promote access.
- Students begin service-learning projects in community agencies. (Technical writing, counseling, and nursing undergraduate and graduate students all participated in service learning.)
- Public Hospitals and agency-based Food Stamp Outreach workers publicize, through their agency and throughout the county, their outreach functions, which are to provide Food Stamp Program information and to assist low-income families gain access to the Food Stamp Program.

As in the months of October and November featured here, each month or two months had a list of the activities slated for completion during those time periods.

What Media Choices Are Best for Campaigns?

The various media choices for campaigns need to be thought out carefully. You want to be able to match the medium with your target audience. In the USDA/DSS Food Stamp campaign, I chose a number of different media. My students and I developed a video about the ease of applying for food stamps now as opposed to the more difficult and labor-intensive application process of the past. In addition to emphasizing ease of application, we also advertised the new Electronic Benefits Transfer (EBT) card, which is now used in several states to purchase groceries. The EBT card is a plastic card not unlike a debit or credit card used by applicants. Such a swipe card used at the supermarket and grocery stores makes buying food with assistance much less visible than when recipients had to count out very obvious stamps that did not resemble printed cash. The stigma of applying for and using food stamps was thus greatly reduced for recipients. See Exhibit 7.2 for the poster featuring the EBT card now used in the state for food stamps/assistance.

Students developed and printed 100 copies of the EBT poster and 100 copies of another poster to hang in DSS offices, free clinics, community centers, volunteer service outlets, hospitals, and other facilities our target audiences visited. We kept the posters simple and plain to maximize impact. All posters included the food stamp hotline toll-free number for SCDSS.

The video we developed featured a scene filmed in a supermarket and showed the swipe card (EBT card) being used to purchase groceries. A narrator explained how the process worked and the ease of use. The video was sent to local television studios for public service announcement spots (PSAs) and is circulating on the closed-circuit televisions set up in large regional hospital systems. We knew that we'd have a large, ever-changing, captive audience within the hospitals as patients and visitors flipped through channels in patient rooms and waiting areas. As in all media choices and development, the key is to make sure the media is applicable for both the message and the audience.

Although we did develop three Web sites during two years of the campaign, one for internal communication among partners and two for social service agencies, our main emphasis was on colorful brochures, simple posters, videos, audiotapes for radio station PSAs, and newspaper ads for small papers, local dailies, and weeklies. As mentioned before, many people in low socioeconomic brackets do not have good computer access, and, according to Cindy Selfe and others, the "digital divide" is increasing (Moran and Selfe 1999, 48). There are studies showing that some people within that bracket will use computers at public libraries to conduct searches. But our primary target audiences were not computer users, except for those who used a particular community center and would also volunteer to be a food stamp counselor within our campaign. We chose instead the media that would most interest our information users. Students in technical writing classes usability tested (field tested) various iterations of brochures to assure clarity and appeal. There was a huge difference between the state-developed brochures (text-heavy, no visual interest, full of jargon and legalese, in eight-point font size) and those developed and tested on real audiences in soup kitchens in the Upstate. Every one of the audience members approached rejected the state-developed brochures in favor of the ones produced by students. When our test audiences found confusing phrasing or graphics that were less than effective, students revised the documents to reflect the advice of members of the target audience. As we have pointed out in several of these chapters, involving the users or audience members in the process is the

Electronic

Benefits

Transfer

Food Stamps
Just Got Easier!

SC Department of Social Services Food Stamp Hotline: 800-768-5700

Exhibit 7.2 Poster for the EBT (Electronic Benefits Transfer) Developed for USDA Campaign in Clemson University Technical Writing Class
Source: Heifferon, 2001

best way to ensure appropriate and useful health materials. Brochures developed by detail-oriented office workers and those not tested with real audiences are not as successful. Because of the myths (and in some cases the reality) of the nightmares of applying for food stamps, it is important that attractive and accessible materials are developed.

What Are Some Possible Strengths and Weaknesses?

There are many important aspects of successful campaigns: adequate resources, skilled leadership, excellent collaboration, good design skills, extensive and accurate research, and well organized communication plans and applications. One public campaign that didn't succeed was Nancy Reagan's "Just Say No" to drugs campaign and its attendant DARE program. Both have failed to adequately address the problem, even though resources, time, and energy have gone into keeping the DARE program going. Although it and Reagan's efforts did address an extensive problem in the culture of the United States, statistics show that the campaigns did not have the impact everyone had hoped for. Although younger children in the upper grades of elementary school seem like a good target audience, the campaign was misdirected. One would have thought that T-shirts carrying the DARE slogans, a curriculum of antidrug slogans and thinking, and much work on the part of teachers and school resource officers (from law enforcement agencies) would have been effective strategies. If you listen to the public service announcements on television in the early part of the twenty-first century, you'll notice that the focus has changed (not that just one focus is adequate for such an extensive health problem). Today television ads target an older age group (adults, not children) with conversations that call attention to the violence and mayhem of the drug-selling lifestyle. It follows a seemingly logical argument that if you buy drugs, even for recreational use, you are connected to the network of horrendous drug violence. The ad is dramatic and noteworthy. Hopefully the networks and the agencies responsible for the ads are conducting assessments to see what kinds of effects this type of advertising and health promotion is having on the consumers who watch television (most adults in this culture).

Another ad, again targeting adults and especially parents of teens, suggests that talking to your children about drugs (even if they seem to ignore you), being involved in their lives, and setting some limits and boundaries (even though teens resent and resist such "interference"), are all effective strategies for keeping teens off drugs. This campaign is based on research that supports such conclusions. Whether or not that approach encourages the behavioral changes in parents who are not already parenting this way remains to be seen. The campaign is a relatively new one.

How Can I Troubleshoot a Campaign?

Campaigns, to be successful, should be subjected to troubleshooting before large investments are committed to implement the plan. Critiquing a campaign and judging these successes and failures does not mean that in conducting a campaign, there are obvious right and wrong ways to do it. Campaigns are difficult to construct and usually need to involve multiple audience approaches, strategies, and media. One way to troubleshoot a campaign before its launch is through a pilot program, which involves field-testing or having focus groups pretest the products, approaches, and strategies in advance of the release. In my technical writing classes, we developed a variety of visual, electronic, and print documents and pretested them in several ways. We have draft or beta evaluation meetings similar to studio "crits." A "crit" in art and architecture studio classes consists of bringing out the artwork or architectural plans in front of classmates, other professors, clients, etc., and receiving feedback on what is working and what is not working. Students, designers, and artists use the critique to revise and improve the product or work. Crits work also in technical communication and/or health communication classes in which students must design visual communication products for a campaign.

Another way to send up "trial balloons" in addition to pilot studies, small-scale releases, and crits is to conduct focus groups to see what campaign components are successful and which are not. Just as focus groups are often recruited in the mall for early release of new products with some incentives (coupons, gift certificates, small amounts of cash), focus groups on health campaign components can be gathered in similar ways.

In addition to the above strategies, campaign planners can also plan usability testing (especially on Web sites and intranets) and/or field-testing, which are less formal usability tests conducted outside of the usability laboratory setting (Gould and Doheny-Farina 1999).[14]

Whatever research or testing methods you use, carefully recording, analyzing, and synthesizing the feedback you receive will enable you and/or your team to set up revision suggestions and strategies. These revisions will be easier to map out once you've got a good sense of the data. Ideally this data can be triangulated; in other words, use several different feedback methods or multiple groups or audiences to try out your health messages. This way you do not get a false reading around which you then design an extensive and expensive campaign.

What Assessment Techniques Should I Use?

In our USDA Food Stamp campaign example earlier in this chapter, much of our assessment data was quantifiable. We set up evaluations based on frequency counts, because our goals and outcomes were based on increasing the number of food stamp applications in seven counties in the Upstate. Unless you operationalize your definition of success, you will not be able to develop an effective assessment plan. In other words, articulate your goals and outcomes first, and then you can develop assessment strategies, which include surveys (quantitative or qualitative), frequency counts of all kinds, and other ways to measure changes in behavior. In the Healthwise example mentioned at the beginning of this text, the informational book clearly outlined whether specific symptoms dictated home healthcare, a visit to the regular doctor, or an emergency room trip or ambulance ride; therefore the measure was to be a reduction in the number of emergency room visits by nonemergency patients. The difficult part of measuring successful outcomes is controlling for variables. These situations cannot be set up as experiments in which variables are controlled, because dealing with human behaviors is not an experiment-appropriate situation.

For example, even if emergency room visits dropped during the year follow-ing the introduction of the Healthwise books in thousands of homes, there could be other factors influencing the numbers. If a large population group relocated away from the area to follow a major corporate employer, if people moved away because of a toxic pollution problem, or if health insurance became more afford-able for poor people using emergency rooms rather than general practice physi-cians, the number of emergency room visits could be reduced without being im-pacted by the health campaign itself.

Here's an example of the outcomes and frequency counts set up for measur-ing the success of the Food Stamp campaign:

Maintaining Records and Reporting to DSS

The FSOCUP project will use the 2000–2001 report forms approved by DSS in order to continue developing a longitudinal and aggregate database. The data from second project year 2001–2002 will be added to the existing Food Stamp Data Base to establish aggregate data. Report forms include the individual activ-ity forms, monthly activity forms, DSS monthly forms, time sheets, and travel sheets. The Food Stamp preassessment forms are coded to indicate which com-munity agency and which department at XXXXX University were involved in completion of the preassessment forms.

Expected Outcomes and Measurements

Goal 1: Need to inform low-income individuals and service providers about the availability, eligibility requirements, and application procedures of the Food Stamp Program.

Outcome 1: Low-income individuals presenting at the ZZZZ Hospital System or XXXX Regional Healthcare System for inpatient or outpatient services will be in-formed about the availability, eligibility requirements, and application procedures of the Food Stamp Program.

Measurements:
- Frequency count of events and educational materials distributed
- Frequency count of number of Food Stamp preassessments completed
- Frequency count of referrals to DSS
- Frequency count of referral sources visited by social worker or other healthcare professional (information and preassessment)
- Frequency count from research-qualified individuals (current, previous applied, previous disqualified, and in-depth interview)
- Frequency count of applications processed by DSS of individuals not requiring preassessments (codes on Food Stamp materials)
- Develop Food Stamp video and obtain approval by DSS for use in hospitals (Done—Tech. writing students)
- Evaluation of outreach activities by service providers

Outcome 2: Service providers and students will be informed about the availability, el-igibility requirements, and application procedures of the Food Stamp Program.

Measurements:
- Frequency count of agency and student outreach workers participating in Food Stamp Outreach activities

- Evaluation of training materials developed
- Frequency count of events and educational materials distributed
- Outreach activity evaluation by service providers and students

Goal 2: The need to enhance access to the Food Stamp Program by assisting low-income individuals in applying for the Food Stamp Program. The need to train service providers and students on how to assist low-income individuals to access the Food Stamp Program.

Outcome 1: Low-income individuals have been assisted in applying for the Food Stamp Program.

Measurements:
- Frequency count of number of low-income individuals assisted in the application process
- Frequency count of number of applications processed in DSS county offices attributable to outreach activities
- Compare patient payer source (Medicaid, Medicare, and self-pay) to county poverty level data and numbers of Food Stamp applicants
- Collect demographic data from in-depth interview (race and zip code)
- Measure time frame from in-depth interview to qualification
- Collect from DSS the frequency of coded in-depth interviews that received Food Stamps
- Collaborate with DSS office regarding the error rate from the GHS in-depth interviews according to federal guidelines
- Compare hospital admission and discharge rates to information and access outreach activities assessment information (maintain 90 percent+ information and activity rates)

Outcome 2: Service providers and students are trained in the Food Stamp Program application process.

Measurements:
- Frequency count of the number of service providers and students trained to assist low-income individuals in applying for the Food Stamp Program.
- Training process evaluation by service providers and students

These are just two of the goals we outlined in our campaign proposal.

If you set up the system from the beginning to measure the outcomes and enter the information into an electronic data bank, you will have excellent data by which to measure your success *and* to apply for further funding. Often funders are more willing to support a campaign for following years after a successful, first-year run.

How Do I Debrief and Learn from a Campaign?

One of the most important last steps in any campaign is to conduct a debriefing. Some campaigns will require several of these, depending on how long your assessment tool is expected to monitor and catalogue results. In other words, you may not realize the effectiveness of the various media you have constructed as a health communicator. Changing the behavior of target audience members takes time. You may reflect on the data you have collected repeatedly over a number of years. It is good, however, to reflect on the initiatory and implementation phases of your campaign, just as you reflected and critiqued your design choices in the mid-course reviews or corrections.

The most valuable reflection for the campaign planning process takes place soon after the campaign is launched, while your difficulties, weaknesses, and successes are fresh in your minds. Often these details are lost over time, so if you take time to record (audiotape or videotape) and write up your reflections on the campaign afterward, you will have a record of campaign planning strategies for the next one you work on. The analysis is particularly valuable to gather both as a team and individually, as different team members remember and/or learned different lessons along the way. In this way, you will reinforce your memory and your findings.

Some grantors and other funding agencies or clients require a "postmortem" or final report that details not just the data you have collected so far, but also a process document that lays out the methodologies you used and your analyses of their successes or failures. You may have review boards or feedback committees complete this process for you, as well, but do your own internal review anyway. Find professional ways of critiquing a colleague who did less than stellar work. If you are in a management position on the media development of the campaign and you document failings in your review process, you have important data for the employee evaluation. Again, recalling and reflecting on these experiences not too long after the launch will keep the necessary details patent.

What Grants and Foundational Support Are Available?

Many grants and foundations will fund outreach efforts and media development that are part of a health campaign. NIH, NIMH, and NSF are obvious governmental granting agency choices. Often, too, it is worthwhile for a large campaign to farm out several aspects of the project to different funding organizations. In other words, cobble together numerous grants to complete a big undertaking. Grantors are more comfortable with an extensive project if they know other funding entities are also involved. Sometimes having another grantor or foundational support will mean the difference between success and failure in obtaining more funding.

Summary

In this chapter we have examined what makes writing, composing, or designing a large health campaign different from some of the other genres of writing we do as health professionals. You have gotten some insight into what criteria to use in choosing a topic that is appropriate for a public health campaign. You've learned that a team of professionals needs to identify a problem for which there is a solution. In addition, for a solution to be effective, it must be one that human behavior can alter. Whether it's a matter of education or information, that new knowledge needs to lead to some action that can be initiated, changed, or stopped by a member of the target audience.

You also realized what ethical considerations are important for a public health campaign. Because you are attempting to change behavior, i.e., manipulate someone to do something he or she is not currently doing, you must examine the ethics of that attempt. In addition to the previous ethical considerations we outlined in an earlier chapter, for campaigns it is important to also consider health policy, laws, and regulations. You do not want to be attempting to ask people to change their behavior if it goes against standing legislation. In that case, you need to refocus the campaign and the target toward those who write and design the policies or laws, not toward the individual behaviors. You also want to determine if your suggestions go against the cultural norms of the target group. If so, the campaign is unethical because it is asking people to change behaviors that may jeopardize their memberships in the support groups and environments that they need to maintain health and happiness. Often there are other approaches that can be used to achieve the same goals without destroying support groups.

You learned the importance of developing good audience analyses, research agendas, and proposals for campaigns. Because those trained in communication studies or public health analyze audiences differently from those trained in rhetoric and technical or professional communication, you learned the different terminology (*audience segmentation*) that is used and how terms such as *audience segmentation* and *sense-making interviews* are defined. Various other methodologies for audience analyses and research were discussed, and a model of a detailed campaign plan was used to familiarize you with the genre.

Media choices in a large campaign are also important, and usually numerous genres are used to achieve success. We emphasized matching the medium to the audience (if multiple audiences are addressed) or using media that was most effective and frequently used by the target audience(s).

We briefly discussed some of the strengths and weaknesses of past campaigns, acknowledging that there are many factors and multiple variables that must be in place to make a campaign work well. One way to heighten the odds of success is to troubleshoot the campaign before its launch and come up with revision strategies appropriate for mid-course corrections. We also suggested studio "crits" of the various media that health writers develop to catch problems early on. Usability and informal field-testing are also effective ways to troubleshoot early iterations of whatever deliverables you are developing.

The USDA health campaign served as an example throughout this chapter to illustrate the various principles, including setting up evaluation and assessment techniques. Although this particular project mostly used frequency counts in determining if the campaign was successful in raising the number of food stamp applications, other types of assessments were discussed. Another strong suggestion was that you assess your own internal performance as a team and as group members after the launch of the campaign. This information provides valuable guidelines for the next campaign you plan, implement, and assess.

The next chapter will examine at length the grants and foundation proposals that lead to successful funding of health campaigns and other projects.

Discussion Questions

1. Find some public health campaigns in the past that were, in your view, unethical. Bring examples to class to discuss. Also, brainstorm some possible campaigns that might be a problem ethically. Discuss, too, the more subtle ethical issues that occur in any number of campaigns. How do campaigners reconcile the greater good versus harm argument?
2. Discuss how you would research health legislation, policy, and cultural norms for your local area. Take notes on this discussion for Exercise #2.
3. Brainstorm or find examples of the positive and negative outcomes of public health campaigns that were effective overall and be prepared to discuss what you have discovered.
4. Discuss various large public health campaigns that used television as one of the media. What audience segments do you think they were trying to reach?
5. Discuss what kinds of questions you would want to ask organizational and institutional personnel who prepare proposals or review them for large projects or media campaigns.
6. What kinds of activities would you set up in a health campaign to achieve the goal of less binge drinking among college students on weekends?

Exercises

1. Find Web sites that address public health problems and are themselves perhaps part of a campaign. Put each target problem and solution through the three criteria to see how they fit. Write a brief, analytical report on why this does or does not lend itself to a public health campaign. Also, detail how the goal could be restructured to then be appropriate for a campaign effort.
2. Based on discussion question #2 above, use your notes to develop a health legislation, policy, and norm research plan. If an electronic discussion board, chat room, or bulletin board is available, post your ideas and plans for your classmates to see. Try some of the avenues of research you have developed and rate which ones are most helpful. Then develop a plan that puts together the most effective research means for your area (you can include state and federal laws, too).
3. Form a team of two to four people and brainstorm how you would address one form of bioterrorism, chemical warfare, or radiation exposure in an educational public health campaign. What combination of genres of documents (print, audio, video, electronic) would you use and why? Document your discussion and then write up a brief initial plan. You may want to map out your

strategies in a visual and/or electronic form (graphics, PowerPoint, Web site, or multimedia presentation).

4. Form a team of two students and discuss a health habit you would like to change. Choose one that lends itself to a campaign and design a skit in which the two of you role-play an approach you think is persuasive to help change a bad health habit. You can choose a sense-making interview for the scene of your one act.

5. Divide into teams of two and pick one of the main segmentation categories to map out for a target audience of your choosing. Come up with several scenarios within your category that illustrate the principle. If there are enough teams, you can choose subcategories under demographics. Present your examples to the class visually, electronically, orally or in written form (use different media).

6. Using the Internet, find other examples of health campaign proposals and compare several of them and report back to your classmates.

7. Visit the sponsored programs or research administration office on your campus or at your health institution. Interview some of the personnel who write, review, or prepare budgets for large projects or media campaigns.

Works Cited

Dervin, Brenda. 1989. "Audience as Listener and Learner, Teacher and Confidante: The Sense Making Approach." *Public Communication Campaigns*, 2nd ed. Ronald E. Rice and Charles K. Atkin, eds. Newbury Park, CA: Sage.

———, M. Nilan and T. Jacobson. 1981. "Improving Conditions of Information Use: A Comparison of Predictor Types in a Health Communication Setting." *Communication Yearbook* 5. M. Burgoon, ed. New Brunswick, NJ: Transaction.

———, T. Jacobson and M. Nilan. 1981. "Measuring Information Seeking: A Test of Quantitative/Qualitative Methodology." *Communication Yearbook* 6. M. Burgoon, ed. New Brunswick, NJ: Transaction.

Frank, R. E., W. F. Massey, and Y. Wind. 1972. *Market Segmentation*. Englewood Cliffs: Prentice Hall.

FSOCUP Partners. 2001. *SC Food Stamp Outreach Community/University Partnership (FSOCUP)*. Greenville, SC.

Gould, E., and Stephen Doheny-Farina. 1988. "Studying Usability in the Field: Qualitative Research Techniques for Technical Communicators." *Effective Documentation: What We Have Learned from Research*. Stephen Doheny-Farina, ed. Cambridge, MA: Massachusetts Institute of Technology Press.

Greenville Hospital System. 1999. "Percentages of GHS System Discharges by County of Residence." *Greenville Hospital System Records*. Greenville, SC: FY1999.

Greenville Hospital System. 2000. "Third-party Payer for Services Provided at Greenville Hospital System." *Greenville Hospital System Records*. Greenville, SC: FY2000.

Haley, R. I. 1985. *Developing an Effective Communication Strategy: A Benefit Segmentation Approach*. New York: J. Wiley.

Heifferon, Barbara. 2004. "The New Smallpox: An Epidemic of Words?" Unpublished manuscript. Under review.

Kotler, P. 1986. *Principles of Marketing*, 3rd ed. Englewood Cliffs: Prentice Hall.

McGuire, William J. 1999. "Behavioral Medicine, Public Health, and Communication Theories." *National Forum* 60: 18–24.

———. "Theoretical Foundations of Campaigns." 1989. *Public Communication Campaigns,* 2nd ed. Ronald E. Rice and Charles K. Atkin, eds. Newbury Park, CA: Sage.

Moran, Charles, and Cynthia L. Selfe. 1999. "Teaching English across the Technology/Wealth Gap." *English Journal* 88, no. 6 (1999): 48–55.

Parrott, Roxanne Louiselle, Mary Louise Kahl, and Edward W. Maibach. 1995. "Enabling Health: Policy and Administrative Practices as a Crossroads." *Designing Health Messages: Approaches From Communication Theory and Public Health Practices.* Edward Maibach and Roxanne Louiselle Parrott, eds. London: Sage.

Skinner, Tom. Quoted in "Survey: Smallpox Misconceptions Linger." http://www.cnn. com/2002/ HEALTH/ conditions/12/20/smallpox. misconception. ap/index.html. December 26, 2002.

U.S. Census Bureau. 2000. "County Age Profile for Persons 65 Years and Over." *Census 2000.* Washington, DC: GAO.

U.S. Census Bureau. 2000. "County Race Profile." *Census 2000.* Washington, DC: GAO.

U.S. Census Bureau. 1998. *Current Population Survey* (March). Washington, DC: GAO.

U.S. Census Bureau. 1997. "Model-based Estimate." Washington, DC: GAO.

Windahl, Susan, and Benita H. Signitzer, with Jean T. Olson. 1992. *Using Communication Theory: An Introduction to Planned Communication.* London: Sage.

Endnotes

1. Although self-exams have been a widely accepted and promoted practice, recent studies have questioned the efficacy of this practice. Conflicting studies present one of those "contested information" situations that are particularly difficult for health consumers to untangle. In other words, the message is no longer clear.
2. Read or review Chapter 2 on Ethics if you have not already done so.
3. Sexually transmitted diseases.
4. Both communication theorists and rhetorical theorists have been influenced by postmodern thought, although there are as many opinions and interpretations of postmodernism as there are individuals.
5. *Alternative* refers to some medical or health practice that is not part of the conventional Western medical system (such as acupuncture). *Integrative* refers to a combination of alternative and conventional Western medical practices.
6. African American women die at a disproportionately higher rate from breast cancer than any other minority group in this country.
7. See endnote 2 in Chapter 2 regarding the Tuskegee experiment.
8. Revisit Bitzer's rhetorical situation and discussion of exigence in the Instructor's Manual if you are puzzled by this term.
9. The design and production of materials worked well, but the grant suffered from incompetent and ever-changing leadership after the first year. Fortunately, some success was achieved, but not as much as some of us had hoped.
10. Direct costs refer to the costs incurred to pay personnel, buy supplies, travel, and fund the activities involved in carrying out the project. Indirect costs are usually a percentage that a university or other institution gets to process paperwork and run the facility at which partners work. Indirect costs are thus similar to overhead costs for businesses.

In the case of this proposal, the institutions waived some indirect costs because of the social service nature of the grant; thus the percentages were much lower.

11. DSS = Department of Social Services.

12. Such campaigns that involve students and faculty working out in the community must submit an Institutional Review or Research Board application.

13. SCDSS = South Carolina Department of Social Services.

14. For more information on focus groups, usability testing, and field testing research, please see the chapter on audience analysis.

CHAPTER **8**

Grants, Proposals, and Government Documents

 ## Overview

In addition to writing grant proposals and answering foundation solicitations in the medical arena, this chapter examines other institutional documents, specifically governmental and insurance letters and forms. Chapter 8 provides both a broad overview as well as some more detailed writing guidelines. Health writers, health public relations professionals, practitioners, and researchers write both proposals and grants for campaigns, for their and others' research, and for projects.

You will learn what's most important in institutional health writing and discover answers to the following questions:

- How do I get started on a grant?
- What are the basics of grant writing?
- Why aren't government documents user friendly?
- How do I rewrite government documents?
- What is the role of documentation?
- How do insurance documents differ?
- How long do records need to be kept?

Grant applications, governmental and insurance documents, like many medical documents (charts, etc.), are all specific genres of writing and communicating that have many formal constraints. As we know from our learning experience in schools in this country, genre norms vary greatly across literature, technical writing, and various professional forms of communicating. In this chapter, these particular genres have very strong conventions and norms; because of this factor, the constraints on writers are many. In all of these documents, you must prepare copy that does not vary much if at all from the requirements mandated by the forms. Of the three genres featured in this chapter (grants, governmental and in-surance documents), the grants are the most malleable.

How Do I Get Started on a Grant?

The first thing any health professional who wants to write a grant should know is that the best grants start with an excellent idea. Most grantors and foundations are interested in new approaches to old problems. Those most successful in garnering grants are not only good writers, which is an important skill, but they are also cre-ative thinkers. Coming up with ideas that no one has tried before, as long as they are not outlandish or impossible to accomplish, is one of the best ways to ensure success in securing a grant. As a health professional, you want to think of ways to solve health problems or educate patients in ways that have not been tried before. Or, if you know of campaigns or media that solved problems and were effective, you can plan various offshoots of such projects. For example, one of the most successful ef-forts ever mounted in a culture on a large scale was the eradication of syphilis in mainland China. While this was done under a regime that gave citizens few choices, syphilis was rampant in previous times and was brought thoroughly under control. Although many cultures do not have such captive audiences or similar apparatus for social influence, we can still learn lessons from such disease control. One of the strategies that was most helpful in achieving this ambitious goal of eradication was the use of "barefoot doctors." This ability to train and disperse large numbers of health personnel was key to a society with limited technological access. By not hav-ing to rely on health professionals like physicians and nurses with lengthy, educa-tional processes, more quickly trained paramedical personnel could be recruited and sent out. So a creative grant idea may be to train local community members to ad-dress a singular, chronic, or acute problem and fund the pilot project with a grant from the National Institutes of Health (NIH).

Other ways to come up with creative solutions is to interview the patients or families of patients with the problem. Often those who are most involved can come up with the most creative and pragmatic ideas. Because they live the experi-ence, in many ways, they are the experts. As we emphasized in the chapter on au-dience analysis and in the strategy of sense-making interviews in the chapter on campaigns, this means of brainstorming creative solutions works. Remember: the first "basic" of the "basics" of grant writing in and for the health professions is coming up with a great idea.

What Are the Basics of Grant Writing?

Searching for grants that welcome applications (rather than those that discourage or will not accept unsolicited applications) is easier now with the Internet. Previously a health writer had to leaf through large volumes listing foundations and granting or-ganizations, often not well indexed or intuitively arranged, to find appropriate grants. Now we have the Internet and good searching techniques, which can find many applicable grants. The most common sources for medical professionals include the following: National Institutes of Health (NIH), National Science Foundation (NSF)[1], National Institutes of Mental Health (NIMH), United States Department of Agriculture (USDA), Department of Health and Human Services (HHS), Department of Defense (DOD), and Duke Endowment, to name a few. Many other foundations and granting agencies will support health-oriented research and will support educational, preventive, and outreach activities. If you live in a rural area, there are many sources for support from grantors recognizing the great need of good healthcare and better health information outside of urban centers. In addition, groups that traditionally are underserved by healthcare facilities are especially good populations to target in grant projects. Research on these populations to determine which messages are behavior changing and encourage good health practices is also readily fundable.

What Is the First Step After I Have Located a Grant?

One of the most important steps in the process after you have developed a stellar idea and located a grant possibility is to very carefully read the RFP. An RFP means request for proposal, but many of the larger health-oriented entities have numerous grants that are available on rolling deadlines and do not call an application an RFP per se. If you work at a health foundation or other granting organization, you know that it is surprising how many applicants do not carefully read or follow the directions in an RFP or grant application. As obvious as it seems, accurately following the directions, submitting the correct forms and formats, answering all the required questions, and giving yourself or your team enough time to adequately develop the application itself are commonsense strategies that some ignore. Those applications are the first rejected. If you cannot follow the directions, your credibility (*ethos*) as a professional is already impaired. If there is a direction you cannot understand, call the granting agency (the larger ones all have toll-free numbers) or email them for a clarification. For questions about the RFP, most Web sites or mail-in application forms have a specific person or persons who answer questions about the RFP itself, especially in the large organizations with multiple grants.

What Writing Style Do I Use in a Grant?

Often a description, narrative, or executive summary is the first part of the grant you will fill out. Don't use complicated vocabulary to impress the grantors; they will not be impressed. Avoid long, convoluted sentences; use your best professional writing style. Often academics write in a very different style from practicing health professionals. Administrators sometimes write in bureaucratic language. Be able to change your style to a more effective one for this genre. Know your audience for the grant. Often grantors are not members of the more focused discourse community of which you are a part. In other words, if you are writing a grant and your specialty is genetic counseling, don't use medical language specific to your area that those outside your discourse community will not understand. In grants it is advisable to carry out the KISS directive: "Keep it simple, stupid" or "Keep it short and sweet."[2] You do want to sound knowledgeable but you don't want to lose your audience among highly specific, scientific verbiage. Some of the members on the grant review board may not even be in the medical field. If possible, obtain a successful grant from the grantor to use as a model or research the members of the review board to determine their fields or backgrounds. Although every component of a grant or proposal must be accurate and persuasive, the descriptive narrative or executive summary is the place where your persuasive and descriptive abilities are most important. Here's where you can sell the great idea you've developed. If the idea is strong, innovative, and feasible, and if you build credibility into the document, you are more than halfway there. Other portions of the grant may be negotiable or may not be perfect, but a good idea that is well presented will be the heavy hitter in your application.

In this persuasive text, you will also want to establish a need for your idea and the expertise of you and your partners. There are at least three main ways to establish a need:

1. **Use statistics.** (These are readily available on-line or in government documents areas and reference desks in libraries.)
2. **Use a literature review.** In this case, a review of the literature will include sources that persuade the reader that your idea is well supported by other experts in the field. You can also use the literature review to prove that previous approaches didn't work; thus your innovative approach will break the mold of recent unsuccessful projects.
3. **Use a pilot study.** You can either propose a pilot study (especially attractive for NIH and NIMH) or use a pilot study to show that your idea will work. As we discussed earlier, the fact that others have financed your good idea as a pilot is persuasive, too.

What About Outcomes and Assessments?

Sometimes the narrative or description includes outcomes and activities. In other grant instructions, you are to prepare a timeline of activities separate from the executive summary or description. Either way, make sure you include detailed outcomes (because they are measurable, clear, and operationalized) and include activities that support and lead to those outcomes. In this way proposals and grants are similar. In the previous chapter on health campaigns, we outlined clear objectives, goals, or outcomes and the steps to achieve them. An example of a grant proposal written this way is:

Outcome 1: Start a Web site for health information most useful to citizens in XXXX County for their Health Department.

Activities:
1. Set up interviews with health department personnel to determine the main health issues they see in their department.

2. Assess staff computer expertise and availability.
3. Assess the computer software, processor speeds, and memory capability available on XXXX County health department computers.
4. Assess the computer capability of county public libraries to determine browser types.
5. Review researched census data for the county to determine computer access. (Some of this research should have been done before the grant was written but not all, especially if the grant asks for solutions to health problems and you propose several.)
6. Make sure the designer can also train or be able to hire a trainer to teach some of the county staff to maintain and update the site once it has been launched.
7. Determine linkage possibilities to other health information sites.
8. Make sure software needed for you and your designers of the site is available or can be ordered through the site.
9. Research the need for and availability of domain names.
10. Determine access to Internet service providers and ongoing funding to support such access or see if the site can be uploaded on government servers already *in situ*. (For more examples, see Chapter 7 section on campaign proposals and assessments. Those examples were written for a USDA grant.)

Because you have set up careful outcomes and activities (plus their timelines) in the grant proposal, you have helped your team develop an assessment tool as well. That's what we mean when we say outcomes are measurable. You can now set up a measuring system based on your various goals or outcomes and activities. You can use both qualitative and quantitative measures or a combination of both. Be sure you include enough detailed information to answer the "w and h questions":

- Who?
- What?
- When?
- Where?
- How?

What About the Qualifications Section?

Avoid listing your qualifications as an expert in the descriptive narrative or executive summary, unless you have completed a pilot project and want to report that study as part of your success rate. You will usually be asked in another section of the grant for your qualifications and those of your partners. Depending on the grantor, this request for qualifications can be met by sending them either a resume or a curriculum vitae.[3]

Your partners will also be asked to include their vitae or resumes. If you are the P.I. (principal investigator) on the grant, make it your responsibility to check and see that everyone sent in updated text, accurate data, and no padding. If vitae or resumes are padded or are not factual, your ability to get the grant, especially if wrongful reporting is caught, has been jeopardized. That's why as a principal investigator and sometimes even as a coprincipal investigator, I check the qualifications section, too. Often, rather than long *vitae,* an organization will ask you to reduce it to three pages. In that case, you cut all but the most relevant and important information to the specific grant you are proposing.

How Do I Prepare a Grant Budget?

Often as health communicators and/or medical grant writers, we don't have accounting degrees and have limited experience in writing financial documents. But budgets are important in grant writing, and there are a number of ways to successfully complete a thorough and accurate budget:

1. Work with the business officer or accounting department of your health facility to get the help you need in preparing a grant budget.
2. Work with the college or university research office to prepare a budget if you are an academician.
3. Use the Internet to examine costs and find examples of good grant budgets.
4. Find grant budget wizards online. [NIH site for grants $250,000 and less: (http://grants.nih.gov/grants/guide/notice-files/not98-178.html); NIH site for grants $500,000 and over: http://grants1.nih.gov/grants/funding/phs398/fp4.doc].
5. Ask the grantor or foundation to send you a model of a successful proposal and make sure it includes a model budget as well.

Although expertise in the area of financial planning on a grant is important, don't underestimate your own abilities; instead, use the opportunity to learn how to design a good budget. Most health communicators are or should be adept at using MSExcel and other tables, graphs, charts, and spreadsheet software. Because the Internet contains such a wealth of knowledge, even if it is a disorganized "wealth," you can readily find information about costs. In time it would be advisable to at least acquire a good working knowledge of grant budgets.

What Are Red Flags in a Budget?

Pay careful attention to the instructions for the budgetary process for the particular grant you are writing. Sometimes the directions will ask you not to include certain costs, such as those for computer hardware or other equipment, travel funds, or another budget category. Make sure you don't include what you are instructed not to, or your application is immediately jettisoned.

Also, even if you follow the directions carefully, you still may be asked to negotiate the budget. Be prepared to discuss possible changes. Often, especially for government grants, indirect costs may be too high for some grantors to pay. Do not come to the "table" (or conference call) unprepared to suggest what part of your budget is flexible and what part is firm. If you worked with an in-house expert during the budget-writing process, bring that person into the negotiations with you. In academic settings, especially, indirect costs and their percentages can be negotiated and often lowered or matched. Matched funds simply mean that you put up your resources (such as normal faculty salary of the principal investigator and coinvestigators, for example) as a match with the grantor. Be aware that there are different ratios of matches, such as: 1:1, 2:1, 3:1, etc. In the state of South Carolina, for example, when Medicaid is considered, the federal government often gives a 3:1 or 4:1 match. In other words, if the state puts up $100,000, the Feds will put up $300,000 for a 3:1 match or $400,000 for a 4:1 match.

Other red flags go up for the grant reviewers if you do not accurately reflect the costs for equipment, supplies, phone calls, or whatever category you are including in the budget. Especially with the Internet, there is no reason not to be able to check many prices on-line and give an accurate estimate, even if you are not an expert. While the government itself is famous for cost overruns, especially the military services and contractors, it's not good for the ethos or credibility of your organization to go over your budget consistently. Such actions usually signal to a savvy grantor that you are not a good planner or project manager. Also, before you send the budget and grant in, make sure all figures are accurate. Use your calculator even if an expert has done the spreadsheets. Anyone can make a mistake, but that's not quite how grantors look at it.

Lastly, make sure the amounts asked for as accurately as possible represent or match the amount of work involved in the grant. You don't want to ask for too large an amount for a relatively small amount of work and vice versa. Review boards generally have wide experience with reading these proposals, and they can usually spot mismatches. They can also tell if people are overpricing themselves (and therefore the grantor), and the boards are again not persuaded of your abilities. Padding budgets is as grievous as padding vitae or resumes. A slight amount of overage is fine, but a consistently large amount signals problems. Make sure you have accounted for as many legitimate expenses as you can possibly foresee. Like many other genres, this one also takes practice to be able to forecast the budgetary needs. Once you have completed several budgets, you will notice how much easier the grant-writing process seems.

What Is a Good Caveat for a New Grant Writer?

As experienced grant writers tell their trainees, make sure your expectations are realistic. Because good grant writing is based on creative and new ideas, it is easy to overdo the dream and end up in difficulty. Although an innovative grant is initiated via inspiration, as the saying goes: "It's 10 percent inspiration and 90 percent perspiration." You need to be both pragmatic and idealistic to be successful in this arena. You need to have the motivation to save the world and dream big, plus you need to be a down-to-earth pragmatist at the same time. Otherwise, you will have problems realizing your goals and will frustrate yourself, your team members, and your institution. In addition, you will develop a bad track record as an initiator of grants with funding organizations. Review the outcomes and activities to ensure that the activities will achieve the projected outcomes successfully.

In the Works Cited section at the end of this chapter, you can find additional resources. An especially good site is Galen II, which is the University of California at San Francisco Digital Library at http://www.library.ucsf.edu/ref/path/grants/reference.html. This site is particularly good because librarians identify and evaluate sources based on their quality and credibility (Fang 2001, 1). Look under "Selected Internet Resources." The site is also updated regularly. Grants on-line are easy to find, and numerous good databases are available.

Why Aren't Government Documents User Friendly?

Government documents represent their own genre in that they are part of a bureaucratic and legal system that determines the discourse of the documents found there. Efforts are being made to change some of the language to make them more accessible, but just as traditional medical language has evolved over centuries to give us the peculiar language in medicine, so has governmental language evolved to give us difficult verbiage. However, some governmental documents are more arcane than others.

Physicians and others often complain about the myriad of forms and tons of paperwork required by the government by such federal programs as Medicare and Medicaid. The filing of numerous forms per eligible patients creates whole staff positions in healthcare facilities and agencies to handle the paperwork alone. In addition to these programs, licensure of health professionals at all levels requires certifications and updating to ensure the quality of healthcare in this country. Often there are both state and federal document requirements, not to mention county and city paperwork, too. Insurance companies, large health systems, and HMOs also require more forms and documents. Pharmaceutical companies must file protocols for the release of new drugs, and much medical research is also highly regulated to ensure the safety and protection of the populace. Like federal grant programs, some of the documentation is now available electronically, but not all of it is digitized. Many complain that even with the increase of on-line systems of reporting, the paperwork stacks loom larger, not smaller. Safety boards must inspect and guarantee many practices that otherwise would put the citizenry at risk. To cover all genres of governmental documents would be impossible; thus a few examples are featured here to help you visualize strategies to make them more user-friendly for readers. Examples include a set of federal guidelines, the new requirement in response to the enactment of HIPAA[4], and a pharmaceutical protocol.

How Do I Rewrite Government Documents?

A governmental project that I've worked on is a revision of the *Guidelines for Design and Construction of Hospital and Health Care Facilities,* a set of guidelines for the construction and renovation of health facilities, which is required in over 40 states and which is a "best practices" document for the rest of the states. Some of the document is difficult to comprehend, but because patients' lives and health are at stake, the lengthy document undergoes a thorough revision every five years to ensure that it is as usable as possible.

The team for this effort consisted of a former cardiopulmonary and special procedures technician turned professional communication/health communication professor, two professors in health architecture, and a number of graduate students in health architecture. Our charge from the Health Federal Guidelines Institute (FGI) was the following:

- Identify confusing, conflicting, and contradictory wording.
- Identify inconsistent use of terms and words that need to be defined.
- Identify dimensions or clearances that are inconsistent, confusing, conflicting, or ambiguous.
- Identify apparent errors, typos, or omissions.
- Determine if departments and supporting services are where one would expect to find them.
- Determine whether there are areas in the text, tables, or appendix that need clarification.
- Without changing the intent, offer suggestions to correct perceived discrepancies.
- Make suggestions for better chapter titles, headings, or subheadings including requirement location, numbering system, and space definition.

Notice the emphasis here on document design, covered in Chapter 3 of this text.

The actual document itself is in a two-column format, and the advisory information is listed in a tabbed, notecard format at the bottom of each column and shaded gray to distinguish it from the normal text. At the beginning of each section the following words appear: "In this edition appendix material appears in the main body of the document; however, it remains advisory only" (*Guidelines* 19). Thus, rather than putting the appendix material at the end of the document, the last revision in 1996 put the material on the appropriate page. Much of the text contains asterisks, and they indicate the appendix material or advisory information at the bottom of each page. One clarification our team suggested is that this notion of "advisory only" be explained. Some of the material is actually clarifying and interpreting, so the idea of labeling all those various rhetorical strategies as advisory seems confusing.

Another confusing inclusion for our team was the term *functional program*. This term was used repeatedly throughout the book, and while health architects and regulators may understand such a term, many hospital administrators and other health professionals may not. Because every aspect of hospital design is covered, a typical entry using the jargon is: "When the concept of swing beds is part of the functional program, care shall be taken to include requirements for all intended categories" (19). We consulted readers and found that indeed it was confusing to them. Also, as technical writers and teachers have long observed, users do not read a manual the same way they read a novel. Therefore, we cannot assume that the definition or explanation in the front of the text has been read by those consulting a later chapter, as they construct or renovate different areas of the hospital or health facility.

Other suggestions for revisions follow and serve as examples of types of revisions health communicators can make in health documents.

- "Additional parking may be required to accommodate outpatient and other services" (19). Change this and other passages from passive to active voice. In this case: Outpatient and other services may require additional parking.
- "Separate and additional space shall be provided for service delivery vehicles and vehicles utilized for emergency patients" (19). Change to: Provide separate and additional space to service delivery and emergency vehicles. This second wording is much clearer and easier to read.
- Again, the example used above: "When the concept of swing beds is part of the functional program, care shall be taken to include requirements for all intended categories" (19). First of all, we have the term "functional program," then the "concept" of swing beds (which are undefined), plus some elusive phrasing that "care shall be taken to include requirements for all intended categories." This language is vague and confusing and needs definitions and rephrasing for clarity and concreteness.
- Again, on this same page, the following language appears: "Maximum room capacity shall be two patients. Where renovation work is undertaken and the present capacity is more than two patients, maximum room capacity shall be no more than the present capacity with a maximum of four patients" (19). If the guidelines simply change the permissible size during construction or renovation from two to four, it needs to say just that.
- Many of the entries include detailed measurements and dimensions about how far beds should be from walls, what clearance is necessary for equipment, and other complicated numerical data written in dense prose paragraphs. We suggested that those entries could greatly benefit from schematics or drawings showing clearance needs, dimensions of the room, and other measurements.

Most government documents do use similar discourse. You will find vague and confusing language, undefined terms, few visuals for the visual learner, passive voice, and roundabout ways of saying what is meant.

What Is the Role of Documentation?

By now, you have a sense of how to revise based on a basic understanding of good technical writing. To make documents more accessible will continue to be important; document design and layout, whether print or digital, will continue to call for the advanced skills and analytical tools of health communicators/writers and communication savvy health professionals. When we ask about the role of documentation in health and medicine, because we have covered patient charting, education, and promotional campaigns in earlier chapters, much of the documentation in this chapter is about regulations, safety guidelines, and recording data for legal, financial, reimbursement, and insurance purposes.

The health industry and its careful documentation is heavily regulated; thus the discourse becomes dense and sometimes difficult to decipher. Plus each area of the health industry has its own norms and genres. One particularly interesting and difficult genre is the professional and technical communication surrounding pharmaceuticals. Because medicines in Western culture are so tightly regulated, often the protocols reflect this heavy emphasis on policy. The Food and Drug Administration (FDA), of course, reflects this reliance on regulation and clinical trials to protect patients, and its documentation serves to ensure that medications have passed through rigorous testing. Most of the time, these regulations work well to keep pharmaceuticals' profit motives from doing damage to the trial and testing period. However, that doesn't mean they're one hundred percent compliant all of the time. Here's where the ethics we discussed earlier in the book come into play. You wouldn't want to complete documentation as a health writer or professional verifying good research methodology and results unless it was true. Given the nature of pharmaceutical work, which involves numerous disciplines, the various team members act as checks and balances on each other.

Stephen Bernhardt and George McCulley argue that instead of waiting until the end process of drug development to write documentation for a new pharmaceutical, when the teams employ writing throughout the process, consensus is reached on substantive issues. Bernhardt and McCulley, as consultants at McCulley/Cuppan LLC, have worked with numerous international pharmaceutical teams to develop this successful documentation process. As we mentioned above, the interdisciplinary team that develops a new drug "require[s] close, well-articulated collaboration across a range of disciplines, from synthetic chemistry to medicine, with major contributions from chemical engineering, analytical chemistry, toxicology, medicinal chemistry, pharmacology, and other well-defined scientific disciplines" (Bernhardt and McCulley 2000, 22).

Pharmaceutical companies with their teams must develop persuasive documents and "labels" to get their new products through the regulatory agencies successfully and to compete with other pharmaceutical companies.[5]

What Pharmaceutical Documents Are Necessary?

The teams must produce NDA (new drug application) dossiers that run from 200,000 to 600,000 pages containing hundreds of different data sets such as:

- Chemical development
- Animal research
- Clinical trials on humans (Bernhardt and McCulley 2000, 22).

(For more information on the background of the various data sets and NDAs, see Rubin 1984; Spilkera 1986; and Bonk 1999.) Because a drug development process can take six to 12 years, often changes have taken place in personnel, so the experience level on writing the NDAs in general and on a particular drug in particular is not high. In addition, few medical writers have been trained to work in pharmaceutical companies (although that, too, is changing given the size and complexity of these documentation projects). Often the scientists themselves (chemists, engineers, toxicologists, physicians and pharmacologists) are the lead authors on such a document (Bernhardt and McCulley 2000, 23). In addition, regulations vary across the globe, and reviewers don't always have the same knowledge sets as the particular scientists, although they do make up an expert audience. The document is too unwieldy to localize and tailor to a particular culture and its worldview and expectations. In short, the rhetorical situation here is daunting. A few professional communicators who have made pharmaceutical writing their specialty are adding valuable input to make this process more efficient and feasible.

Many of the scientists prepare laboratory reports and articles for publication. Neither of these processes is similar to the development of NDAs. Instead of rhetorically analyzing the audience, context, and genre conventions of the particular writing challenge, the scientists lapse back into their own patterns of writing. The NDAs demand a very different hierarchy of organization. As a health professional who has learned to conduct these analyses, you will be able to more quickly learn new ways of writing. For researchers who cannot develop such strategies, Bernhardt and McCulley created a heuristic[6] and then a simple matrix to better structure the writing task and get various team members on the same page. The reviewers didn't want to know what journey the scientists took to discover and refine the new drug; they instead wanted to know answers to the following questions:

- Is it safe?
- Does it work?
- Does it meet a real need?
- Is it worth the money? (Bernhardt and McCulley 2000, 23–24)

Bernhardt and McCulley developed what they call a "seed document," and among other helpful strategies it offers a simple matrix and organizes information not around long narrative histories of each study, but instead around the key questions and issues surrounding the development of the new product within its clinical context. The "seed document" is divided into four columns with the following labels: *Issue, Response, Rationale, and Support* (25). As a health writer or consultant to the team of scientist writers, using such a martix can help them develop the arguments they need to be most persuasive to their reviewers. As mentioned above, you can now see why it is important to employ these strategies early in the process, before the document has already taken on the usual pattern of writing from the familiar genres of the scientists. Without such successful argumentation, products can be delayed from reaching the market to enhance the health or save the lives of patients.

How Do Insurance Documents Differ?

Another important example of a major genre of documents impacting health writing is health insurance documentation. The government programs of Medicare and Medicaid form a national health insurance plan in many ways, with Medicare targeting senior citizens and Medicaid targeting the impoverished. Because catastrophic illnesses and their attendant treatments can cost thousands of dollars, both governmental programs and other forms of health insurance have become essential. Unfortunately, even with these programs, and state programs as well, many individuals and families with low economic status cannot afford health care.[7] It hasn't been easy to estimate how many of the 300 million Americans do not have healthcare (Darr 2002, 31). The most recent estimates are that one-quarter of the population (approximately 75 million people) is covered under the governmental programs of Medicare, Medicaid, or military medicine (Goldberg 2002, A17). The current population survey (CPS) has been a source for health insurance information and has increased its accuracy in the last three years, but is not a health insurance survey per se (Darr 2002, 32). However, the CPS estimate of the uninsured was 41.2 million in 2001 (Goldstein 2002, A1).

Many hospitals in this country are operating in the red, especially those that treat patients who cannot afford to pay. "The American Hospital Associated estimated that American Hospitals provided $21.6 billion (6 percent of total hospital expenses) in uncompensated care in 2000" (Thrall and Scalise 2002, 38). The American Medical Association (AMA) estimated that physicians "provided 35.38 billion in uncompensated care" (Darr 2002, 32). Some health facilities turn patients away when they cannot prove ability to pay immediately upon entering the emergency rooms or admissions offices. "It was somewhat shocking when the prestigious Mayo Clinic recently announced that it will no longer accept Medicare at its clinic in Jacksonville, Florida" (Darr 2002, 31). More recently, some physicians are refusing to accept Medicare assignments; some will not treat Medicaid recipients; and some limit the number (ratio) of Medicaid patients they will service.[8]

The problematic financial situation has numerous causes: high cost of well developed and technologically advanced treatments; high costs of malpractice insurance, aggressive lawyers, and large settlements; large gaps between the wealthy and the poor; expensive and extensive research necessary for development of new treatments; and others. Add to all those variables the extensive and mind-boggling documentation demands within the insurance agencies themselves, and you have another daunting system to learn and sort through.

How Does Insurance Communication Work?

Many of the intricacies of this health insurance system are code-dependent; thus often the coding you see on the charts is tied to regulated business practices developed in tandem with Medicare, Medicaid, and insurance companies that are billed for health services. Many management experts and consultants with large hospital systems are attempting to help streamline the process. To understand a bit about the documentation involved, we need to gain an overall perspective of the communication structure of the health insurance system. Some of the processes are generic to all insurance practices. Lloyd Taylor and Don Sheffield conducted a case study of a Texas-based claims processing company, operating since 1993 with a good client base, which helps us gain an understanding of a representative system.

The bulk of this claims-processing company serves 44 physicians, one hospital, and four clinics (two urgent cares and two primary care facilities) in a medium-sized town.[9] As a result, most of their claims come from office visits (Taylor and Sheffield 2002, 13). Some claims do come from the hospital visits, which vary significantly in form and language from the office visit claims. The company processes about 24,000 claims per month and raises its revenue through preset, flat-fee billing practices. The staff at the company is divided into teams that handle one or more clients, and Taylor and Sheffield's case study looks at one team, which they maintain is representative of other teams at this particular insurance company as well as at other claims-processing operations (13).

If a patient visits the doctor's office or clinic, one "professional component claims form" is submitted (Taylor and Sheffield 2002, 14). If the patient stays at a facility, two forms are submitted: one "professional form" and one "technical form." The more common professional form"uses the Centers for Medicare and Medicaid Services CMS-1500 form " (Taylor and Sheffield 2002, 14). At most facilities, staff types the information into the computerized system and the software formats the information to fit the CMS-1500.

The CMS-1500 form is divided into two sections, as are many professional health forms:

- Demographics
  Patient's Name
  Address
  Insurance Information
- Visit Details
  Purpose of the Visit
  Diagnosis
  Results of the Visit

In general, current procedural terminology (CPT) and Healthcare Common Procedural Coding System codes describe the procedure performed during the visit, and codes from the *International Classification of Diseases,* 9th ed., are used to convey diagnosis. All three sets of codes are distinct and complex. The American Medical Association distributes CPT codes and incorporates code additions and modifications. (Taylor and Sheffield 2002, 14)

Regardless of what happens during the office visit, the correct coding is mandatory if the provider wants to receive proper reimbursement. Because the CPT codes are changed somewhat every year, physicians are expected to keep up with the changes in order to get their maximum reimbursement. Given heavy patient loads on doctors, some of them have no time to keep current; thus they don't receive the maximum credit for their services. At the end of the visit, the health practitioner dictates or makes notes for a staff person that include the CPT code for the situation. New, inexperienced, or overworked staff members do not always double-check the codes, although a good staff member will also attempt to stay abreast of the latest CPT code changes. If an outdated code goes in on the CMS-1500, the claim is compensated at a lower level or rejected outright. It's important to have enough well-trained office staff to keep current on CPT codes in order to double-check and correct the physicians. Also, doctors' notes must be comprehensive enough that the staffer can select the correct codes.

How Long Do Records Need to Be Kept?

Because of the nature of the health professions, medical records need to be kept *ad infinitum*. Laws forbid destroying medical records, although many larger health systems put old records on microfiche to save storage space. Some lawsuits and investigations may go back many years in order to determine and bring certain evidence into court. Dental records are especially valuable for identification purposes even much later in life, although infant footprints and other forms of identification are used, too. A large number of the medical witnesses called to testify in medical cases have backgrounds in medical and dental records research. Even with statutes of limitations in place, it's still best to keep records intact. When in doubt, *don't* throw them out!

Summary

In this chapter you have considered several new genres of health documentation. You've looked at grant proposals and applications, government documents, regulatory guidelines, pharmaceutical research and promotional documentation, and health insurance forms.

When you studied health grant applications, you learned that successful grants begin with an outstanding and creative idea. You must have the insight and foresight to discover an innovative way to solve a health problem or initiate a new health behavior. You understand now that you must read the RFP (request for proposal) carefully and not miss any of the steps required for the application process and the project itself. In addition, you realize that the discourse in grants is more persuasive than descriptive, and the claims you make in your argument for funding your project need to be based on research (review the literature), statistics, or perhaps a pilot study or your own experience. You need to keep the language simple (KISS), because not everyone on the health grant review board will be an expert in your area, and it's not necessary to try to impress a review board with big words they may not understand. There is also a maxim: The good news is you got a health grant; the bad news is you got a grant. Make sure you have enough time and resources to do what you promised to do for the grantors.

In addition to learning how to write the prose part of the health grant, it's important to articulate your outcomes, goals, and assessment tools for health grants. If you outline the steps in the health grant that can help you achieve your goal, whether they are called for or not in the formal RFP, you can organize the grant work so that you will be able to better evaluate what you need to request in the budget and timeline.

You read about some of the documents used by the government in addition to grant requests, especially guidelines for building and renovating hospitals and other health facilities. You learned how to clarify wording and identify vague or confusing passages and how to format such documents, in case you must interpret, use, write, or revise such documents in your health workplace (almost all of us end up working with some form of government regulations or federal and state documents).

Next you turned your attention to pharmaceutical writing and learned about the lengthy reports and the long development and documentation cycle that is part of the pharmaceutical invention process. Experts from various areas of science write the documents but the employment of a matrix developed by professional communicators can help ensure that the documents are persuasive when they compete in the marketplace and are reviewed by the regulators. Often scientists and regulators are looking at different issues and write in very different ways; thus the matrix, with its identification of issues, helps solve problems when used early in the process of the drug development and documentation.

Lastly, you examined the process and the most common reporting form used in the health insurance industry. You discovered that this writing process also demands a team approach and that the process is very much directed and controlled by intricate and complex coding systems. You learned a little about the context in which the health insurance dramas take place on a daily basis and some statistics about the uninsured and financial losses to hospitals and physicians.

Discussion Questions

1. Discuss as a class what you think fundable health and safety problems would be. How and why are these granting opportunities? What criteria are you using? What health and safety problems would not be fundable and why? Make sure in your discussions you articulate the reasons for the difference between the two types.

2. Within the class break up into two-person teams and discuss one grant opportunity you would like to put forward as a class project. Together discuss making an argument for your project being fundable and present your argument to the rest of the class. At the end of the class, have everyone vote and rank their preferences for a granting opportunity based on the success of the arguments presented.

3. Choose a fundable health problem from those you discussed as a class in Discussion Question #1. As a class or in small groups, brainstorm outcomes and

activities to achieve those goals. Have someone in class record them via notes or a computer and distribute them to class members as a model of what this important component of grants (campaign and other proposals) looks like.

4. Discuss the formation of pharmaceutical teams. What kinds of problems can you imagine with such a diverse and extensive group? What strengths does such an interdisciplinary team bring to the process? What is your experience with different disciplines in your studies and in your technical writing classroom if it is not a discipline-specific section?

5. After reading the section on pharmaceutical documentation, discuss possible matrices you could set up within your health specialties in order to better meet the documentation requirements, especially those that are of a regulatory nature. What other possibilities can you design to organize other documents within your particular workplace?

Exercises

1. With a partner from your class, research grants on the Internet within an agreed-upon time frame. In other words, agree as a class to spend between one and three hours (in or out of class depending on what you decide) finding a number of health-oriented grantors or foundations. Prepare a list of your findings. If you have a shared electronic workspace, post your findings and generate a document or Web page that lists all of these possibilities. If you choose not to work electronically, prepare a printed list to keep in your files for grantors in the health field.

2. Interview someone who writes grants for local hospitals or other facilities or someone who writes grants for departments of nursing, nutrition, public health, or other health-related fields. Ask detailed questions about their process, including information on choosing grants, past experiences, budgets, writing styles, success rates, etc. In addition to preparing a transcript of the interview, also make a visual aid (graph, flowchart, or other form) that helps you and your classmates see this particular person's grant-writing process.

3. Search the Internet or go to your library and check for government health or medical documents (most libraries have a special section or floor). Find a document you can copy and share with your classmates. Analyze the document design and layout, usability, and style. Suggest changes that you think might make the text more user-friendly. If you have a print copy, scan it into your computer, and using the MSWord editing tools (track changes and insert comment), illustrate the changes you have suggested.

4. Gather several drug inserts and compare them. Work up an outline or style sheet so that you can become familiar with the norms of the genre. What sections are required in such a document; what graphics are normal for the insert? How do they compare across companies? Develop a brief report on your findings. Include some on-line research on the background of the pharmaceutical

company. Notice the financial reports to stockholders or earning reports on their Web sites and add that to your "background check."

5. If you have any of the disciplines on pharmaceutical development teams listed in the text above on your campus, set up interviews with them about what they see as their perspectives and what they could contribute to such. Interview several pharmacists and physicians, too. Divide up the interviews so everyone in class has an interview to conduct. Present the interviews to the class via Power-Points to share your information with all class members.

6. Design a communication flowchart that illustrates the path of some document that you are responsible for generating in your health workplace. If you are new to a field and have not yet had the experience to rely on, visit a health site and ask permission to observe and ask questions to determine the document flow.

7. With a partner or partners, find the CMS-1500 forms online (see Works Cited page for Web site address) and look up the coding systems mentioned in this chapter. Fill in as much of the form as you can with plausible (if hypothetical) information. Design a how-to sheet that includes a list of the information you need to fill in such a form and where to find that information.

Works Cited

Bernhardt, Stephen A., and George A. McCulley. 2000. "Knowledge Management and Pharmaceutical Development Teams: Using Writing to Guide Science." *Technical Communication and IEEE Transactions on Professional Communication.* (February/March).

Bonk, Robert. 1999. *Medical Writing in Drug Development: A Practical Guide for Pharmaceutical Research.* Binghamton, NY: Haworth Press.

CMS (Centers for Medicare and Medicaid Services) Web site. 8/28/03. August 28, 2003. http://cms.hhs.gov/providers/edi/1500info.asp.

Darr, Kurt. 2002. "The Uninsured: A Reality Check." *Hospital Topics: Research and Perspectives on Health Care* 80, no. 4 (Fall): 31–34.

Facilities Guidelines Institute. 2001. *Guidelines for Design and Construction of Hospital and Health Care Facilities.* Dallas: FGI.

Fang, M. Emily. 2001. Information Services Librarian at UCSF. *Grant Resources in the Health Sciences: A Selected Bibliography.* San Francisco: The Regents of the University of California, 2001.

Galen: Digital Library of University of California at San Francisco. 8/28/03. August 28, 2003. http://www.library.ucsf.edu/collres/reflinks/grants/

Goldberg, Robert. 2002. "Choice, Not Health Confusion." *Washington Times.* (17 December): A17.

Goldstein, Amy. 2002. "Health Coverage Falls." *Washington Post* 30 (September): A1.

National Institutes of Health Web Site. 8/4/03. August 5, 2003. http://www.nih.gov

Rubin, Alan A., ed. 1984. *New Drugs: Discovery and Development. NY: Marcel Dekker, Inc.*

Spilker, Bert. 1986. *Guide to Clinical Studies and Developing Protocols.* NY: Raven Press.

Taylor III, Lloyd J., and Don Sheffield. 2002. "Goldratt's Thinking Process Applied to Medical Claims Processing." *Hospital Topics: Research and Perspectives on Health Care* 80, no. 4 (Fall): 13–21.

Thrall, Teresa Hudson, and Dagmara Scalise. 2002. "America's Uninsured: Rethinking the Problem That Won't Go Away." *Hospitals & Health Networks* 76, no. 11: 30–32, 34, 36, 38, 40.

Endnotes

1. NSF (National Science Foundation) will no longer accept print versions of applications. All applications must be written in the on-line form and must be submitted electronically over their Web site.
2. KISS: for example, in the featured grant, one of the goals is "Assess staff computer expertise and availability." This is a direct and brief statement. A statement that violates the KISS principle is: We will determine how the computer skills of the staff are to see if they need further training and if there is adequate computer equipment for the staff.
3. A *curriculum vitae* is Latin for the course of your life and is the lengthy academic document that you use instead of your resume. It contains publications, education, courses taught (usually but not always), presentations, and previous grants.
4. HIPAA is the Health Information Portability and Accountability Act. See Chapter 2 on Medical Ethics for more information on this legislation.
5. The "label" refers to the prescribing information for physicians, contraindications, drug interactions, and other information that makes up the fairly extensive package insert. In addition, it plays an important role in the marketability of the product and its journey through the regulatory process. Bernhardt and McCulley write, "In a cross-functional team environment, it is a challenge to get the team working together to align the data with the strongest possible arguments for drug approval. The approved label provides marketing leverage and can determine the success or failure of the new drug product" (22).
6. A heuristic is simply a series of questions that helps you elicit and organize the important information within a body of work.
7. Darr makes an argument that many low s.e.s. individuals and families do not take advantage of the federal and state programs available to them. However, because of the many cultural and social factors that feed into why people do or do not seek healthcare from the American health system, it's important to examine the issue historically and with extensive research so that an author does not come across as "blaming the victim."
8. Doctors do not want to take on Medicare assignments because of "the paperwork, bureaucratic complications, claims denials, and low payment" (Darr 2002, 31).
9. One anesthesiologist, who works at the hospital, also bills through this company.

CHAPTER 9

Multicultural and International Medical Writing

 ## Overview

In Chapter 1 you learned to analyze your audience. In this chapter, that analysis will become more focused on multicultural and international medical writing. This chapter goes into more detail both in examining how to write for various cultural groups found in the diverse U.S. society and for other cultures, too. "Patterns of communication behavior, to say nothing of values, are deeply rooted in language-culture complexes. Though technology [may] be in place, there are cultural obstacles[1]" (Ulijn and Campbell 2001, 77). Because different areas of the world practice medicine and healthcare in other ways, we will briefly examine what U.S. citizens refer to as alternative or integrative medicine, although in other sites around the world, our Western biomedicine would be the "alternative" medicine. We will feature examples from African American, Hispanic, Asian, and Native American healthcare and writing for those audiences in this country and abroad. "The diversity of the United States population continues to change at a rapid pace" (Nelson et al. 2002, 209). In 2000 an estimated ten percent of the U.S. population was born in a country other than the United States (U.S. Bureau of the Census 2001). English is a second or even third language for many of those born outside the United States (Gudykunst and Mody 2002). "Even though much of the world's business is conducted in English, there are often problems of interpretation between one variety of English and another, such as British English as opposed to U.S., Australian, Indian, or Nigerian English. More potential for difficulty exists where English is the second or third language of both parties to communicating" (Ulijn and Campbell 2001, 78). It is impossible to cover all cultural groups, but the methodology described in the chapter on audience analysis, as well as the analysis of differences and sensitivity to cultural norms discussed in this chapter, will provide a basis for composing and designing health documents for diverse groups.

You will learn about the question that continues to perplex scholars in international professional communication as well as learn the answers to the following questions:

- Should messages be local or global?
- How do I find particular cultural data?
- What research methods are best for cultures?
- What are "high-context" and "low-context" cultures?
- How do I avoid offending demographic groups?
- What about translation issues?
- How will computers impact other cultures?

In the past it may have been easier to write and design visuals for health audiences because we were not able to connect with our distant neighbors so easily. Obviously our context has changed dramatically; technology now makes it possible to see live coverage from anywhere in the world via satellite television feeds, and to view another culture and initiate contact through Web sites, email, and instant messaging. Internet cafes can be found in most large cities around the world. Cell phones and palm pilots also provide easy access. Although scholars have broadly critiqued the term "global village," we should be aware that to be culturally sensitive, we should use the term *global* carefully. That term connotes to many people around the world the idea that the United States is attempting to impose its culture on others. Large demonstrations have taken place around the world against the idea of *globalization,* as some countries perceive that their own cultural artifacts and constructs are being lost in the wake of the overwhelming influence of Western culture, a.k.a., *Westernization.*[2]

Should Messages Be Local or Global?

This question is at the center of a debate in international professional communication, so we need to consider it here in international health communication and medical writing. Gail Hawisher and Cynthia Selfe argue that the "global village" is a myth. They suggest that in fact our technology maps some problematic Western cultural assumptions onto unsuspecting and unwitting cultures that are not as technologically developed as we are. In an earlier article, Selfe argues that in fact the global village narrative "is shaped by American and western cultural interests" (Selfe 1999, 292). The representation of the world as a global "village," Selfe maintains, "has been criticized by technology studies scholars who question its accuracy and point to the specific national and cultural interests" of the more highly developed technological countries (2). In addition to seeing this myth much less positively than U.S. citizens, people of other cultures interpret even "the global expansion of the Web within the historical context of colonialism" (9). Hawisher and Selfe suggest that we need to look at the individual cultural contexts to discover the "culturally specific literacy practices," thus calling into question the idea of a globalization that shuts down individual differences by using the global village metaphor.

In addition, Hawisher and Selfe suggest that the global village myth "provides a convenient and ideologically effective way of making efforts to expand free-market economic development, provide active support of fledgling democratic political efforts and intervene militarily in the affairs of non-western countries" (Hawisher and Selfe 2000, 9). To buy into the myth may place more emphasis on the financial gain than the benefits or the consequences cultures face by being a part of the "global village." Issues like culture, literacy, and accessibility become factors for many countries that may not be as economically and/or technologically advanced as Western cultures.

The other side in this debate is represented in another collection of essays. In Carl Lovitt's book on international professional communication, Jane Perkins does not advocate a global village approach per se, but she offers a way of using generic lenses within professional communication to construct cross-cultural communication. She acknowledges that in fact professional communication will continue to "benefit from 'inter' [intercultural or between cultures] research that helps us understand what confuses and what offends—verbally, visually, proxemically, socially, and so forth" (Perkins 1999, 17).[3] While she does not suggest we use the global village metaphor, she does suggest that "Professional communication teachers and researchers need to consider additional metaphors, new ways of thinking about international communication; we can begin from an understanding of some current changes in corporations—changes toward borderless, yet multicultural, professional communication" (18).[4]

To clarify our terms:

> The terms *local* and *global* are increasingly used to refer to the problems emerging from the fact that variations in professional communication conventions and expectations derive from linguistic and cultural differences. Indeed, the term *localization* has been used to refer to the strategy of a company to adjust its documents to the local culture of the market area. However, LOCALIZATION is considered an unsophisticated strategy that has largely been ignored in practice and in research. It is no surprise then, that much of the current research in the past few years has looked only into how primarily American strategies are translated to accommodate local cultures, their conditions and expectations. (Yli-Jokipii 2001, 105) (See also Hofstede 1991; 1994; Hofstede et al 1990; Trompenaars and Hampden-Turner 1998; Weiss and Stripp 1998; and Hall 1983).

Given the debate, then, how are we as health communicators and message designers for international and multicultural audiences supposed to respond? Instead of seeing these positions as "either/or," it might be better to find a balance and approach the question from a "both/and" viewpoint. We need to be as culturally sensitive, as specific and localized as possible to persuade our target audience to employ or cease certain health behaviors. Our goal is the overall health and quality of life of individuals. We can never design any materials that will please everyone or meet each individual's exact needs. But we can and must be sensitive to issues within the populations we hope to reach. Otherwise, if we do not attempt a culturally specific analysis and design process, we risk alienating or offending the very people we had hoped to reach. We need to realize, too, that other cultural medicines and medical treatments may be as valid as our own, and we need to respect them and the people who believe in what may seem to be unusual procedures for us. Realistically speaking, time and resources prevent us from being able to customize each particular application to a specific audience member. Some of the choices we make cannot help but be based on a more generalized view of someone else's culture or of cultures overall. One learns through the process of trial and error or looks back and reviews what someone else tried and failed to do in a different cultural context. With that ethical and theoretical lenses in place, then, let's look at how we learn to design for another culture.

How Do I Find Particular Cultural Data?

In other words, how does a writer in the health professions learn about another culture? Of course the ideal way to get to know another culture is through immersion in that culture: living in it, learning its language, and getting to know its people one-on-one. This ideal is difficult to achieve in most instances, unless you are one of the fortunate few who can live and work in the new culture for which you write. Most of us have to find other ways to develop an insider's view or at least to be culturally knowledgeable. Fortunately, we have a wealth of articles published in many journals. (See Works Cited for examples of relevant articles.) Healthcare theorists, providers, and other professionals realize the need for "[e]levating the awareness of practitioners about cultural differences in communication, and on increased training on how to create culturally appropriate messages" (Nelson et al. 2002, 209; Gudykunst and Mody 2002). However, that doesn't mean we don't need more research into particular groups; we do. Reading articles will help, especially if they are based on good and extensive *experiential* research. Note if articles and books are written by members of the culture and/or include interviews and firsthand accounts. You'll want to avoid stereotypical, third-person accounts that remove the data far from the subjects of the studies. Ideally, the subjects should be involved and have major roles in determining the direction of projects and initiatives.

From the beginning of your international project, no matter what media format it takes, you want to find out about the audience's preferences, trigger points, and views of medicine. For example, some Latinos do not like to be referred to as *Latino* or *Hispanic,* but would like to be referred to according to their country of origin, such as *Mexicano, Chilean, Costa Rican, Guinean, Puerto Ricano,* etc. "In a Gallup poll in mid-2001, Americans of Spanish origin overwhelmingly said they prefer *Hispanic* rather than *Latino,* by 67 to 13 percent" (Valdés 2002, 7). However, "[p]resently, Hispanic consumers in some areas such as California or Texas tend to prefer *Latino*" (7). Although people think of *Hispanic* as a racial category, it is not; people from South and Latin America can be any race (6). Some African Americans prefer to be referred to as Haitian, Jamaican, or Brazilian. Native Americans often have labeling issues, too. To all of these groups, lumping them together or not respecting their names smacks of bias and prejudice. Native Americans in Southern Arizona once called *Papagos* do not like to be referred to by this moniker. *Papago* in Spanish means "bean eater" and was introduced by the arrival of *conquistadores* on the shores of the Pacific on the Mexican coast. The correct term to show respect for this Native American tribe is *Tohono O'odham,* which loosely translated means "people." The Navajos also refer to themselves as "the people" or, in their original Athabascan language, *Diné.*

Some may reject this rhetorical strategy, thinking it sounds like "political correctness." However, in order to understand and reach another culture, it is important to call people by the name that they prefer. To alienate or offend an audience from the beginning will only cause your health message to fail.

What Research Methods Are Best for Cultures?

In Chapter 1 you learned about the following methods, which you may want to review:[5]

- Library research
- Internet searches
- Field studies and case studies
- Surveys (email, telephone, person-to-person)
- Interviews
- Focus groups
- Usability studies
- Field-testing

In that chapter we also briefly mentioned ethnographies. The most valuable, indepth research done on a culture is ethnographic research. Janice M. Lauer and J. William Asher defined ethnographies as a "kind of qualitative, descriptive research, [which] examines entire environments, . . . [and] derives primarily from phenomenology, anthropology, and sociology" (Lauer and Asher 1988, 38). Sociologists, anthropologists, and phenomenologists regard this qualitative method as "a window on culture," and so do those who study various cultures' views on health (Lauer and Asher 1988, 38). As a researcher, you employ *thick description,* as you did in case studies in Chapter 1, and *triangulate* your data. Ethnographies need much time in order to be completed—ideally from several months to several years. You can position yourself as a participant-observer who takes part in the culture while taking notes and studying the members of the group or as an observer who does not get involved as a cultural member. Remember, though, it would be better to involve the members of the culture in ways that let them be more of an active part in the research, design, and implementation processes.

Objectivity is not possible in whatever role you choose for yourself, so we must recognize that all of our impressions of a culture's values and health practices are viewed through our own cultural lenses. Also, it's important to remember that because this research method is a qualitative one, it will not be generalizable. At most you can generate some interesting conclusions and form some further research questions that you may want to employ based on what you discovered in the ethnography. However, if you also have the members of the cultural group critique your health designs and messages throughout the process, you can enable more effective communication.

For example, suppose that you were to design a campaign for another culture that had tapeworm infestations among its population, and yet the health practices and medication you were responsible for administering was resisted. By studying the culture, you discovered that this cultural group believed that boiling potatoes and drinking the potato water was a cure for tapeworms. In order to research this one health practice, you could set up interviews or survey people (probably informally) to find out the details and belief system underlying their practice. You could also find out about why people were resistant to the medicine and health practices other professionals had tried in this location. You'd need this type of in-depth research and cultural understanding to create some persuasive ideas for getting residents to try the effective practices and medications. Especially for hard-to-change habits, it is important to find ways to work with the cultural belief system rather than against it; get the people's ideas for how they might be persuaded to change an earlier practice by listening closely and recording their input.

The best cross-cultural medical ethnography that I've read is *The Spirit Catches You and You Fall Down: A Hmong Child, Her American Doctors, and the Collision of Two Cultures* by Anne Fadiman. In this nonfiction story, the author "explores the clash between a small county hospital in California and a refugee family from Laos over the care of Lia Lee, a Hmong child diagnosed with severe epilepsy. Lia's parents and her doctors both wanted what was best for Lia, but the lack of understanding between them led to tragedy" (Lannon 1997). The medical communication lessons here are extensively illustrated with Fadiman's prose, and it should probably be held up as a gold standard for other cross-cultural medical ethnographies. Each culture should have such a guide to the belief systems and communication issues about the body. As recent immigrants to the United States from northern Laos (less than 30 years ago), many parents of the children seen in California clinics have a very different knowledge base than that of the U.S. culture. One doctor reports, "The language barrier was the most obvious problem, but not the most important. The biggest problem was the cultural barrier. There is a tremendous difference between dealing with the Hmong and dealing with anyone else. An *infinite difference*" (Fadiman 1997, 69).

Dan Murphy [physician] said,

> The Hmong simply didn't have the same concepts that I did. For instance, you can't tell them that somebody is diabetic because their pancreas doesn't work. They don't have a word for pancreas. They don't have an *idea* for pancreas. Most of them had no concept that the organs they saw in animals were the same as in humans, because they didn't open people up when they died, they buried them intact. (Fadiman 1997, 69)

The little Hmong girl with epilepsy was well cared for and loved in her family. In fact she was given special status because of her "falling down" sickness. Hmong believe the spirit takes the child's soul, and the seizure manifestations mark the child as a very special and honored person. But their belief system and culture were such that they could not administer medications that could keep her alive as her seizures increased in severity. After many visits and years of management efforts in conjunction with the local hospital, the child died. Efforts were made by both sides to bridge the cultural gap before her death, but it was still too wide to reach across. Some of these efforts included the help of a social worker, who took on the Lee family as a crusade. She worked hard with an interpreter who was well liked in the Hmong community to get to know the family and the culture. In time she was able to effect some small change; she printed out charts to try to regulate the home care of Lia Lee, the patient. She even had them printed up in the "swirly Lao script," but then discovered that no one in the Lee family spoke or read Lao (Fadiman 1997, 115). Because the Hmong culture was an oral one, in which all knowledge one needed was passed down verbally from tribal members and parents, there was no need to read and write within their own culture. One thing the social worker did learn and pass on to the doctors at the hospital where the Lee family brought their daughter when she was having seizures was that the best way to get cooperation was by asking permission for everything that needed to be done. Although the parents did bring their child in when she had *grand mal* seizures, they still were frightened of the hospital and often thought the doctors were making her worse and not better.

> Lia Lee's pediatrician said, at one point: "The parents had to go along with a lot of stuff, an oxygen mask, lots of IVs, bloodwork, an arterial line to measure the oxygen and carbon dioxide in her blood, real invasive stuff" (116). [Her father] remembered this time as the time when Lia "had a lot of plastic all over her." He and Foua [the mother] slept by Lia's side every one of the fourteen nights she spent in the hospital. He recalled, "The doctors made Lia stay so long in the hospital, and it just made her sicker and sicker." (116)

The realities of modern healthcare and social work often make it impossible to work with families from other cultures one-on-one. If you do have a particular cultural group or groups in your healthcare practice, an effort to understand the culture on the part of any and all health practitioners dealing with those groups will help some. But often, given the reality of managed healthcare that is both expensive and time-limited, we must approach a culture via a group approach.

M. Isabel Valdés refers to her practical approach to Latino marketing, which involves audience segmentation based on cultural belief systems, as "in-culture" and provides faster ways to get information, as conducting or even reading ethnographic studies takes such a long time. She is, fortunately, a participant-observer, and recognizes that Latino culture sees certain stages of life as separate from others, and somewhat differently from other societies (Valdés 2002, 77). An example of her in-culture analysis is her age segmentation of audiences that she conducted for a March of Dimes health campaign to increase the consumption of folic acid among pregnant women. JMCP, a Latino advertising agency, developed a Spanish-language Web site as part of its communication strategies (www.nacersano.com). To find out more about the campaign as a case study, check the site at www.incultureapproach.com;[6] check www.todobebe.com for other examples of in-culture Web sites developed for healthy babies. In addition, to the "in-culture" approach, certain fundamental belief systems of cultural groups can be more easily understood if you divide them into what are called "high-context" and "low-context" cultures.

What Are "High-Context" and "Low-Context" Cultures?

Anthropologists have suggested that cultures fall on a continuum between "high" and "low" contexts. See Exhibit 9.1 for the characteristics that define high- and low-context cultures. You'll quickly realize that the culture of the United States is a very low-context culture. Our emphasis on open and informal communication privileges individualistic attitudes of competition. Private enterprise and "making it on your own" are strong cultural messages, even if they don't quite reflect the economic reality of our interconnectedness. Our history of coming as exiles or entrepreneurs to a vast, sparsely populated land and a democratic vision of equal opportunity created the scene for a low-context culture to take root and thrive. Populations from various classes, groups, and countries with different languages mixed together and survived based on this shared yet individual-oriented vision. Of course, some groups were also brought here by force, and others immigrated to this country and received less than enthusiastic welcomes.

Other high-context cultures evolved more slowly over long periods of time in which the culture was fairly isolated from other influences (except by attempts to conquer them, which caused further group cohesion for survival purposes). Members of these cultures needed each other, and therefore they constructed strict rules of conduct and indirect, formal methods of communication to keep the culture from being weakened from the inside. Too much internal strife would make the society more susceptible to external enemies. The indirect communication style remains one of the largest difficulties when high- and low-context cultures must work together to reach common healthcare goals. Often high-context cultural members are reluctant to express requests and commands directly and also are reluctant to refuse requests. "It has been said that Japanese don't have a word for *no*. While this is not true, it is true that the Japanese, like many other people, are reluctant to give a direct refusal" (Thrush 2001, 34).

Many cultures emphasize oral not written language. In particular, ". . . Arabic and Latin American [cultures], place a high value on personal and oral communication" (Lewis 1996, 20). On the other hand, some cultures are as direct or more direct than U.S. citizens, for example, the German culture. "German business people . . . prefer factual detail in documents and are likely to thoroughly read and absorb written documents. Consequently, Germans may be likely to tolerate a substantial amount of textual material [even] in web documents" (Zahedi, Van Pelt, and Song 2001, 86). Other countries have problems with the Western insistence that everything be put in writing. In their view, the negotiation stops once this happens. Because writing is often regarded as a formal process, there are often fewer notions of "a working document." In writing up a case study about Czech society, Emily Thrush presents an interesting dialogue that she observed, which illustrates this cultural difference well.[7] In the conversation that follows, "H" is Helen, the U.S. citizen, and "D" is the director of the Czech program.

> H: As mentioned before, the American participants really need to get some response from you to feel sure that they've been accepted for the program.
>
> D: You know the problem—it's very expensive to mail materials to the United States.
>
> H: But the program is very inexpensive for Americans. You could easily increase the fees a little for Americans to cover the postage costs.
>
> D: Well, I don't know if we can do that.
>
> H: Will you think about it and see if it's possible?
>
> D: Yes, I'll think about it.

Both Helen and the director knew that the answer to Helen's request was "No." When Helen heard the director hesitate, she gave the director a way to avoid giving a direct

High and Low Context Characteristics[8]

	Low Context	High Context
Approach to Disagreement	Resolution via conflict	Resolution via consensus
Business Affiliations	Based on monetary benefits	Based on tradition & personal relationships
Conflict Resolution Style	Face-to-face	Through liaisons & intermediaries
Discourse Groups & Communities	Belonging to many communities & groups	Belonging to and identifying with one predominant group
Interpersonal Relationships	Informal	Formal
Life Goals	Achievement up to the individual	Achievement up to the group or in the hands of fate
Motivation	Individual goals	Group goals
Play & Work	Kept separate	Integrated
Relationship to Nature	Separated from nature & the environment	Integrated with or a part of nature
Self Concept	Thinks of self as a single individual	Thinks of self as a part of a group
Status in Relationships	Egalitarian	Hierarchical
Time	Future orientation	Past & present orientation
Trust Level	Open	Closed except to group members
Usual Behavior	Independent, oriented toward self	Dependent, oriented toward group
Values	Active—doing	Passive—being

Exhibit 9.1

refusal by asking her to think about it. This gave the director a chance to give a positive response so that both she and Helen would feel good about the exchange. It worked in this case because Helen understood that she had been told that the fees would not be raised and information would not be mailed to the Americans.

This sort of subtle, indirect communication is easier in oral interactions, so people in high-context cultures often prefer this mode of communication to written documents. This explains, to some extent, why the Czechs did not understand the need for written information and did not want to communicate with the Americans that way. Once things are set down in writing, they are no longer open to negotiation [the Czech view]. (Thrush 2001, 34–35)

How Do I Avoid Offending Demographic Groups?

No matter how well you know the culture and how culturally sensitive you are, there are no guarantees that you will always be able to prepare appropriate and inoffensive health messages in a culture other than your own. Frankly, there are even problems in targeting and marketing areas of the U.S. culture, because of our diverse nature. But you can build in as many efforts as you can to keep problems from being insurmountable. Althought we mentioned it before, it bears repeating: involve members of the culture at all levels as designers and consultants, as participants on review panels, and as usability test participants. Develop focus groups and ask them to review your materials, too.

At one point in the process of developing materials for the Department of Social Services (DSS) for the food stamp project at my university, technical writing students were to test food stamp application recruitment brochures designed by the previous class. In the previous class, we had segmented the audience and designed various materials for targeting different segments. We had a large and diverse review made up of many different university and government officials. One of the segments we targeted in our food stamp materials was African American families. Students had used different graphics to depict these families. Several African Americans on our review board objected to the graphics as pictured, because they portrayed all African Americans as being of one hue. They pointed out to us that of course African Americans have a variety of skin tones, not just one. This feedback was the type we had hoped for, because the class's assignment was to test and redesign (revise) the materials before they were sent to the printer for exactly that kind of cultural faux pas. The students spent much time manipulating graphics and changing skin tones. Then they decided to avoid the issue and just picture food graphics on the materials. Those graphics were then field-tested at soup kitchens and Wal-marts in the area. The reactions were all positive, with some materials receiving very enthusiastic responses, especially when compared with the densely written bureaucratic phrases and confusing materials put out by DSS itself (not correctly targeted toward people of less education than the DSS administrators). An appropriate way to picture a mix of people representing a Spanish-speaking audience would be to feature a number of subtle facial and hair differences, because Spanish speakers can be Caucasian, Native American, and African American.

What About Translation Issues?

Even translators aren't always able to bridge the culture gap, as was the case with the Hmong family earlier in the chapter. But in some cases, they do provide tremendous services and are able to increase the effectiveness of diagnosis, treatment, and a successful outcome. As writers in the health professions, we will be mostly involved with written translations (as well as visual and electronic ones). Bruce Maylath's case study, "Translating User Manuals: A Surgical Equipment Company's 'Quick Cut,'" provides us with a case study of the issues often involved in such translation projects. His study involves a hypothetical company that manufactures heart patches. Although the case is fictionalized, it is based on typical experiences of large manufacturing firms that work with outsourced translation companies to translate manuals accompanying the products. In this situation, the firm is "Cordipatch," a company that makes synthetic heart patches to implant in perforated hearts. The research questions for the case study are:

- How can technical [or health] communicators ensure that their documents in translation are accurate and usable—in other words, of high quality?
- Will surgeons reading the translated manuals be able to attach the heart patches correctly? (Maylath 2001, 65)

These questions, when placed in a healthcare context as they are here, are no longer a matter of technical accuracy, but involve patient lives; such translations and their translators are under tremendous pressure to perform the kinds of translations that are unequivocally accurate. As a health communicator, you may be involved in having to translate or assist in translating documents or verbal instructions at your workplace. One of Maylath's conclusions is that we as health communicators and U.S. citizens need to be fluent in at least one language other than English in order to understand the issues involved. He cites the Coca-Cola and Ford example of replacing U.S. workers and even CEOs in other countries with native speakers and urges us to attain fluency in another language, whether we live in this country or another (78).

> Many translation companies report that, while the largest segment of their work is for documentation headed in and out of the United States, the fastest growing segment of their business is for documentation staying within the United States. Part of the growth is a result of the North American Free Trade Agreement [NAFTA]. Go to any store in the United States, and you'll see many package labels in English, French, and Spanish for the joint U.S., Canadian, and Mexican market. (Maylath 2001, 79)

Within our own country, then, translation issues will come up in our workplaces. Right now the segment of our population most in need of translated health materials and translators are our Spanish-speaking populations. Immigration from Latin and South America is still growing, and this population, with its attendant high birth rate, is the least able to afford health insurance. The underinsured arrive at our emergency rooms in dire need of medical attention and often unable to understand English.

If you are in a health facility or organization that outsources its translation needs to a translation company, like Cordipatch in the case above, you will need to be aware of the kinds of processes that should be in place in that company before you sign a contract with them. Let's look at both sides of the translation equation because many employees are involved on both sides. On the manufacturing or product side, you will have a staff of:

- Designers
- Engineers
- Line workers
- Project managers
- Technical communicators

On the translation company side, you will have a staff of:

- Project managers
- Translators
- Editors
- Reviewers
- Document designers

With a best-case scenario, your translators will be on-site in the countries where the translation will be used; they will also have medical experience (in this case be a surgeon). While this scenario of a translator/surgeon does happen, it is not always possible. If the translator is not also the end user of the surgical product, you need to be able to have the surgeons who will use the translated manual available to review and comment on the manual before it is released for use with the product. Ideally the company you work with will have in-country personnel known as vendors (Maylath 2001, 66). The project manager for the translation company should match both language ability and content knowledge with the project. The vendors should be able to understand and use the technical language called for in the document. See Exhibits 9.2 and 9.3 for a flow chart of the documentation process when a translation company is used.

While the case study is centered on large translation companies, you can also adapt the models for your own smaller projects, even if they are in-house. By having experts in subject areas and in the language review the documents, by revising carefully and consistently, and by then retesting the translations, you can be reasonably assured that the information will be comprehensible and accurate. One problem my students and I ran into when we developed our bilingual Spanish-English diagnostic and triage system was that, just like there are many forms of "English," there are also many forms of Spanish. Often the dialects and vocabulary are different enough to justify localizing the documents even more. However, time constraints and budgets often limit how much time and effort you can spend on localization. When you use the above model, you will see early in even a small, in-house project that time constraints will impact your translation project in a major way. By testing the document with a good, representative sample of your audience, even if you only have time to do a small focus group or conduct informal usability tests, you will have a document that works for most. Also, the case study above is a bit misleading. It sets up a hypothetical scenario in which a

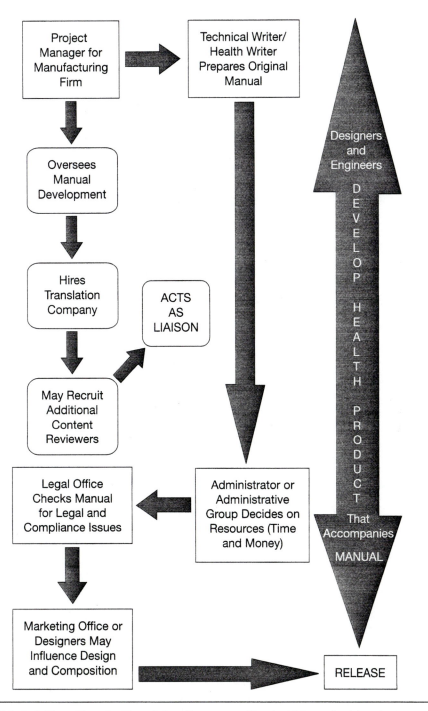

Exhibit 9.2 Flow Chart for Translation Document (Surgical Manual)—Manufacturer

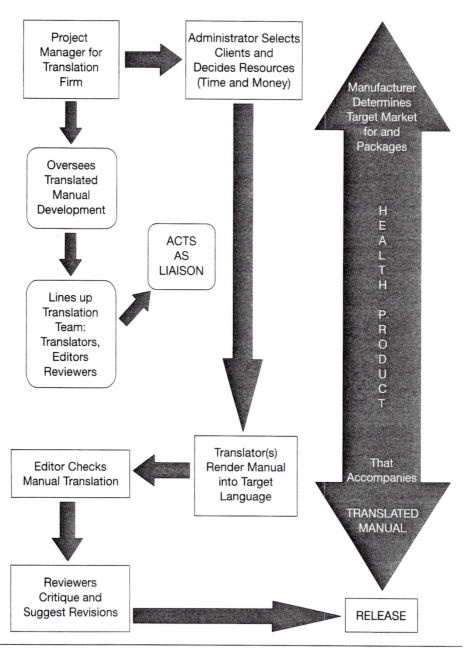

Exhibit 9.3 Flow Chart for Translation Document (Surgical Manual)—Translation Co.

surgeon is ready to do his second heart patch operation and having difficulty with the translated manual. In reality, in such crucial areas as open-heart surgery, much more time is spent studying the procedure and working with someone who has done the procedure more often.

There were times as a cardio-pulmonary technician that new machines would come in, and I would need to set them up and figure out the instructions. These machines weren't used on patients until I had figured out the instructions, rewritten them, given them to my supervisor and medical director (cardiologist), revised them, took them to the ICU and/or CCU, taught the nurses how to use the new machine, and given them the rewritten instructions. Then I was available twenty-four hours a day, every day, to come troubleshoot any difficulty they had in setting it up, calibrating it, or operating it. Even though this translation involved translating technical English to nursing English, it required many steps.

While we are talking about translating and communicating in a language that is not our native language, it's important to make clear that it isn't always the patients or people in other countries who need translating help. Harvard medical school, for example, has many medical students from other cultures both outside of and within the United States. You may find yourself working with a resident physician from another country and culture. You might even find yourself explaining or clarifying an English idiom (our language is replete with them) or even explaining whole documents. Also, because of our present nursing shortage (which actually occurs cyclically in our culture), we are training and hiring many health professionals, especially nurses from other countries. Here, too, there are both language and cultural differences that as a good communicator you need to be aware of in the workplace.

> Transnational[9] mobility and the movement of people across borders have impacted the rhythms of life in local contexts. These global relocations are accompanied by the circulation of multiple discourses resulting in new configurations of culture and identity (Giddens 2000). The repercussions of this world-in-motion are felt at the level of everyday experiences of immigrants in their world of work and personal relations. The *transformations caused by globalization force a rethinking of fixed notions and cultures.* (My emphasis, Hegde and DiCicco-Bloom 2002, 90)

In other words, in the United States you may have patients with English as their second language who are treated by a nurse from still another country, who is being supervised by a resident from still a third country. While this scenario presents a rich, cross-cultural exchange, it could also present some major miscommunications both via language differences and cultural differences. To solve some of the communication problems and cultural differences, my students and I in our translation project relied on graphics or visuals that could illustrate a symptom in an almost universally understood manner. Again, though, visuals are also used differently across cultures and, without testing the individual graphics, you can't be sure whether you are on target or not. Still another language barrier or tool in making health communication more effective (it depends on how well it is designed and how it is used) is electronic communication. In fact, if you localize the interface (Web site, etc.), some of your translation issues will be solved, because how you design the site for a different audience will dictate some of the text you can and must use (Yli-Jokipii 2001, 111).

How Will Computers Impact Other Cultures?

This question is a very important one. Because Computer-Mediated Health Communication (CMHC) is still new for some countries, changes are ongoing. With computer-aided health communication emerging from so many different cultures, we still are not clear about all the pros and cons of CMHC. The most recent example of a world health crisis, SARS (Severe Acute Respiratory Syndrome), has been more quickly controlled because of the ability to pinpoint world patterns of contagion, as well as to update potential travelers to prevent a further and more disastrous spread of the disease via international communication. The World Health Organization (WHO) was able to gather data and communicate quickly with greater impact simply because of our ability to communicate directly with any country in the world.

Many health and other professional communicators are speculating on the changes ahead as the technology develops further and faster. Ulijn and Campbell suggest, "The new communication technology will force us, more than ever, to adapt messages both linguistically and culturally to the needs of culturally distinct audiences (see Ulijn 1996, 69). The rhetoric of English as the international language of technical communication will contrast increasingly with culture (see St. Amant 1999, 297; Campbell 1998, 1)" (Ulijn and Campbell 2001, 78). Because our health practices, medical systems, and beliefs are so diverse, mediating health communication via computers will present even more challenges and opportunities.

Even though the technological changes are recent, a number of theoretical articles, empirical studies, and hypothetical studies have been published. There is a significant need for more research of this kind in health communication. In a report on international, electronic business writing with Belgians, Finns, and U.S. citizens, Verckens, deRycker, and Davis claim that ". . . understanding patterns of communication behavior evidently can be facilitated by technology . . . However, technologies impose their own patterns" (Ulijn and Campbell 2001, 78). Just as the changes in medical technology have reconfigured Western medicine into an arena of increasing specialization, the attendant communication and culture have reflected those changes in the discourse communities. The technology itself can force a mold or pattern on a new culture that seeks to use it. For example in B. L. Thatcher's study, ". . . the experience of a North American organization in imposing standard accounting rules in its South American operations . . . may be emblematic of this kind of problem" (Ulijn and Campbell 2001, 78).

Yli-Jokipii's empirical study on factors that differentiate English and Finnish Web sites (topic, substance, and visual presentation) reveals a rhetorical strategy of customizing or tailoring a Web site based on the knowledge base and motivations for viewing the site of the cultural members (Ulijn and Campbell 2001, 78). This study, then, speaks to the major research question raised in the first part of this chapter: Should a health communicator attempt to compose messages that are more globally focused based on what she perceives as a "universal" basis, or should she design messages that are more locally based? Yli-Jokipii's study argues for a more localized approach in a "global" or international, electronic medium (the Worldwide Web). In this example of computer-mediated communication, the electronic interface offers a synthesis of both local and international aspects of culture, although Yli-Jokipii maintains she is presenting a counter argument, "reversed angle" or "counter-mainstream framework" (2001, 105).

Other electronic impacts include emails, listservs, and networking programs. You will learn more about the electronic realities and possibilities in the last chapter. Emails, listservs, discussion boards, and chat rooms (both asynchronous and synchronous) present other challenges for health professionals in one culture communicating to health professionals in another culture. Emails and chats and other media in this genre tend to encourage informal and very fast responses. The areas for stepping on cultural toes are enormous. Here's where one first learns (usually) to adapt discourse to the culture he or she is addressing. People from the United States, as a low-context culture, are not as polite or formal as members of a high-context culture. In order to address the needs of our audience, we as U.S. health professionals will have to learn and respect the different cultural norms that will apply to various cultures.

Of course how computer-literate cultures are will make all the difference in terms of any mass cultural health appeals and attempts to link medical researchers. Some countries do not yet have the infrastructure to support the kinds of electronic communication other countries already have in place. For example, although Finland is a small country with a small market and language area, the Finns are well known for their state-of-the-art technology. In fact, "at the end of 1999, Finns had Europe's densest coverage of Internet networks, and the world's second densest, with 121 computers with Internet interfaces per thousand inhabitants. The U.S. held the highest figure at the same point of time, with 134 computers per thousand inhabitants" (Yli-Jokipii 2001, 105). Japan had the third-greatest amount of Internet uses in 2002, behind the United States and Germany. Other countries have large areas not yet part of the electronic world, so the impact on their cultures will be much less. However, if delivery of healthcare becomes more and more computerized, such impoverished countries will suffer even more obstacles to providing medical treatment.

Ulijn and Campbell raise interesting questions regarding computer-mediated communication (CMC) for future research, which we can adapt to computer-mediated health communication (CMHC) but will only be addressed briefly here:

- How will different cultures and languages develop with the [growth of CMC]?
- What will be the impact of CMC on Asian cultures and their interactions with others?
- To what extent can CMC replace FTF [face-to-face]?
- Is electronic healthcare messaging that efficient? (2001, 79)

Ulijn and Campbell point out that the Internet disrupts both the physical environment (through laying fiber optic cable) and the cultural environment. Smaller cultures especially feel overwhelmed by the technological capabilities of larger, more developed countries, although the Australian aborigines are using the Internet as a way to save their culture (2001, 79).

Because of the large differences between Eastern and Western cultures, a number of studies have examined the impact of CMC on Asian nations (see Ramati 2000, Schwarz 1992, and Triandis 1995). Ramati argues that CMC will change the tight social group structure (high-context culture) in places like Malaysia, having a much stronger impact on those types of societies as opposed to the low-context, loosely organized cultures like that in the United States and Australia (Ulijn 79). The high-context cultures, then, can expect to feel the impact of electronic media, and care must be given to include them in ways that will access their needs.

In many cultures, the proxemics, face-to-face, personal contact is extremely important. In healthcare the face-to-face contact is even more key, and video-conferencing is not going to work that well in many health situations. There are exceptions, however, such as video-conferencing for learning surgical techniques and other distance-learning approaches. However, many cultures are dependent on the visual, nonverbal, and face-to-face information we get from one another. Latin American people rely heavily on body language and eye-to-eye contact to built trust. Also, even European and other Western cultures receive much cultural and individual understanding via the face-to-face interactions; thus the chance of face-to-face (FTF) replacing CMC and especially CMHC is doubtful. At Clemson we have a new nursing suite for the nursing department with mechanized "patients," who can be programmed with various heart sounds, etc. Educational researchers need to study that type of programmed patient as an interface for learning purposes.

The last question Ulijn proposes is the efficiency one. Is CMC or CMHC as time-saving as we might think? The answer here is "yes and no" or "yet to be determined." Some facets of CMC are definitely more efficient, because the speed with which people can access health information alone is phenomenal. However, especially cross-culturally and in large health systems, there are drawbacks. Often there is considerable lag time in email response or listserv response by those who are already overwhelmed with such electronic communication tasks. Also, the larger the company or health concern, often the more difficult it is to get through. In some instances the CMC or CMHC is time-saving. But we've also all had the experience of our computer systems being down, and no work via computer could be done.

Summary

In this chapter, you learned that although a cultural analysis is part of an audience analysis, it is much more difficult to conduct than the other segments, such as race, gender, age, etc. The debate among those who design multicultural and international health communication is between designing a universal or more localized health message. You also learned that the term *global* could be problematic; in many countries it means an overwhelmingly Western culture overtaking smaller and less technologically able cultures. There are pros and cons to each approach. Sometimes localization has been heavily critiqued, and other times the universal approach has been rejected. Some researchers have developed syntheses or bridges between the two extremes, using a universal medium and adapting the specifics to a particular culture.

You learned that the research methodology for understanding other cultures is time-consuming, lengthy, and extensive. It involves *triangulating* data and, if you do ethnographies or case studies, using *thick description.* But there are also other ways to understand cultures, by working directly with those people who are well versed in or native to the culture and including them in the design process. In addition, the Instructor's Manual features a broad hypertextual approach to cultural research, through which, in the absence of time and resources, you can at least get a feel for another culture visually and textually.

One way to distinguish between cultures, at least on a macro level, is to examine them using the categories of *high- and low-context.* The United States is a low-context culture, while many Asian cultures are more formal and high-context cultures. See Figure 9.1 for more ways to comprehend this concept. While no culture can be pigeonholed absolutely in this way, when used as a tool, it does provide some ways to get insight into other styles of thinking and communicating. In order to deliver health messages to another culture, it's most important not to offend the members of the culture and lose the ability to communicate with and persuade them. Before print or electronic health messages are sent out, if possible field-test them or have a consultant look at the materials.

Besides the cultural analyses for international and multicultural audiences whose health practices are deeply engrained in their societies, translations present more challenges. In-country or in-community translators are ideal; for big projects health companies use translation companies. Consult the models in this chapter for ways to set up an effective management system for translations. With good writing and management processes (revisions, feedback, good editing), your translation projects will be accurate and effective. The translation software itself is still not sensitive enough to successfully translate important medical documents.

We can't underestimate the impact of electronic media on cultures, and that impact varies from culture to culture. The technologies themselves have the ability to dictate changes in language, graphics, and ways of perceiving and conceptualizing identities. Cultural identities will change vis-à-vis their adoption or adaptation of electronic media in healthcare; traditional health practices may be lost or changed. As health professionals, we need to be sensitive to these changes and dislocations. Rather than impose our Western medicine on cultures, our audiences would be better served if we took their cultural needs into consideration and let them use the technology in ways it works for them. More research needs to be done into health and electronic media across cultures. Especially in epidemics and with bioterrorism, the need for better and faster communication across the world could be key.

Discussion Questions

1. Find an article on the Internet on international healthcare. Bring it in to discuss and share with your class.
2. Have you ever been in a situation in which you were the "other" person or the only one from a particular race, class, gender, ethnic group, or culture? How would you want your health issues addressed if you were from another culture?

3. As a small class team, spend a half-hour outlining a brief research plan for an ethnographic, cross-cultural health project your teacher assigns you and then discuss the pros and cons of your research plan in class.

4. Discuss high-context and low-context cultures using examples from cultures that you know. Make a list on the board of your own culture first before you discuss other cultures.

5. After reading a text (book or article) on a cross-cultural, health-oriented case study or ethnography of another culture, discuss the various problematic communication situations (rhetorical situations) featured in the work. Identify and discuss these culture gaps in class and what options were available to change the outcome.

6. Make a list of the various high-tech communication systems we use now in medicine. Picture those systems being deployed in various cultures. How do you think those systems will impact the society? How could technology for healthcare and for communicating that healthcare mold or change a culture?

7. If you had to design health messages for a culture like the Hmong in the United States, what are some of the research and strategies you would plan?

8. Discuss how you would want to be addressed as a receiver of a health message designed by some other culture. What sensitivities would you want the designer and communicator to know? What would offend you? What would persuade you?

9. Using different cultures as examples, discuss the various impacts you think CMC and CMHC would have on these societies.

Exercises

1. Find two health Web sites from two different countries. What similarities do you see on the two sites? How are they different? Design a PowerPoint to show your classmates how the two Web sites are different. Based on the two health Web sites (small case study), can you speculate what those differences in visual design and the use of the technology might indicate about the two different cultures?

2. Invite several international students on your campus to your class and ask them to talk about healthcare in their countries. Ask them also what difficulties and confusions they and others from their counrty have about Western medicine. Listen carefully, take notes, and ask for suggestions of ways that you as a health communicator can make healthcare more accessible and understandable for them and their families.

3. Each student in the class should read and prepare a report on an article such as those listed in the chapter. (Make sure you each have one targeting a different cultural group.) At the end of your report, draw some conclusions about what issues you would want to be aware of when designing health messages for your particular group. Condense some of your insights and advice into a report, PowerPoint presentation, or on a Web site that you design. Share them with your classmates.

4. Using the Internet, try to find Web sites that speak to a different culture and in their language, particularly about health issues. Then compare these Web sites with U.S. health Web sites. Note the differences. It's especially good if you know another language besides English. Also, see if you can find Web sites for other cultures that are in English or give you options in the opening page of a Web site to choose other languages. Notice if the designs stay the same or not as the languages change; also write down a list of advisory suggestions for a Web designer who must develop a Web site for that culture.

5. With a team of your classmates and classmates from a language class, develop a script based on cultural issues (between high- and low-context societies) in which you attempt to address a health problem. Write a 15-minute script for your dialogue and then perform it for the class with a group discussion scheduled after the performance.

6. With a team made up of several members of your class plus several students from a language class, research attitudes toward healthcare in the other country (on-line or in interviews). Now, write up what you think would happen with the introduction of new communication media. Each team could explore a different technical medium and a different culture. (For example, a Spanish-language class could compare healthcare Web sites in Mexico with those in the United States or could explore the use of electronic charting in Madrid hospitals.)

Works Cited

Bosley, Deborah S. 2001. *Global Contexts: Case Studies in International Technical Communication.* Boston: Allyn-Bacon.

Campbell, C. P. 1998. "Rhetorical *Ethos:* A Bridge between High-Context and Low-Context Cultures." *The Cultural Context in Business Communication,* 1–47. S. Niemeier, C. P. Campbell, and R. Dirven, eds. Amsterdam: Benjamins.

Centers for Disease Control (CDC). 2001. *CD Cynergy.*

Cooper-Patrick, L., J. Gallo, J. J. Gonzales, H. T. Vu, N. R. Powe, C. Nelson, and D. E. Ford. 1999. "Race, Gender and Partnership in the Patient-Physician Relationship." *JAMA* 282: 583–89.

Dushay, R. A., M. Singer, M. R. Weeks, L. Rohena, and R. Gruber. 2001. "Lowering HIV Risk Among Ethnic Minority Drug Users: Comparing Culturally Targeted Intervention with a Standard Intervention." *American Journal of Drug and Alcohol Abuse.* 27: 501–24.

Fadiman, Anne. 1997. *The Spirit Catches You and You Fall Down: A Hmong Child, Her American Doctors, and the Collision of Two Cultures.* New York: Noonday Press/Farrar, Straus and Giroux.

Giddens, A. 2000. *Runaway World.* New York: Routledge.

Gudykunst, William B., and B. Mody. 2002. *Handbook of International and Intercultural Communication* 2nd ed. Thousand Oaks, CA: Sage.

Hall, E. T. 1983. *The Dance of Life: The Other Dimension of Time.* New York: Doubleday.

Hall, E. T., and M. R. Hall. 1987. *Hidden Differences: Doing Business with the Japanese.* New York: Doubleday.

———. 1990. *Understanding Cultural Differences: Germans, French and Americans.* Yarmouth, ME: Intercultural.

Hawisher, Gail E., and Cynthia L. Selfe. 2000. *Global Literacies and the Worldwide Web.* London: Routledge.

Hegde, Radha S., and Barbara DeCicco-Bloom. 2002. "Working Identities: South Asian Nurses and the Transnational Negotiations of Race and Gender." *Qualitative Research Reports in Communication* 3, no. 4: 90–95.

Heifferon, Barbara A., Rebecca Barnett, Angela Davis, Jason Durham, Andrew Edds, and Kaushal Seshadri. 2000. "Adopting or Adapting New Technology in an International Arena." Unpublished manuscript.

Helton I. R. 1995. "Intervention with Appalachians: Strategies for a Culturally-Specific Practice." *Journal of Cultural Diversity* 2: 20–26.

Hofstede, G. 1991. *Culture and Organization: The Software of the Mind.* New York: McGraw-Hill.

———. *Culture's Consequences: International Differences in Work-Related Values.* 1980. CA: Sage.

———. *Uncommon Sense about Organizations: Cases, Studies and Field Observations.* 1994. London: Sage.

———. B. Neuijen, D. Ohayv, and G. Sanders. 1990. "Measuring Organizational Cultures: A Qualitative and Quantitative Study of Twenty Cases." *Administrative Science Quarterly* 35: 286–316.

In-culture Approach Web Site. © 2002. Isabel Valdés. August 29, 2003. www.incultureapproach.com

Lannon, Linnea. 1997. Book jacket end comment. Excerpted from *The Detroit Free Press.*

Lauer, Janice, and J. William Asher. 1988. *Composition Research: Empirical Designs.* New York: Oxford University Press.

Lewis, R. D. 1996. *When Cultures Collide.* London: Nicholas Brealey.

MacNealy, Mary Sue. 1999. *Strategies for Empirical Research in Writing.* Needham Heights, MA: Allyn Bacon.

March of Dimes Spanish-Language Web Site. 8/29/2003. August 29, 2003. www.nacersano.com.

Maylath, Bruce. 2001. "Translating User Manuals: A Surgical Equipment Company's 'Quick Cut.'" *Global Contexts: Case Studies in International Communication.* Deborah S. Bosley, ed. Boston: Allyn-Bacon.

McLaughlin, L. A., and K. L. Braun. 1998. "Asian and Pacific Islander Cultural Values: Considerations for Health Care Decision Making." *Health Social Work* 23: 116–27.

Nelson, David E., Ross C. Brownson, Patrick L. Remington, and Claudia Parvanta, eds. 2002. *Communicating Public Health Information Effectively: A Guide for Practitioners.* Washington, DC: American Public Health Association.

Perkins, Jane. 1999. "Communicating in a Global, Multicultural Corporation: Other Metaphors and Strategies." *Exploring the Rhetoric of International Professional Communication: An Agenda for Teachers and Researchers.* Ed. Carl Lovitt with Dixie Goswami. Amityville, NY: Baywood.

Pinker, S. 1994. *The Language Instinct: How the Mind Creates Language.* New York: William Morrow.

Ramati, N. 2000. "The Impact of Cultural Values on Computer Mediated Work." *Cultural Attitudes Toward Technology and 2000,* 257–74. F. Sudweeks and C. Ess, eds. Murdoch, W. Australia: University School of Technology.

Schwartz, S. 1992. "Universals in the Content and Structure of Values: Theoretical Advances and Empirical Test in 20 Countries." *Advances in Experimental Social Psychology* 25: 1–65. M. Zanna, ed. NY: Academic.

Selfe, Cynthia L. 1999. "Lest We Think the Revolution is a Revolution: Images of Technology and the Nature of Change." Gail E. Hawisher and Cynthia L. Selfe, eds. *Passions, Pedagogies, and the 21st Century Technologies.* Logan, Utah: Utah State University Press.

St. Amant, K. 1999. "When Culture and Rhetoric Contrast: Examining English as the International Language of Technical Communication." *IEEE Transactions on Professional Communication* 42, no. 4: 297–300.

Tanjasiri, S., S. P. Wallace, and K. Shibata. 1995. "Picture Imperfect: Hidden Problems Among Asian and Pacific Islander Elderly." *Gerontologist* 35: 625–50.

Thatcher, B. L. 1999. "Cultural and Rhetorical Adaptations for South American Audiences." *Technical Communication* 46, no. 2: 177–95.

Thrush, Emily A. 2001. "High-Context and Low-Context Cultures: How Much Communication Is Too Much?" *Global Contexts: Case Studies in International Communication.* Deborah S. Bosley, ed. Boston: Allyn-Bacon.

TodoBebe Web Site. © 1999-2003-Derechos Reservados. August 29, 2003. www.todobebe.com

Triandis, H. C. 1995. *Culture and Social Behavior.* New York: McGraw-Hill.

Trompenaars, A., and C. Hampden-Turner. 1998. *Riding the Waves of Culture: Understanding Diversity in Global Business.* 2nd Ed. New York: McGraw-Hill.

Ulijn, Jan M., and Charles P. Campbell. 2001. "Technical Innovation and Global Business Communication: An Introduction." *IEEE Transactions on Professional Communication* 44, no. 2 (June): 77–82.

Ulijn, Jan M. 1996. "Translating the Culture of Technical Documents: Some Experimental Evidence." *International Dimensions of Technical Communication,* 69–86. D. Andrews, ed. Washington, DC: STC.

U.S. Bureau of the Census. 2001. *Statistical Abstract of the United States. 2001.* Washington, D.C.: U.S. Census Bureau.

Valdés, M. Isabel. 2002. *Marketing to American Latinos: A Guide to the In-Culture Approach.* Ithaca, NY: Paramount Market Publishing, Inc.

Verckens, J. P., T de Rycker, and K. Davis. 1998. "The Experience of Sameness in Difference: A Course on International Business Writing." *The Cultural Context in Business Communication* 247–61. S. Niemeier, C. P. Campbell, and R. Dirven, eds. Amsterdam: John Benjamins.

Weeks, M. R., J. J. Schensul, S. S. Williams, M. Singer, and M. Grier. 1995. "AIDS Prevention for African-American and Latina Women—Building Culturally and Gender-Appropriate Intervention." *AIDS Education and Prevention.* 7: 251–64.

Weiss, S. and W. Stripp. 1998. "Negotiating with Foreign Business Persons: An Introduction for Americans with Propositions for Six Cultures." *The Cultural Context in Business Communication,* 51–118. S. Niemeier, C. P. Campbell, and R. Dirven, eds. Amsterdam: John Benjamins.

Yli-Jokipii, Hilkka. 2001. "The Local and the Global: An Exploration into the Finnish and English Websites of a Finnish Company." *IEEE Transactions on Professional Communication* 44, no. 2 (June): 104–13.

Zahedi, Fatemeh "Mariam," William V. Van Pelt, and Jaeki Song. 2001. "A Conceptual Framework for International Web Design." *IEEE Transactions on Professional Communication* 44, no. 2 (June).

Endnotes

1. Culture, as Hofstede has suggested, is something like "software of the mind," the operating system that allows human individuals to share and make sense of experience. Such operating systems consist partly of language systems and partly of tacit rules. Such rules govern all aspects of how humans relate to each other—for example, how

they show deference to people higher in the social hierarchy, conduct courtships, acknowledge their bonds of identity through ceremony, even how close they stand to one another on a bus (see, for example, Hall; Hofstede; Trompenaars and Hampden-Turner; and Weiss and Stripp).

2. I witnessed the demonstrations in Genoa, Italy, when the World Trade Organization scheduled a meeting there. Demonstrations have occurred in Washington, D.C., and Seattle in this country and in many, many cities around the world. They all center around the large international political and funding organizations, such as the World Trade Organization (WTO), the International Monetary Fund (IMF), the World Bank, etc.

3. "Proxemically" refers to distance or space between and among people. If you have known someone from another country who stands closer to you than you are comfortable with, this is a matter of proxemics. People from the U.S. typically feel more comfortable with a larger personal space.

4. Perkins claims that the boundary metaphor is problematic.

5. In addition, see MacNealy 1999 and Lauer and Asher 1988 for good, concise discussions and parameters of research methods.

6. See the Instructor's Manual for a list of excellent on-line sources for Latino research.

7. Thrush says, "[w]hile the situation described below actually happened, the names of the individuals involved have been changed. Also, the documents represent a compilation from several people over the seven years of history of this agreement" (2001, 27).

8. Obviously these characteristics are applicable only in some cases. Gender, a person's unique personality, age, and many other factors can alter any of the above.

9. *Transnational* may be a better term than *globalization*, although these authors use both terms. With the term *globalization* being problematic (see the first part of this chapter), try to choose *transnational, international, intercultural,* etc., instead.

CHAPTER **10**

Presenting Written Materials Visually

 ## Overview

Although this book is about medical *writing,* some attention needs to be focused on the visual presentation of material, because written texts often migrate to other forms and health professionals are often called upon to deliver data visually. Also, much written material now is heavily dependent on graphics, and many of the younger people of our culture are more visually oriented than textually oriented. Because we now have a vocabulary of design terms and principles, introduced in Chapter 3 on document design, we can build on that as we discuss visual communication. This chapter then flows naturally into the last chapter on electronic formats.

In this chapter, you will learn answers to the following questions:

- How long has visual communication been used?
- How do I decide whether to use illustrations?
- What about creating illustrations using computers?
- What do I need to know about photographs?
- Which visuals are most important?
- How do I construct tables?
- How do I decide whether to use tables?
- How do I format tables?
- What kind of information is best for graphs and charts?
- How do I format graphs and charts?
- What about flowcharts and molecular graphics?

This chapter includes discussions on the most frequently used visual communication tools in healthcare: medical illustrations, graphs, tables, and charts, while the next chapter addresses PowerPoint presentations and other electronic media.

How Long Has Visual Communication Been Used?

Historically, medicine has been passed down as oral history or what we call the "oral tradition." Knowledge of medicines and treatments was passed from one or more older people in a tribe to younger members of the tribe, so, in times of warfare or epidemics, healing remedies were not lost with the elderly.

In addition to communicating verbally, some cultures also used symbol systems, such as hieroglyphics (picture writing) and other drawing systems to record medical treatments, healing plants, epidemics, and the number of lives lost in battle. "Ancient Egyptian writing is among the oldest recorded language in history, appearing during the late pre-dynastic period and continuing into the Ptolemaic period, when it was superseded by Greek" (Brewer and Teeter 1999, 123). This time period means that "'Egypt has the longest recorded history of any of the earth's languages: over 4500 years" (Ray 1994, 51). "The Egyptian hieroglyphic language combined ideograms and phonetic signs and functioned like a rebus: symbols representing sounds of actual objects were combined to form more abstract words" (Brewer and Teeter 1999, 117, 119; Heifferon n.d., 8).

Many centuries later, Leonardo da Vinci's drawings of the human body marked the beginning of a lifting of the taboo against autopsy and the exploration and subsequent drawing of organs and body systems in the Western world.[1] Although da Vinci didn't publish his anatomical drawings, completed between 1505 and 1510, researchers believe he influenced Vesalius, who wrote *De Humanae Corporis Fabrica* in Venice and Padua in 1542, to whom the history of English medicine owes a great debt (Thebeaux 1997, 186).[2] In the book Leonardo hoped to finish and publish, his illustrations would be dominant, with little attendant text (the text that is there is anatomically descriptive only and doesn't prescribe herbs or treatments) (O'Malley and Saunders 1952, 15).

Da Vinci's contributions changed medical drawings significantly. During the Middle Ages, no one was drawing parts of the human body in such animated and multidimensional ways; his visuals became the standard for medical illustrations (Thebeaux 1997, 186). His contributions include:

- Illustrating different systems of organs in logical succession;
- Illustrating the skeletal system—he was the first to show accurately the bones of the hand [Belt];
- Illustrating the relationships between the musculature and the skeleton system;
- Showing the relationship between internal organs and the body's surface;
- Using cross sections to show the topographical relationships between the bones, muscles, and nerves;
- Using exploded views of the skull to show the relationships among its parts;
- Using three-dimensional drawings to illustrate structure and function of anatomical parts [Herrlinger 1970, 70–72; O'Malley and Saunders 1952, 15–18]. (Thebeaux 1997, 186).

Thebeaux also noted in her history of technical writing that during the Renaissance in England, self-help medical books were one of the largest categories of texts (14). Some were illustrated with drawings, especially line drawings of plants for medicinal usage. In his study of Renaissance medicine, Paul Slack said that medical books comprised "only an eighth of titles, but a quarter of all editions and nearly a third of works of 'pocket-size'" (Slack 1997, 247). "Books for non-medical professionals dominated the list of medical bestsellers between 1485 and 1604" (247).[3] The most popular book for midwives during this same period was a translation of Roesselin's book *The Birth of Mankynde,* which enjoyed 10 editions between 1540 and 1604 (Roesselin 1598).

Medical illustrations are just one form of graphics often used in the health professions. Except for preparing health education materials and medical reports, you, as a health professional, may not do much drawing. However, there are also medical illustrators whose careers center on this visual work. If you do not have an illustrator on staff, you will then outsource to medical illustrators, i.e., professional artists. You will need to purchase permissions on all copyrighted materials in order to use them (unless you use drawings from freeware and shareware, clip art, etc.). You will need to know how to embed illustrations so that they are appropriate and fit well with the text you are developing. You may develop some simple drawings and logos, especially if you have some visual talent.

As a document planner, you will need to be aware that preparing illustrations takes time. Therefore, as soon as you know what your project entails (whether it is print or electronic) and whether or not your illustrations are photographs, drawings, X rays, etc., you will need to build in time for their creation and fine-tuning. You'll want to make sure you have the proper personnel to prepare the illustrations for you. Also, make sure you understand how definitions are used in your workplace. Sometimes illustrations include graphs and charts; other times illustrations refer more to drawings and photographs. For our purposes here, we separate tables, graphs, and charts from illustrations, thinking of illustrations more as line drawings, drawings, and photographs.

The most frequently used graphics in medicine are tables, graphs, and charts. These are often the visual genres used most by scientists, and this traditional norm has migrated from science into our science-based medicine. Edward Tufte is known in professional communication circles as the Guru of Graphics, particularly when it comes to charts, tables, figures, and graphs.

In a well-known example in professional and technical communication courses, Tufte discusses the health and safety issues in the Challenger space shuttle incident. In his text, *Visual Explanations,* Tufte argues that the reason the Challenger was not grounded and instead was sent up in freezing weather was that the visual message given via charts and graphs did not adequately communicate the safety message that the O rings would not function well in the cold. He shows how the data was presented to NASA visually and also how it could have been presented more effectively through a data matrix and a scatter plot that he developed.

In another example in *Visual Explanations,* Tufte examines the cholera epidemic in London in 1854. John Snow, a chronographer of those times, describes it here:

The most terrible outbreak of cholera which ever occurred in this kingdom, is probably that which took place in Broad Street, Golden Square, and adjoining streets . . . . The mortality in this limited area probably equals any that was ever caused in this country, even by the plague; and it was much more sudden . . . . The mortality would have been much greater had it not been for the flight of the population. . . . in less than six days from the commencement of the outbreak, the most afflicted streets were deserted by more than three-quarters of their inhabitants. (Snow 1855, 38; Cholera Inquiry Committee 1855; Scott 1934; Tufte 1997, 27.)

In this case, Tufte argues that because Dr. John Snow created statistical graphs that revealed the data rather than obscured it as in the Challenger case, he could find the cause of the outbreak and stop the epidemic (Tufte 1997, 27). Snow hypothesized that the public water pump well at Broad and Cambridge streets in central London was contaminated, even though he did not see obvious problems in the water at first (Snow 1855, 39; Tufte 1997, 28). Using a map and a list of 83 cholera deaths from the General Register Office in London, Snow plotted out the cholera victims; the map revealed "a close link between cholera and the Broad Street pump" (Snow 1855, 39–40). He went door-to-door and talked to the inhabitants to determine if his data were correct. He was able to talk to all but six of the victims' families, and he confirmed his data that indeed they had used the well. A few (another six) claimed that the victim had not had well water. Snow went to the Board of Guardians for St. James Parish and convinced them to remove the pump handle. The epidemic was ended almost as quickly as it had come (Tufte 1997, 28). According to Tufte, Snow was able to make good causal connections and intervene because in addition to his causal theory, he had a good method of communicating through visual rhetoric. Snow showed "a shrewd intelligence about evidence, [and] a clear logic of data display and analysis" (29). We have to remember, too, that this was in a time when most medicos thought cholera was an airborne disease.

Here are the constructs for good visual communication that Tufte gleaned from the cholera example, which illustrate why Snow was successful in exhibiting his data:

- "Placing the data in an appropriate context for assessing cause and effect." A textual display here with long descriptive passages (more routine form of reportage at the time) would not have shown the causal relationship, nor would just the timeline on which he recorded the deaths. Instead, Snow developed a cluster analysis methodology using a map, still used today to track outbreaks.
- "Making quantitative comparisons." Snow also mapped out the cholera deaths in relation to sizeable populations in the same area that did not have any deaths: the brewery and the workhouse. In interviews he discovered that indeed these locations both had wells on their sites and didn't use water from the Broad Street pump, even though it was close by. Also, the brewers almost never drank water, but were allowed to drink a certain quantity of the malt liquor they made.
- "Considering alternative explanations and contrary cases." Sometimes researchers are reluctant to report examples that are contrary to their main

thesis (the news media often carries such stories today) and therefore present the data without the counterarguments. Snow followed up with the families of several cholera victims who lived at some distance from the Broad Street pump and found that they had purchased the water from Broad Street that day. Also, one of the distant victims went to a school that used Broad Street water.

- "Assessment of possible errors in the numbers reported in graphics." Snow carefully backs up his data with a thorough discussion of his method. He drew a dot map, marking each cholera victim's death. Death rates were not shown, and if the number had been too high, the map could have been cluttered with detail and not present the evidence well. On the other hand, he avoided the problems sometimes seen with aggregate maps. The data would have been presented even better if he had done a map that showed the death *rates* as well, because we can't see on his graphic how many people died compared with those who lived. Mark Monmonier, a modern mapmaker and analyst, also worked with Snow's data in producing aggregations in graphic format. He points out by using three different aggregate maps how the data could have been misinterpreted (Monmonier 1991, 142–143; Tufte 1997, 29–37).

Snow is still widely credited with the discovery that cholera was indeed waterborne and not airborne.

This methodology is still used today, and with the recent examples of anthrax, SARS, and monkey pox, our skills in presenting data causally to stop certain epidemics are important.

Another visual theorist in addition to Tufte and Monmonier we should consider is John Berger. Like those who teach English composition to students and ask them to make their writing as persuasive as possible, in *Ways of Seeing,* Berger also stresses the *composition* of the visual image to emphasize its persuasive power. "The compositional unity of a painting contributes fundamentally to the power of its image" (Berger 1972, 13). This composition or arrangement and organizational strategies of visual images helps us plan the work carefully, using our design vocabulary to consider *chunking* like items together either by *proximity* or *similarity*.[4] We also need to consider *white space, saturations, hues, values,* and all the aspects of good design. The more difficult the task to change health behaviors, all the more necessary it is to be able to create persuasive and clear images. Whether an image is combined with extensive text (not a good idea for posters, flyers, Web pages, and PowerPoints, but fine for journals, reports, and books) or combined with just a few words, "the meaning of an image is changed according to what one sees immediately beside it or what comes immediately after it" (Berger 1972, 29). Sometimes visual communicators refer to this view as "Gestalt Theory," even though it has a different origin. *Gestalt* in German means wholeness or the whole picture. Originally used by Fritz Perls, a German psychologist, to re-create the whole picture for patients, it applies to the completeness or cohesion of a visual document as well (Perls 1973).

Up until reproduction of artworks and images became technologically simple, artists in the upper class and/or guild members created images. Now, with easier access to digital cameras, scanners, and computer drawing programs, almost anyone can create images and use them for whatever purpose she desires. In healthcare, these images can be used to explain conditions and educate patients; they can also be used to better illustrate the results of certain diagnostic tests as well as make diagnosis possible in itself (through CAT scans, X rays, sonography, and magnetic resonance imaging, or MRIs, for example). The process of creating images is no longer as class-based as it was in the past. Graphics has become a new language in which many of us can speak. "If the new language of images were used differently, it would, through its use confer a new kind of power . . . what matters now is who uses that language for what purpose" (Berger 1972, 33). As we know from our ethical awareness as health professionals, with that power comes responsibility.

How Do I Decide Whether to Use Illustrations?

Just as you will find specific criteria for creating tables, graphs, and charts, you will also find that the use of illustrations is governed by certain criteria, too. You do need to keep in mind that in health document design, illustrations are the most expensive for the publisher to add to publications (if they are print-based). Therefore, like the other genres in visual communication used in the health professions, these more expensive formats will be carefully reviewed. Illustrations should serve one or more of these needs:

- Evidence
- Efficiency
- Emphasis

Evidence means using an illustration to document a health claim or argument that you are trying to make. For example, if you said that impetigo and monkey pox on the skin look very similar, you would, of course, back up that claim with an illustration of each disease manifestation side-by-side. If you are making a particular argument or documenting a procedure that is difficult for a person to accept, by using illustrations as evidence, your points will be more persuasive.

Efficiency in illustrations means that it is more efficient to use an illustration than to explain something in long phrases of text. Readers will get lost in long explanations, but a drawing or photograph will be much more quickly grasped.

Emphasis is using an illustration to emphasize some information that you want the reader or viewer to remember and keep in mind. Often when you are delivering a presentation, the most effective means of getting your audience's attention is through visual means vis-à-vis a PowerPoint presentation, illustrations on a Web site, a flip chart, poster, photograph, or slide show. Be aware, though, that "[e]mphasis is the reason least likely to be accepted by the editor for use of illustrations" (Huth 1999, 152). Again, be careful not to overuse illustrations, especially if they do not strongly meet the criteria. Your editor will most likely want fewer than you have planned, so assign priority to the illustrations in your own mind as you prepare them or send them to someone else for preparation. "If you can anticipate how the editor is likely to judge their usefulness, you may save some time, effort, and money by not having to discard illustrations" (151).

If you're illustrating your own health facility's materials, do not rely on sizes of letters and widths of lines as they show up on the monitor, because in print they often look very different. Instead, see how they will look once they are reduced to their printed size. After reduction of the illustration, you can figure that the original is twice as big as the printed illustration. That means to get the standard two-millimeter-high capital letters, the capital lettering on the original needs to be twice as high. In other words, since the typical size of the capital letters should be two (2) millimeters when printed, put them on the original at four (4) millimeters in height. Make sure, too, that the x- and y-axes and the trend lines shouldn't be any "wider than the width of the lines making up the letters" (Huth 1999, 158).

Sometimes it's appropriate to resize and reformat your illustration. The most frequently used format is a vertical rectangle, but sometimes a horizontal rectangle or a square is more appropriate. For example, a kidney X ray is a vertical rectangle, but it's possible that a square cutout of the area showing a large renal stone would be a better choice than showing the entire field of the film.

Color illustrations cost much more than black/white or grayscale images if they are to be printed professionally. Be sure to inquire within your department or consult your publisher's guidelines and budgets to find out if you are permitted to use color illustrations and what the limitations are. Most publishers will prefer color transparencies (slides) or negatives from color prints. For many professionally printed genres, you will be asked to prepare your black-and-white illustration using a photo studio software program to copy them. Of course, using desktop printers with less professional-looking color copies (although rapid improvements are being made) is much more affordable.

Don't forget to make sure you have correctly labeled the illustration itself and any parts within the illustration with callouts, etc. Although the text needs to refer to the illustration, too, it should be recognizable for the most part without having to consult the text too much.

What About Creating Illustrations Using Computers?

Because our capability to do desktop publishing has increased exponentially over the last several years, many of us trained in more traditionally print-based programs are now doing much of our document design, photography, and illustrating via computer drawing programs, newsletter templates, digital photography software, etc. We also have increased capability for visually interesting multimedia productions. We will discuss many of these skill sets and their potential in the health professions in the next chapter. But it's important to keep in mind that illustrations that we use for print documents do not always translate from monitor screen to book or other print format when digitized. Make sure you know what you're doing, have the equipment you need, and have talked to your publishers or journal editors about what platforms their own printing equipment supports. If you have the time, talent, and access to the appropriate facilities and software, then you can save both money and "translation time" (putting your ideas of what needs to be done in some other designer's head) in your overall project budget and time line. (See Chapter 3 for project management tips.) Make sure you can do a professional job, because your credibility with readers/viewers and other professionals will suffer with poorly done illustrations. Huth suggests consulting *Preparing Scientific Illustrations* and *Visual Data* for more specific guidelines (Huth 1999, 153). I would also add such sources as Edward Tufte, Charles Kostelnick, and David D. Roberts.

Line illustrations (black ink on white paper) can range from very simple to very complex. Again, know what you want up front and communicate that vision if you need to outsource the line illustrations. Simple line drawings are fast and easy to complete, and are actually less time consuming than working with photographs. However, if you need very complex line illustrations (for example detailing steps in a new surgical procedure), these types can be time-consuming, labor-intensive, and expensive. If such illustrations are required, think about limiting some of the other illustrations, photographs, tables, charts, and graphs to keep costs down, unless you have a large or unlimited budget.

Electrocardiograms (EKGs) and other medical tracings are often necessary to embed in a report or other document. You can copy them photographically or with a copy machine, placing the tracing on a white background. If the tracing lines are thin, though, the person preparing the camera-ready copy or doing the copying "may not be able to reproduce the fine detail" (Huth 1999, 154). In that case, you can trace the waveform onto a piece of paper, create a facsimile on another piece of paper, and photograph that EKG tracing. However, this does have to be very exact work, so only attempt this if you can make sure that you are faithful to the original (154). For any reproductions you attempt, make sure you use an excellent laser printer, not inkjet or dot-matrix (if there are still some of those in your workplace) to get the cleanest and sharpest copy you can.

What Do I Need to Know About Photographs?

You will need a photographer to create good prints of X rays, clinical views, and photomicrographs (these cannot be done with line drawings). Such illustrations are referred to as *continuous-tone illustrations.* When working with photographs or continuous-tone illustrations, keep in mind the copyright and patient privacy issues involved. Especially in photographs of patients themselves, you must ensure their anonymity. Remember to have patients sign permission slips before you photograph them. (See Chapter 2 for more specific information.) Rather than the black strip[5] across the face, you can also crop pictures so that patients can't be identified. You do need to include enough visual information so that the reader/viewer knows what particular area you are photographing. For example, if you are taking a picture of a bad lesion on a forearm, include the hand and elbow so that the viewer can identify where the lesion is located.[6] In roentgenograms, echocardiograms, scans, and other sonograms, don't worry about including extra information if the reproduction or copy is to be read by experts. You might need to identify right and left; if the pictures are shot with different dates, those should be noted on the copy, too. For lay readers, you could include a line drawing on top of the scan to show the location of abnormality.

If you need to crop a photo, you have numerous choices:

- Mark the photo with a wax pencil with crop lines. The downside to this method (and the advantage) is that the marks are easily wiped off.
- Put tissue over the photograph and lightly mark the crop lines there.
- Use the cropping tools in photo and drawing programs. Send it to your printers or other sources electronically.
- Copy the photograph and mark the lines on that sheet of paper rather than on the photo itself.
- Make a sketch of the area you want shown in the final report. (Huth 1999, 158)

Whatever method you choose, do make sure the directions are clear to your collaborator.

The text accompanying your photographs is also an important consideration. You may need to put some labels and arrows directly on the photograph to identify an area of interest. Again, in drawing and photo software programs, you can easily do this by inserting text and arrows or inserting text boxes with a transparent or appropriate background color if you want to label areas of interest that way. You can also put the pressure-sensitive letters on the prints (these materials are available at art and university book stores) (Huth 1999, 159). Remember, as on Web sites, Microsoft PowerPoint presentations, and other graphic genres, use light letters on dark backgrounds and dark letters on light backgrounds so that the words are not lost. Basically, as you learned in the earlier chapter on visual communication, keep figure-ground separation and contrast in mind as you compose. In addition to these ways of orienting your readers and yourselves when using photographs, you can also use the drawing tools and other graphics in Microsoft Word to draw in arrows and labels to identify parts of the photograph. Always remember to have the photographer (or you) make an additional backup copy of all your illustrations and photos, especially if the subject photographed is particularly rare. If you're working digitally, of course remember to make backups regularly. If size is important, do insert the centimeter-millimeter rule so that readers/viewers can better comprehend significance.

In electron micrographs and photomicrographs, "[a] professional photographer will use lighting and color filtration to ensure the best images but will need your guidance on what fields are to be shown and whether low-power views are to be used to orient the reader on locations of high-power views" (Huth 1999, 156). You and the photographer (if outsourced) should keep detailed data about the power at which each individual picture was shot and what size the original image was. For photomicrographs, you will have to include scale.

If you need to submit your illustrations to a publisher or other outsourced project-team member, you can do so electronically if your photos and illustrations are digitized. To move print formats to electronic formats, you need to scan the images and take them into a photo or drawing program. In the most recent versions of the commonly used software program, Adobe PhotoShop, you need to save the illustration as a copy; you have many options, including .tiff formats, the most appropriate for printing. You will need to send these via zip disks or burn them onto CDs because they are memory heavy. Also, if you are preparing these illustrations in color rather than grayscale/black and white, do note the CYMK (cyan, yellow, magenta, black) printers' color scale (see Appendix B). If you are sending hard copy illustrations:

- Label each one on its back with the figure number
- Title the print
- Add names of project directors or other identifying information
- Write on a pressure-sensitive label to apply to the back, or gently write on the backs of photos or other illustrations after placing them on a hard surface (without pressing so hard as to create lines that will appear on the front of the print)
- If necessary, put guidelines on the print indicating top in addition to right or left
- For photomicrographs, the indication of "top" is especially important.

In addition to labeling your work, you also need to consider the wording in your legends. That information should be consistent, i.e., use the same font and point size, figure numbers, or letters (7.1, 7.2 or 7.a, 7.b, or 7.i, 7.ii, etc.) and language usage for titles. If you title one figure with the descriptive label "Patient 412's Left Lower Lobe Lung Lesion," don't label the next one with a title such as "This is the suspicious lesion on patient 128's upper right lung." Keep the syntax of each legend consistent. Also, include technical details; if you are using micrographs and photomicrographs, include the stain and magnifications. If you have a publisher printing your work, submit the legends for the illustrations on a separate sheet. Submit your illustrations according the specifications of the publisher or organization with which you work. If you are sending electrocardiograms (EKGs), echocardiograms, or other tracings, most people will ask you to send a scanned or photographed version of them. Again, make sure you have backup copies (or the originals) of everything you send.

When you look at the larger project and its layout, consider grouping your illustrations. Huth suggests, "Photographs of closely related subjects are sometimes more effectively presented as a single, multipart illustration. An example is a general view of the cut surface of a kidney sliced longitudinally accompanied by several close-up photographs of details of the cortex, a pyramid, and the pelvis [all parts of a kidney]" (Huth 1999, 159).[7] If you are doing a Web site, PowerPoint presentation, or oral presentation, you can prepare a board or poster using a plotter, a drawing or photo software program, a page paste-up program, and layout programs such as Quark, Adobe PageMaker, and Adobe InDesign. If you send the poster to a professional printer, find out in advance what platforms the printer supports so you don't design a spread in a program that is not compatible with your printing choice.

If you need to work with hard copy or are sending the graphic to a publisher or having someone else assemble it, make sure you label the parts of the bigger project so that it can be composed correctly. You'd want to label the separate figures such as "10.1A, 10.1B, 10.1C," etc. You can also provide a sketch that shows where the various components go.

We've discussed the importance of imagery in illustration and what Leonardo da Vinci, as one of the first Western medical illustrators, contributed. You know that as health professionals many of us are less apt to make the illustrations ourselves, although we need to know how to use them in our compositions. You've learned about many of the specific composition principles and design features that are important in working with illustrations. We are more apt, however, to be called upon to design graphs, charts, and tables to illustrate findings of various research studies, compare demographic groups, evaluate the seriousness and range of epidemics, and communicate many other health and medical issues.

What Visuals Are Most Important?

Graphs, charts, and tables are often used in medical writing, and they are often the least described and taught. This area is Tufte's strong point, and, as pointed out in Chapter 3, he was the originator of the term *chartjunk*. To review briefly, this term refers to the overloading of graphs, charts, and other visual representation with too much unnecessary data or too many distractions (Tufte 1997, 65). Here are some of the aspects to keep in mind as you design charts, graphs, and tables to illustrate or represent your data.

- Sequence the data logically, e.g., from youngest to oldest patient, from mildest to most severe, males grouped together and females grouped together if gender is an important variable in the study, etc.
- If previous issues or reports have double-column pages and your table is a single-column one, your table width should not be greater than 60 characters (counting spaces, too).
- If the table needs to be a double-column width, then the character count should not be greater than 120 (including spaces and row headings).
- If the table needs to be even wider for your data (several columns wide), consider asking the publisher or institution if you can put tables on facing pages, although this format can be less reader-friendly. If this is the best choice, do repeat the row headings so that readers don't have to relocate them across from the facing page on the left.
- Also, if you have a large table with case studies, you might want to divide it up into two tables by similarity. For example, divide it up into test findings and clinical features or other logical divisions.

Charts, graphs, and tables are subject to the same kind of preparatory analysis as any text or image: who is the audience and what is the purpose? Most reports and research articles prepared for your peers and/or other health professionals will require tables for the data. Be sure to be consistent throughout your report or article from one graph to another or one table to another. "Use consistency in symbols and abbreviations and in the style of citing references in all tables, graphics, and text for a single manuscript. When abbreviations are used, define or explain them (in parentheses or footnotes, respectively)" (Evans 1994, 39).

How Do I Construct Tables?

You need to know what the parts of a table are called for outsourcing a document to graphic designers; for teaching, transcribing, or dictating data to be set up in tables; and for communicating with publishers, reviewers, and members of foundations or granting agency staffs. These are the parts of tables:

- The *title* contains the number and briefly labels content.
- The *field* is all the space or area for the numerical data, descriptive terms, or brief phrases.
- The content of the field is arranged in
  Horizontal *rows*
  - Each *row heading* identifies the kind of data and descriptions aligned in the horizontal row to its right ("stub" is a term for the group of row headings).
  Vertical *columns*
  - Each *column heading* identifies the type of data and descriptions vertically underneath it.
  - The column headings for the field are also known collectively as the *box heading.*
  - The column heading for the row headings beneath it is called the *box heading for the stub* (see row above).
- The *footnotes* explain details of the content of the table (see Exhibit 10.1 for an example of a descriptive table).

Tables are usually one of two types: either descriptive or numerical. One of the complaints that the users of the *Guidelines for Design and Construction of Hospital and Health Care Facilities* have made for the revision process in Exhibit 10.1, which features a descriptive table, is that there are too many footnotes in this particular table. There are 20 footnotes for the one-and-three-fifths-page document.

> Avoid having too many exceptions, i.e., too many footnotes, long explanations, and descriptions. You will lose your reader and the intent of the table. Remember that the original type size of tables and other graphics will probably be reduced in the printed publication, making such information difficult to read unless it is clear and well spaced. (Evans 1994, 39)

Unlike the descriptive table in Exhibit 10.1, when you prepare a numerical table for a report of an epidemiological study, laboratory or clinical research, or drug trial, you generate and then analyze quantitative (numerical) data. If you

Table 7.2
Ventilation Requirements for Areas Affecting Patient Care in Hospitals and Outpatient Facilities[1]

Area designation	Air movement relationship to adjacent area[2]	Minimum air changes of outdoor air per hour[3]	Minimum total air changes per hour[4,5]	All air exhausted directly to outdoors[6]	Recirculated by means of room units[7]	Relative humidity[8] (%)	Design temperature[9] (degrees F/C)
SURGERY AND CRITICAL CARE							
Operating/surgical cystoscopic rooms[10,11]	Out	3	15	--	No	30-60	68-73 (20-23)[12]
Delivery Room[10]	Out	3	15	--	No	30-60	68-73 (20-23)
Recovery Room[10]	--	2	6	--	No	30-60	70-75 (21-24)
Critical and intensive care	--	2	6	--	No	30-60	70-75 (21-24)
Newborn intensive care	--	2	6	--	No	30-60	72-78 (22-26)
Treatment room[13]	--	--	6	--	--	--	75 (24)
Trauma room[13]	Out	3	15	--	No	30-60	70-75 (21-24)
Anesthesia gas storage	In	--	8	Yes	--	--	--
Endoscopy	In	2	6	--	No	30-60	68-73 (20-23)
Bronchoscopy[11]	In	2	12	Yes	No	30-60	68-73 (20-23)
ER waiting rooms	In	2	12	Yes[14, 15]	--	--	70-75 (21-24)
Triage	In	2	12	Yes[14]	--	--	70-75 (21-24)
Radiology waiting rooms	In	2	12	Yes[14, 15]	--	--	70-75 (21-24)
Procedure room	Out	3	15	--	No	30-60	70-75 (21-24)
NURSING							
Patient room	--	2	6[16]	--	--	--	70-75 (21-24)
Toilet room	In	--	10	Yes	--	--	--
Newborn nursery suite	--	2	6	--	No	30-60	72-78 (22-26)
Protective enviornment room[11,17]	Out	2	12	--	No	--	75 (24)
Airborne infection isolation room[11, 18]	In	2	12	Yes[15]	No	--	75 (24)
Isolation alcove or anteroom[17, 18]	In/Out	--	10	Yes	No	--	--
Labor/delivery/recovery	--	2	6[16]	--	--	--	70-75 (21-24)
Labor/delivery/recovery/postpartum	--	2	6[16]	--	--	--	70-75 (21-24)
Patient corridor	--	--	2	--	--	--	--

(continued)

Exhibit 10.1 Facilities Guidelines Table 7.2

Table 7.2 (continued)
Ventilation Requirements for Areas Affecting Patient Care in Hospitals and Outpatient Facilities[1]

Area designation	Air movement relationship to adjacent area[2]	Minimum air changes of outdoor air per hour[3]	Minimum total air changes per hour[4,5]	All air exhausted directly to outdoors[6]	Recirculated by means of room units[7]	Relative humidity[8] (%)	Design temperature[9] (degrees F/C)
ANCILLARY							
Radiology[19]							
X-ray (surgical/critical care and catheterization)	Out	3	15	--	No	30-60	70-75 (21-24)
X-ray (diagnostic & treatment)	--	--	6	--	--	--	75 (24)
Darkroom	In	--	10	Yes	No	--	--
Laboratory							
General[19]	--	--	6	--	--	--	75 (24)
Biochemistry[19]	Out	--	6	--	No	--	75 (24)
Cytology	In	--	6	Yes	No	--	75 (24)
Glass washing	In	--	10	Yes	--	--	--
Histology	In	--	6	Yes	No	--	75 (24)
Microbiology[19]	In	--	6	Yes	No	--	75 (24)
Nuclear medicine	In	--	6	Yes	No	--	75 (24)
Pathology	In	--	6	Yes	No	--	75 (24)
Serology	Out	--	6	--	No	--	75 (24)
Sterilizing	In	--	12	Yes	No	--	--
Autopsy room[11]	In	--	10	Yes	--	--	--
Nonrefrigerated body-holding room	In	--	10	Yes	--	--	70 (21)
Pharmacy	Out	--	4	--	--	--	--
DIAGNOSTIC AND TREATMENT							
Examination room	--	--	6	--	--	--	75 (24)
Medication room	Out	--	4	--	--	--	--
Treatment room	--	--	6	--	--	--	75 (24)
Physical therapy and hydrotherapy	--	--	6	--	--	--	75 (24)
Soiled workroom or soiled holding	In	--	10	Yes	No	--	--
Clean workroom or clean holding	Out	--	4	--	--	--	--

Exhibit 10.1 (*continued*)

Table 7.2 (continued)
Ventilation Requirements for Areas Affecting Patient Care in Hospitals and Outpatient Facilities[1]

Area designation	Air movement relationship to adjacent area[2]	Minimum air changes of outdoor air per hour[3]	Minimum total air changes per hour[4,5]	All air exhausted directly to outdoors[6]	Recirculated by means of room units[7]	Relative humidity[8] (%)	Design temperature[9] (degrees F/C)
STERILIZING AND SUPPLY							
ETO–sterilizer room	In	--	10	Yes	No	30-60	75 (24)
Sterilizer equipment room	In	--	10	Yes	--	--	--
Central medical and surgical supply							
Soiled or decontamination room	In	--	6	Yes	No	--	68-73 (20-23)
Clean workroom	Out	--	4	--	No	30-60	75 (24)
Sterile storage	Out	--	4	--	--	(Max) 70	--
SERVICE							
Food preparation center[20]	--	--	10	--	No	--	--
Warewashing	In	--	10	Yes	No	--	--
Dietary day storage	In	--	2	--	--	--	--
Laundry, general	--	--	10	Yes	--	--	--
Soiled linen (sorting and storage)	In	--	10	Yes	No	--	--
Clean linen storage	Out	--	2	--	--	--	--
Soiled linen and trash chute room	In	--	10	Yes	No	--	--
Bedpan room	In	--	10	Yes	--	--	--
Bathroom	In	--	10	--	--	--	75 (24)
Janitor's closet	In	--	10	Yes	No	--	--

[1]The ventilation rates in this table cover ventilation for comfort, as well as for asepsis and odor control in areas of acute care hospitals that directly affect patient care and are determined based on healthcare facilities being predominantly "No Smoking" facilities. Where smoking may be allowed, ventilation rates will need adjustment. Areas where specific ventilation rates are not given in the table shall be ventilated in accordance with ASHRAE Standard 62, Ventilation for Acceptable Indoor Air Quality, and ASHRAE Handbook —

HVAC Applications. Specialized patient care areas, including organ transplant units, burn units, specialty procedure rooms, etc., shall have additional ventilation provisions for air quality control as may be appropriate. OSHA standards and/or NIOSH criteria require special ventilation requirements for employee health and safety within healthcare facilities.

[2]Design of the ventilation system shall provide air movement which is
(continued)

Exhibit 10.1 (continued)

generally from clean to less clean areas. If any form of variable air volume or load shedding system is used for energy conservation, it must not compromise the corridor-to-room pressure balancing relationships or the minimum air changes required by the table.

[3]To satisfy exhaust needs, replacement air from the outside is necessary. Table 7.2 does not attempt to describe specific amounts of outside air to be supplied to individual spaces except for certain areas such as those listed. Distribution of the outside air, added to the system to balance required exhaust, shall be as required by good engineering practice. Minimum outside air quantities shall remain constant while the system is in operation.

[4]Number of air changes may be reduced when the room is unoccupied if provisions are made to ensure that the number of air changes indicated is re-established any time the space is being utilized. Adjustments shall include provisions so that the direction of air movement shall remain the same when the number of air changes is reduced. Areas not indicated as having continuous directional control may have ventilation systems shut down when space is unoccupied and ventilation is not otherwise needed, if the maximum infiltration or exfiltration permitted in Note 2 is not exceeded and if adjacent pressure balancing relationships are not compromised. Air quantity calculations must account for filter loading such that the indicated air change rates are provided up until the time of filter change-out.

[5]Air change requirements indicated are minimum values. Higher values should be used when required to maintain indicated room conditions (temperature and humidity), based on the cooling load of the space (lights, equipment, people, exterior walls and windows, etc.).

[6]Air from areas with contamination and/or odor problems shall be exhausted to the outside and not recirculated to other areas. Note that individual circumstances may require special consideration for air exhaust to the outside, e.g., in intensive care units in which patients with pulmonary infection are treated, and rooms for burn patients.

[7]Recirculating room HVAC units refers to those local units that are used primarily for heating and cooling of air, and not disinfection of air. Because of cleaning difficulty and potential for buildup of contamination, recirculating room units shall not be used in areas marked "No." However, for airborne infection control, air may be recirculated within individual isolation rooms if HEPA filters are used. Isolation and intensive care unit rooms may be ventilated by reheat induction units in which only the primary air supplied from a central system passes through the reheat unit. Gravity-type heating or cooling units such as radiators or convectors shall not be used in operating rooms and other special care areas. See Appendix A for a description of recirculation units to be used in isolation rooms.

[8]The ranges listed are the minimum and maximum limits where control is specifically needed. The maximum and minimum limits are not intended to be independent of a space's associated temperature. The humidity is expected to be at the higher end of the range when the temperature is also at the higher end, and vice versa.

[9]Where temperature ranges are indicated, the systems shall be capable of maintaining the rooms at any point within the range during normal operation. A single figure indicates a heating or cooling capacity of at least the indicated temperature. This is usually applicable when patients may be undressed and require a warmer environment. Nothing in these guidelines shall be construed as precluding the use of temperatures lower than those noted when the patients' comfort and medical conditions make lower temperatures desirable. Unoccupied areas such as storage rooms shall have temperatures appropriate for the function intended.

[10]National Institute for Occupational Safety and Health (NIOSH) Criteria Documents regarding Occupational Exposure to Waste Anesthetic Gases and Vapors, and Control of Occupational Exposure to Nitrous Oxide indicate a need for both local exhaust (scavenging) systems and general ventilation of the areas in which the respective gases are utilized.

[11]Differential pressure shall be a minimum of 0.01" water gauge (2.5 Pa). If alarms are installed, allowances shall be made to prevent nuisance alarms of monitoring devices.

Exhibit 10.1 (*continued*)

[12]Some surgeons may require room temperatures that are outside of the indicated range. All operating room design conditions shall be developed in consultation with surgeons, anesthesiologists, and nursing staff.

[13]The term trauma room as used here is the operating room space in the emergency department or other trauma reception area that is used for emergency surgery. The first aid room and/or "emergency room" used for initial treatment of accident victims may be ventilated as noted for the "treatment room." Treatment rooms used for Bronchoscopy shall be treated as Bronchoscopy rooms. Treatment rooms used for cryosurgery procedures with nitrous oxide shall contain provisions for exhausting waste gases.

[14]In a ventilation system that recirculates air, HEPA filters can be used in lieu of exhausting the air from these spaces to the outside. In this application, the return air shall be passed through the HEPA filters before it is introduced into any other spaces.

[15]If it is not practical to exhaust the air from the airborne infection isolation room to the outside, the air may be returned through HEPA filters to the air-handling system exclusively serving the isolation room.

[16]Total air changes per room for patient rooms, labor/delivery/recovery rooms, and labor/delivery/recovery/postpartum rooms may be reduced to 4 when supplemental heating and/or cooling systems (radiant heating and cooling, baseboard heating, etc.) are used.

[17]The protective environment airflow design specifications protect the patient from common environmental airborne infectious microbes (i.e., Aspergillus spores). These special ventilation areas shall be designed to provide directed airflow from the cleanest patient care area to less clean areas. These rooms shall be protected with HEPA filters at 99.97 percent efficiency for a 0.3 µm sized particle in the supply airstream. These interrupting filters protect patient

rooms from maintenance-derived release of environmental microbes from the ventilation system components. Recirculation HEPA filters can be used to increase the equivalent room air exchanges. Constant volume airflow is required for consistent ventilation for the protected environment. If the facility determines that airborne infection isolation is necessary for protective environment patients, an anteroom should be provided. Rooms with reversible airflow provisions for the purpose of switching between protective environment and airborne infection isolation functions are not acceptable.

[18]The infectious disease isolation room described in these guidelines is to be used for isolating the airborne spread of infectious diseases, such as measles, varicella, or tuberculosis. The design of airborne infection isolation (AII) rooms should include the provision for normal patient care during periods not requiring isolation precautions. Supplemental recirculating devices may be used in the patient room, to increase the equivalent room air exchanges; however, such recirculating devices do not provide the outside air requirements. Air may be recirculated within individual isolation rooms if HEPA filters are used. Rooms with reversible airflow provisions for the purpose of switching between protective environment and AII functions are not acceptable.

[19]When required, appropriate hoods and exhaust devices for the removal of noxious gases or chemical vapors shall be provided (see Sections 7.31.D14 and 7.31.D15 and NFPA 99).

[20]Food preparation centers shall have ventilation systems whose air supply mechanisms are interfaced appropriately with exhaust hood controls or relief vents so that exfiltration or infiltration to or from exit corridors does not compromise the exit corridor restrictions of NFPA 90A, the pressure requirements of NFPA 96, or the maximum defined in the table. The number of air changes may be reduced or varied to any extent required for odor control when the space is not in use. See Section 7.31.D1.p.

Exhibit 10.1 *(continued)*

function as the writer/designer, then the researcher, the medical doctor, or the epidemiologist might have already crafted preliminary tables for the data and given them to you to polish for a final copy. If you or your health professional is writing a review of someone else's work, you may have crafted tables to "help you pull together concise summaries of what you had read" (Huth 1999, 139).

Some researchers, chemists, and physicians work from tables. But as a designer and polisher, you have to decide which tables are necessary and which data would be better illustrated in graphs. Some tables can be summarized and omitted once the conclusions have emerged. This decision making in regard to the graphics, then, becomes part of the writing process. These decisions also depend on which draft you are on or which draft the researcher has given to you. Is it a preliminary, first, second, or final draft? If it's a first draft, you will need to indicate where tables go and what they will contain. If it's a final draft, the tables should be in their final form. It's always better to include more tables than you need and eliminate some as you revise the report or study to its final form. Deletion of unnecessary tables is always easier than creating new ones for final copy, although if you have a copy editor, that person may request more information to make your report clearer.

How Do I Decide Whether to Use Tables?

Some pharmaceutical companies, government agencies, hospital systems, and academic settings may have a specific form for you to use to report data. Sometimes even grants or institutional review boards have specific requirements. Others do not. If a table is not required, and if you are not sure whether your data is best served by a table, you should determine if a table meets one of the following five criteria:

- Presents precise numeric values rather than just proportions or trends
- Presents large numbers of related data
- Summarizes information clearer in a tabular form than in running text
- Presents more data than can be summarized in a few sentences[8]
- Presents complex information more clearly than in running text or a figure (Huth 1999, 139).

One of the main keys here, when deciding whether or not to use tables, is the quantity of data you have to showcase (the second phrase in the bulleted list above). If you decide you only need to show percentages and research results in a general way (on the macro or more global level in which you lump numbers into larger categories), then you would choose a graph instead of a table.[9] When the detailed results need to be shown and you have a large spreadsheet of data, then it is more appropriate to use a table. *Don't* use a table in the following situations:

- If the data can be summarized in prose form in a few sentences
- If the relations of data to each other or over time can be shown more effectively in a graph (Huth 1999, 142).

A good use of tables is for reports of several or numerous cases. Remember the information on case studies in Chapter 4? The data gathered in those studies can often best be reported in tables that show the commonalities of patients' cases rather than by describing one case after another in detail and writing the same information over and over. Of course, each case will most likely have some clinical variations; the numerical data such as age, temperature, weight, blood pressure, and laboratory results can be embedded in a table. You can also create a descriptive table (more complex than the one in Exhibit 10.2) for symptoms, common findings (such as a history of cigarette smoking, for example), diagnostic testing results, etc. You should include at least one full case report to give the syndrome or disease a context. This helps your reader relate the table to the case studies. Also, you can describe the various uncommon or differing findings in prose passages.

If you plan a teaching text or lecture and want to use tables, "you can emphasize important points by listing in small tables the main features of a disease or syndrome, symptoms and signs of adverse effects, and differential diagnoses . . . . [I]nclude the frequency or percentage of occurrence for each item; these additional data help to make clear the relative importance of the listed items" (Huth 1999, 142–43). In Appendix B we talk about grouping by proximity and similarity. If you group the data by chronology, body order, physiological systems, symptoms, or some other very clear grouping, the reader will quickly grasp the relevant information. Also, in addition to addressing the viewers' schema in that way, you also need to sequence the data in a logical order:

- By descending order of frequency
- By body systems
- In chronological order
- Or through some other clear basis.

In addition to grouping and organizing clearly within tables, the tables themselves should be correctly numbered and referred to within the text. You want to make sure your reader can easily find and grasp the tables themselves. One way to ensure that you have ordered the tables themselves in a logical order in relation to the other tables is to read from table to table in the report or presentation (ignore the text in between for now) and make sure that sequencing makes sense. Pay careful attention to how you title the tables, too. "In many clinical papers the title of the first table may adequately identify the main subject of the paper, with shorter titles for the following tables" (Huth 1999, 143). The first table, for example, in a review of 16 cases of gunshot wounds of the heart, might be titled "Gunshot Wounds of the Heart: Clinical Features." The next might then be titled "Operative Findings and Postoperative Course." The titles of the tables should make logical sense in their sequence.

If you decide to use tables, one of the first tasks you want to complete is to find out if the journal, grant, report, or government document has a set number of tables it allows for the length of the text (Huth 1999, 140). Check the submission guidelines first to see if they state the maximum number allowed. The usual rule of thumb is one table per 1000 words of text (or about four double-spaced manuscript pages). You could also estimate the ratio in a recent or model document by counting the tables and number of pages. The same limits may apply to illustrations, charts, graphs, and the combined total of the four different genres. It's simple enough to employ the formula so that you have a general idea of how many visuals are appropriate:

1. Estimate the number of words in a sample article or report (exclude footnotes, endnotes, works cited, and bibliography)
2. Count the visuals in the document
3. Calculate the number of tables and illustrations per thousand words
4. Do a word count in your word-processing program on your document (exclude footnotes, endnotes, works cited, and bibliography)
5. Round off your word count to the nearest thousand
6. Multiply the calculated limit of numbers of tables and illustrations per thousand by the estimate of your text length in thousands
7. This is your maximum number of visuals.[10] (Huth 1999, 140)

Remember that tables are more expensive for publishers to print than regular text. Therefore, they are more apt to tell you to cut tables and illustrations than they are to ask you to add them.

If you have given an oral presentation on your data before preparing your report or if you plan to present your findings orally to a foundation, grantor, group of researchers, or other audience, you don't want to use the same data-heavy tables you prepared for print. Don't, for example, put detailed tables into your PowerPoint presentation. Small text is difficult to read on-screen, and audience members will not be able to comprehend the data if you read it aloud to them either. If you need the tables to refer to and if you can't more succinctly summarize the data on your PowerPoint or while speaking, then prepare your tables on printed handouts for the audience members. Take one example from the table to follow through with the audience so they can grasp the general way to read the table (unless you have an audience of peers or experts on the same level you are on).

However, the opposite advice applies if you are using a very simple table. Then do put it in a PowerPoint slide and don't use it in a print version that will be published. It's certainly permissible to use it in an informal report or document for lay audiences. An example of a simple table is Exhibit 10.2.

In print you would simply say that two patients out of twelve had nasal irritation as a result of the flu vaccine in spray form. The table is too simple to put into print, and an editor would be sure to ask you to delete it.

A colleague once prepared a video in which he also used PowerPoint images of very data-heavy tables. The audience members (corporate board members of Microsoft, IBM, Bank of America, Kearney Consultants, and others) were outraged with the poor quality and design of the video and roundly chastised the professional communicator for his poor attention to visual communication and multimedia design. More discussion about designing your visuals in concert with oral presentations will follow toward the end of this chapter. Tables showing simple results such as Exhibit 10.2 are good when you only have small amounts of data to display in an oral presentation.

Patient Code Identification #	Nasal Irritation after Nasal Spray or Flu Vaccine	No Nasal Irritation after Spray	Other Minor Symptoms
Patient 1	Yes—mild		
Patient 2		None	
Patient 3		None	Sneezing
Patient 4		None	
Patient 5	Yes—severe		Slight Headache
Patient 6		None	
Patient 7		None	
Patient 8		None	Complained of "Dry Nasal Passages"
Patient 9		None	
Patient 10		None	
Patient 11		None	
Patient 12		None	

Exhibit 10.2 Simple Table Showing Nasal Irritation in Patients in a Clinical Trial for Nasal Spray Flu Vaccine[11]

How Do I Format Tables?

In published print medical materials, you frequently don't see many lines or grids in a table, although you can outline your table in a box or with borders. You also should include a line under the column headings and sometimes a vertical line immediately after the row headings. Sometimes publishers and institutions don't specify how tables should be formatted. If that is the case, you can call them and request guidelines or check previous issues of journals (if your tables are for a journal publication) or reports generated for the same institution.

Even titles are important. The title should let the reader know what to expect without going into too much detail.

Consider a table summarizing in three groups of columns the admission diagnosis and roentgenographic and autopsy findings in 20 fatal cases of pulmonary embolism. Its column headings identify the nature of the findings as "Admission Diagnosis," "Chest Film," and "Autopsy," and its row headings run down from "Patient 1" to "Patient 20." The table title can be simply "Cases of Fatal Pulmonary Embolism" rather than "Twenty Cases of Fatal Pulmonary Embolism: Admission Diagnosis, Chest Film, and Autopsy Findings." (Huth 1999, 144)

Just as in tables, much of the formatting and logical structure of medical graphs and charts share the same general features of medical tables.[12] The next section will cover how charts and graphs are different from tables.

What Kind of Information Is Best for Graphs and Charts?

Graphs and charts differ mainly by the kinds of information you need to present (content). If you find yourself with information that emphasizes the importance of known or potential relationships between values rather than the importance of the values themselves, those relationships are best illustrated in graphs or charts of patients' clinical courses, epidemiological maps, such as John Snow's mentioned above, and other graphic genres.

When visually comparing trends or relationships in sets of data, a graph/graphic is more appropriate than a table. Use a graph/graphic when trends or relationships are more important than exact values, when the meaning of the data must be expressed quickly, and when hidden relationships need revealing. For example, an S-shaped curve is more apparent from a graph/graphic than from a table. (Evans 1994, 38)

Such data would include but is not limited to:

- Data on two related variables: a dependent variable whose values are determined by an independent variable, such as maximum systolic blood pressure after different doses of epinephrine, or maximum blood levels of alcohol after different doses of whiskey.
- Data on one or more variables changing through time, such as clinical data like temperature, blood pressure, leukocyte counts for a patient during a hospital stay.
- Data important to the reader for the extent of their differences and how these differences might be related to unknown factors, such as differences in mortality rates for stomach cancer in the individual states of the United States. (Huth 1999, 142)

If you think of charts in general, they're patterned and are used as measures for charting patient information onto a grid of normals, averages, or other comparisons. For example, an eye chart is a patterned poster that simply tests eyesight at different distances. Pediatric charts measure infants' and children's growth patterns (height and weight) on a grid that compares them with other infants and children to reveal into which percentile their weight and height fall. Charts can also be line graphs, so there's no hard and fast rules or definitions regarding these various genres.

How Do I Format Graphs and Charts?

Usually a line graph is drawn (or imported from a computer program) with the independent variable on the horizontal (x) axis and the dependent variable on the vertical (y) axis. Numerous types of graphs are possible in addition to the line graph: two-dimensional or three-dimensional columns, bars, pie shapes, area, x/y scatter, doughnut, radar, surface, bubble, stock, cone, cylinder, and pyramid. You can also use various icons to illustrate your graphs. For example, if you were comparing numbers of people who are chronically ill in various countries, you could have each person stand for 1000 or more people (depending on how many people there are). If you are preparing to outsource your graphs and charts and if you have a particular print, Web, or PowerPoint format in mind, let the designer know. If as an in-hospital publications manager you have a variety of ways to display graphs and charts, you might want to show the author several formats. An alternative process is to format the data in several different ways and ask the client which one he or she prefers. Microsoft Excel lets you try out data in numerous formats with their chart and graph wizard.

Consider the following design suggestions when you are working on your graphs:

- Avoid cross-hatching, striping, or dots in the bars of your graphs or pie wedges, which are difficult to discern once they are printed in black and white and in a smaller format.
- On a line graph, make sure the lines or curves that show your results are darker and thicker than the x- and y-axis lines and grid lines.
- With three-dimensional graphs and graphics, make sure any patterns that you use to differentiate types of data (like checkering in a three-dimensional pie wedge) does not misappropriately represent the data (make that wedge look disproportionately larger than others of the same size).
- Make sure that the software you use to create graphs produces good quality resolution for the media you choose.
- Remember if you plan to choose graphs and charts for a print interface, use the CYMK color designations to save your graphs and RGB for the Web or other electronic mediums.
- Make sure you have chosen colors that don't clash or are not difficult to read. Develop a color palette that is appropriate to your subject (see Chapter 3 and Appendix B of this text).
- Print out or launch your draft or beta of the graphs you will use to make sure they are of good quality before you send them off to the printer or Web master.

Whether it's best to use a graph or chart also depends on the medium you choose. If you are preparing a text for publication and if your reader can more easily understand the information in prose, don't use a graph or chart. In other words, don't clutter up a text with unnecessary graphs, just the necessary ones. However, when you are giving an oral presentation, it's often desirable to show those same data with some graphs and charts on easels, in PowerPoint, or through other electronic media. In oral presentations, the audience can more quickly grasp your point visually. In prose, the reader can go back and reread the passage, whereas audience members within range of your voice cannot.

> A simple bar graph comparing deaths from lung cancer in men and women between ages 50 and 70 might effectively emphasize a big difference in mortality, but the same point could be stated just as efficiently, or more efficiently, in the text. Unless the difference in mortality is the main conclusion of the paper and the editor agrees that this point merits visual emphasis, you will probably be asked to drop the bar graph. (Huth 1999, 151–52)

What About Flowcharts and Molecular Graphics?

Flowcharts are often used to show organizational structures or processes (sequence of procedures). Some software can give you a flowchart template, but even in the earlier versions of Microsoft Word you could assemble the necessary parts to make a flowchart. To make a flowchart, embed shapes and arrows to indicate a process or organizational infrastructure using the *Drawing* and *Forms* toolbars. You will find these toolbars under *View*. Make sure you understand the steps of the process or procedure before you begin your flowchart.

Molecular graphics have become more common in graphic presentations because of the tremendous growth of genetics and molecular biology. Because these graphics are highly specific, we will not cover them here. However, if you are called upon to prepare or if you want to generate some molecular graphics, the most helpful sources for good guidance are:

- *Cell* (journal)
- *Genetics* (journal)
- *Nature* (journal)
- H Briscoe's *A Researcher's Guide to Scientific and Medical Illustrations*

As you design your graphs, charts, or graphics, keep in mind the following design considerations:

- Whenever possible, construct tables and graphs so that they can be read in the same direction as the text, i.e., in a "portrait" direction. Having to turn the manuscript to read a table or graph in the "landscape" position disrupts the reader's thought. Also, if space allows, the direction of text lettering around the table or graph should be in the same direction as the data.

- Notice the "shape" of graphics. If the nature of the data suggests the shape of the graphic, follow that suggestion; otherwise, design horizontal graphics about 50 percent wider than tall.
- Be conscious of proportion and scale, rule (line) weight, and lettering. Graphic elements look better together when their relative proportions are in balance and they have an integrated quality.
- "Play" with several table and graph/graphic design ideas before deciding on one. Select the briefest, clearest, and most parallel design.
- Both tables and graphs/graphics should tend toward the horizontal, greater in width than height. The eye naturally detects deviations from the horizon, and good graphic design uses that fact to advantage.
- Use white space around the type to allow the data to "breathe," but minimize blank areas.
- Select both upper and lower case type. (Evans 1994, 40)

"Ophthalmology has disclosed that the more letters are differentiated from each other, the easier reading is. Without going into comparisons and details, it should be realized that words consisting of only capital letters present the most difficult reading because of their equal height, equal volume, and, with most, their equal width" (Albers 1975).

Summary

In this chapter, we have closely examined the visual presentations of writing for the health professions. The early history of pictographic writing was done in areas of the world other than in our traditional Western countries. You learned briefly about the Italian Renaissance, a time of flourishing medical discovery in Western Europe, and about the innovations in medical illustration spearheaded by Leonardo da Vinci. With the help of John Snow's epidemiological map to pinpoint the source of a cholera epidemic in the 1850s in London, and Mark Monmonier's aggregate maps, you discovered the key role tables, charts, graphs, and maps can play in medicine. You were reminded, too, of gestalt theory and its importance in visual communication. (It is hoped you had already read or reviewed Chapter 3 of this text before you began this chapter.)

We then moved on to the practical designs of and applications for illustrations, tables, graphs, and charts for use in today's visual medical communication. You discovered there are many good principles to keep in mind as you either prepare illustrations for reports, articles, journals, or other media yourself or outsource this work to designers. While using drawing software makes illustrating easier than it used to be, illustrations themselves are still somewhat difficult to do well unless you have had training in art or medical drawing. You do need to be able to recognize which medical illustrations communicate well visually and which do not. The same is true of any graphic or visual art that you use in designing work for any audience (whether for health educational purposes or for peer and expert audiences).

Tables, graphs, and charts share similarities in their design and application needs, but the content can vary for each. In addition, how you design one of these forms depends on whether it will be featured in print or in an oral presentation. You learned that these graphic genres are subject to the same analysis of purpose and audience that our written communication is subject to. The various suggestions were divided between those regarding content appropriate for tables, graphs, and charts and formatting them.

You can use this chapter as a checklist as you begin to prepare illustrations, tables, graphs, and charts. Many of the same principles apply in working with these genres electronically. In the next chapter, we will look specifically at the electronic applications that you might be working with as a writer for the health professions.

Discussion Questions

1. In addition to the Challenger disaster, what other instances do you know of where technical or medical miscommunication caused great harm?
2. How are ethics, visual representation, and verbal communication intertwined? What kinds of ethical dilemmas can you think of that are created through improper use of graphs or charts?
3. Comparing Leonardo da Vinci's drawings (consult the Internet), modern medical illustrations, and three-dimensional computer imaging of the body, how do you think our technical advances in drawing the human body will change our medical practices? Is this development as profound as da Vinci's contributions? Why or why not?
4. What kinds of steps and processes could be put into place in an organization such as NASA to ensure that safety issues are communicated more efficiently and more ethically? What would you do if you had to present a drawing to convince your superiors that there was a flaw that could jeopardize the safety of the shuttle crew?
5. Do you think we have a more successful rate of understanding and mapping epidemics today? What recent examples can you think of that were plotted in a causal method? Do you think being connected worldwide via technology, including air travel, is an advantage or a drawback? Why?
6. Take the composition analogy (between writing and art) further. Is there a grammar of images? What kinds of schemas exist in the audience for both processes? Are there subtexts in each? Discuss as many comparisons, similarities, and differences that you can think of.
7. Your instructor will bring in medical data sets (or ask you to bring in some). Discuss which genre should be used to display the data (illustrations, charts, tables, graphs, or other types of visuals) and why.
8. Discuss the details for preparing illustrations, charts, tables, and graphs and why these principles and suggestions are useful or not useful in specific projects that you have done in the past, are working on now, or will be planning and implementing in the future.

Exercises

1. Using the Internet, find examples of ethical problems created through unreliable or politically motivated visuals or verbal misstatements. You might want to research falsifying results from experiments with new pharmaceuticals, not reporting side effects of drugs, etc.
2. Find modern medical illustrations and Leonardo da Vinci's medical illustrations and compare the two in a PowerPoint presentation that you show to the class. Most of the PowerPoint should feature the differences and similarities between medical illustration then and medical illustration now.
3. Add to the PowerPoint presentation you created in Exercise 2. Add today's three-dimensional, digital drawing capability that we now have. How are these similar to and different from print-based anatomical drawings and medical illustrations?
4. Search the Internet for some aggregate maps of medical and health data. (Hint: Try the CDC.) Also, find some medical data that you believe would be appropriate for dot maps, disease clusters, and aggregate maps. Using the statistics and MSWord Excel or a drawing program, see if you can map out some of the data on the various types of maps.
5. Using the Internet, print-based materials, or both, find a work of art and images for health education materials that address the same or a similar topic. (For example, choose a painting of the Christian artwork about the visit between Elizabeth and Mary, featuring a very obviously pregnant Elizabeth, and compare that with images pertaining to prenatal healthcare.) Construct a detailed analysis of how these materials are different and how they are alike. What are their messages? What are they trying to educate or persuade the viewers about?
6. Find examples of medical illustrations and photographs, both those in print and those on-line (two of each). Do an analysis in which you show either why the data on each are better served by a different genre or why the data are best served in the present form. Ideally, find examples of both well-designed and poorly designed visuals. Use the criteria developed above for your analysis.
7. Find examples of medical tables, charts, and graphs, both those in print and those on-line (two of each). Do an analysis in which you show either why the data on each are better served by a different genre or why the data are best served in the present form. Ideally, find examples of both well-designed and poorly designed visuals. Use the criteria developed above for your analysis.

Works Cited

Albers, Josef. 1975. *Interaction of Color.* New Haven: Yale University Press.

Archives of Dermatology. Vol. 139, no. 12 (December 2003). © AMA (American Medical Association). Accessed December 16, 2003. http://archderm.ama-assn.org/

Berger, John. 1972. *Ways of Seeing.* London: British Broadcasting Corporation and Penguin.

Brewer, Douglas J., and Emily Teeter. 1999. *Egypt and the Egyptians.* Cambridge: Cambridge University Press.

Briscoe, M. H. 1990. *A Researcher's Guide to Scientific and Medical Illustrations.* New York: Springer-Verlag.

————. 1995. *Scientific Illustrations: A Guide to Better Posters, Presentations, and Publications,* 2nd ed. New York: Springer-Verlag.

The Cholera Inquiry Committee. 1855. *Report on the Cholera Outbreak in the Parish of St. James's, Westminster, During the Autumn of 1854.* Presented to the Vestry. London.

Council of Biology Editors. 1983. *CBE Style Manual,* 5th edition. Bethesda, MD.

Evans, Dale. 1994. "Tables and Graphs." *Biomedical Communication: Selected AMWA Workshops—A Practical Guide for Writers, Editors, and Presenters of Health Science Information.* Phyllis Minick, ed. Bethesda, MD: American Medical Writers Association.

Facilities Guidelines Institute. 2001. *Guidelines for Design and Construction of Hospital and Health Care Facilities.* Dallas: FGI.

Heifferon, Barbara A. 2000. "Analyzing the Rhetorics of Ancient Egypt." Unpublished manuscript.

Herrlinger, R. 1970. *History of Medical Illustration from Antiquity to 1600.* NY: Editions Medicina Rara.

Huth, Edward J., M.D. 1999. *Writing and Publishing in Medicine,* 3rd ed. Baltimore: Lippincott, Williams & Wilkins.

Kostelnick, Charles, and David D. Roberts. 1998. *Designing Visual Language: Strategies for Professional Communicators.* Boston: Allyn and Bacon.

Monmonier, Mark. 1991. *How to Lie with Maps.* Chicago: University of Chicago Press.

O'Malley, C., and C. M. Saunders, trans. and eds. 1952. *Leonardo da Vinci on the Human Body.* NY: Schuman.

Perls, Fritz. 1973. *The Gestalt Approach and Eye Witness to Therapy.* Ben Lomond, CA: Science and Behavior Books.

Ray, John. 1994. "Literacy and Language in Egypt in the Late and Persian Periods." *Literacy and Power in the Ancient World.* Eds. Alan K. Bowman and Greg Woolf. Cambridge: Cambridge University Press.

Roesslin, E. 1598. *The Birth of Mankynde, Otherwyse Named the Womans Booke.* T. Raynauld, trans. London.

Scott, H. Harold. 1934. *Some Notable Epidemics.* London.

Slack, Paul. 1997. "Mirours of Health and Treasures of Poor Men: The Uses of Vernacular Medical Literature in Tudor England." *Health, Medicine, and Mortality in the Sixteenth Century,* 237–74. Cambridge: Cambridge University Press.

Snow, John. 1855. *On the Mode of Communication of Cholera.* London.

Thebeaux, Elizabeth. 1997. *The Emergence of a Tradition: Technical Writing in the English Renaissance, 1475–1640.* Amityville, NY.

Tufte, Edward. R. 1997. *Visual Explanations: Images and Quantities, Evidence and Narrative.* Cheshire, CT: Graphics.

Vesalius, Andreas. 1543. *De Humanae Corporis Fabrica, Libri Septem.* Basil: (June).

Whiting, David A., M.D., FRCP (Edin). "Histopathologic Features of Alopecia Areata: A New Look." *Archives of Dermatology.* 2003: 139: 1555–1559. © AMA (American Medical Association). Accessed December 16, 2003. http://archderm.ama-assn.org/cgi/content/full/139/12/1555/FIGDST30032F3.

Endnotes

1. Some authors still assert that da Vinci stole cadavers from graves, as did earlier artists and scientists. However, this notion is disputed in many circles now.
2. Translated into English by Nicholas Udall in 1553.
3. Physicians usually served only the upper class in England at this time. In America, during the colonial period, Puritan ministers served as healers of both body and soul, while midwives birthed babies and ministers' wives and other lay people grew herb gardens for medicinal remedies.
4. If these terms are unfamiliar to you, please consult Chapter 3 and Appendix B of this text.
5. Medical photographs are well known for the black band or stripe across someone's face. While those photos are useful, they do lead readers to see the human body unattached to what makes the individual human being unique. Some photographers are moving away from such old habits of photography in medicine.
6. See http://archderm.ama-assn.org/ for an example of photographs that are recognizable but conceal patient identity.
7. For an excellent example of photographic and illustrative grouping, see Figure 3 at http://archderm.ama-assn.org/cgi/content/full/139/12/1555/FIGDST30032F3.
8. If you describe large amounts of numerical data in text, the reader may have trouble processing this data. If you give the reader the same information in a table, the reader can more easily process it. Your reader can visually grasp the material in this way much more quickly. See Tables 1 and 2 at: http://archderm.ama-assn.org/cgi/content/full/139/12/1555.
9. See Figures 1 and 2 at http://archderm.ama-assn.org/cgi/content/full/139/12/1555.
10. In other words, if you found out a 5000-word report had 5 graphs, your ratio is 1:1. If your report is 7000 words long, you could have up to 7 visuals. If you get a fraction for the ratio, once you plug in the number of words for your document, round the figure up to the nearest whole number. If, for instance, you get a number of 1.3 visuals per 1000 words in the model and you have 6000 words, 6 1.3 = 7.8. Round up to 8 the maximum number of visuals you can insert for your article.
11. This table uses hypothetical data.
12. For even more detailed and specific information on tables, see Huth's *Writing and Publishing in Medicine*, 3rd ed., and the Council of Biology Editors' *CBE Style Manual*, 5th ed., especially Chapter 31.

CHAPTER

Electronic Medical Writing

 Overview

Most of us are aware that writing in electronic formats is quite different from writing in other formats. Earlier in this text, you learned some information about digital composition in the health professions. For example, in the chapter on audience analysis you learned about conducting research via the Web for different demographic groups to whom you might not have easy access. Another good resource for researchers in the health professions is the multitude of excellent medical and health databases. Many of those are listed in that chapter, too. In addition, you understood the need for sensitive approaches to cultural and disability issues in multimedia. We also discussed bulletin boards for support groups as good research sources and how the Digital Divide in computer access continues to plague people in the lower economic groups. The section on usability testing also suggested pretesting electronic interface design.

In the chapter on ethical approaches to medical writing, copyright issues for using graphics and drawing software were key. You learned that the new computer capabilities gained through photo manipulation and other drawing programs also come with some responsibility for keeping patients anonymous, citing other peoples' graphics that you use in your work, and not manipulating photos and drawings in ways that do not accurately represent the truth or do harm to some person, persons, or organization.

In the next chapter on document design and project management, you learned that you needed to develop storyboards for laying out a Web site or other project involving graphics. Many of the document design issues explained in that chapter are an important background for designing in electronic genres. A small example of font styles that can cause miscommunication is that using all capital letters in an email comes across as shouting. You have now learned much of the design vocabulary in that chapter, which you can rely on for design in electronic interfaces.

In the chapter on charting, a discussion of on-line charting helped you realize that using computers in the hospital room can save valuable staff time and save lives. It's easier and faster to send an electronic chart and/or specific information about a patient than it is to transfer long, handwritten charts that are difficult to read.

In still another chapter, the one on health education materials, you learned about composing on-line for genres that attempt to teach the public about a health issue. You learned that health education materials include videos, audiotapes, radio and television public service messages, interactive electronic interfaces, graphics on posters, Web sites, and others. Even most newsletters, posters, and brochures are no longer pasted up by hand as in the older days of journalism, but now software programs provide the layout platform for your content. We looked at reliable health Web sites and determined what the criteria were for choosing ethical and accurate sites for getting health information.

Those chapters helped prepare you for this last chapter on electronic medical writing. No text fully devoted to electronic medical writing could cover all the bases in these times. Healthcare technologies impact all aspects of Western biomedicine: preventing, diagnosing, and treating disease. Such technologies now include pharmaceuticals and their protocols and a multitude of medical and even surgical operations, which are recorded or tracked via some form of electronics. For resources on mainstream medical technologies, you can start with the following organizations:

- Cochrane Collaboration
  http://hiru.mcmaster.ca/cochrane/default.htm
  This is a large, ongoing, international initiative that reviews evidence from clinical trials of various medical interventions.
- ECRI
  http://www.hslc.org/emb (610) 825-6000
  ECRI is a nonprofit agency that evaluates and prepares reports on medical devices and other health technologies.
- Food and Drug Administration (FDA)
  http://www.fda.gov (301) 827-6242
  The FDA is in charge of regulating drugs and medical devices in the United States.
- National Library of Medicine, Health Services/Technology
  http://text.nlm.hig.gov/ftrs-v3/gateway
  Many forms governing technology are available from this site, including the Agency for Health Care Policy and Research Guidelines.[1]
- Office of Medical Applications of Research (OMAR)
  http://text.nlm.nih.gov/nih/uploadv3/About/OMAR/OMAR.html (301) 496-1144
  OMAR is part of the National Institutes of Health (NIH). It holds consensus conferences to evaluate biomedical technologies. These statements are available through the OMAR Web site.
  (Gastel 145–146)

With the background you already have gained from the earlier chapters in this text, you are ready to learn analysis of Web sites and other design and audience issues when writing for the Web. In addition to practical and health-specific advice, you will also be provided with further resources for Web design. Software documentation is another area in which some medical writers work. Medical videos are designed in many large hospital systems for use with patients. They are also used in public health campaigns and in telemedicine. Audiotapes are also designed for patients and used in public health campaigns. Doctors also rely on this method for dictating their histories, physicals, and diagnoses, so health professionals need to be aware of some of the writing and transcription issues here. Multimedia writing includes some of the above as well as other media such as PowerPoint, interactive CD-ROMs, etc. Because PowerPoint is now so popular, we need to give some of our attention to the "do's" and "don'ts" of composing for PowerPoint presentations.

These are some of the questions you will answer in this chapter:

- How long have hospitals used computers?
- How are Web sites used in medicine?
- What kinds of designs are best?
- How do I write software documentation?
- How are audiotapes used?
- How do I design multimedia?
- How are videos used in healthcare?
- What about PowerPoint presentations?

How Long Have Hospitals Used Computers?

Hospitals and health facilities have been using computers for over 30 years and, along with the military, were among the earliest users of computers. In the early seventies, personal computers (PCs) were still on the somewhat-distant horizon. I first used computers in the clinical laboratory of a large university hospital and medical school complex in 1973. In those days, the computer needed a room of its own because it generated much noise and heat. The computer room in the clinical laboratory was about the size of a small bedroom or large bathroom. The computer itself was larger than the typical wall unit that holds your television set and other media. It was at least eight feet tall and probably ten feet wide. It functioned via very large external tape drives, slightly smaller than the size of auto steering wheels. There were two of them, of course: a reel-to-reel operation with probably miles of tape between them.

In days when the computer tape drives malfunctioned, I would sometimes walk into the computer room to see a large pile of tape on the floor. The tangle and mess were daunting; yet those tapes contained valuable information that needed to be recovered. Next to the computer was the line printer. It was the size of a small organ, pianoforte, or harpsichord. It printed out data on very large, thin, green-lined paper; the clattering of the line printer when it printed was loud.

If the door to the room was left open, the lab technicians and medical technicians screamed for the door to be shut. The computer was hooked up to several modern desk monitors and keyboards and also old-fashioned teletype machines stationed at intervals around the laboratory.

This university hospital was one of the first, in 1973, to computerize its lab results, requests for phlebotomy, the scheduling of blood drawing, and billing for lab tests. We had several data clerks who entered all the specimens to be analyzed, from tissue samples to kidney stones to sperm samples to blood. You name it, the clerks assigned it a number, and the computer had been programmed to call up the appropriate form to record the results. We also employed several data clerks to give results when doctors or nurses called to request them. In addition to me as supervisor of the program, we had a computer expert who took care of setting up and maintaining the programs. On the real hardware questions, the computer company technicians would be called in for repairs. The system functioned amazingly well for the size and newness of its approach. When we had breakdowns of computer equipment, it was sometimes difficult to switch back to a paper system.

Since those early times, everything is now computerized. The calculations I used to have to make on a calculator for pulmonary functions tests (an average two hours of computation for every one hour of testing) are now almost instantaneous. Ditto with the monitors and recorders of intercardiac pressures during heart catheterizations. No more stopping the film (cineangiogram) to measure the left ventricle to calculate pre- and post-contraction stroke volumes. Computers are now used in every phase of measurement and virtually in every phase of medicine. You will note variations depending on hospital size and location in whether or not there are computers in every patient room. In some hospitals (usually urban, large, and/or teaching) nurses regularly record all vitals, scan every product's barcode every time they open a piece of equipment in the patient room (whether it's a gauze, 4 x 4 sponge, or a Foley catheter kit), and record input/output, etc. In other hospitals the charting, equipment usage, and other functions are still recorded by hand. The same is true of medical records. Although there are usually computers involved in all admissions work and billing, some medical records operations are more computerized than others. This system may change more completely when the voice recognition software is foolproof and unfailingly accurate. Then doctors could dictate without needing transcriptionists. Again, some systems are more computerized than others. Small operations often find it difficult to afford the initial layout for extensive computer systems, fearing they will not be able to recap their expenses in a reasonable amount of time.

Some doctors now respond to patients via email; some physicians now have their own Web sites. The surgeon who removed my gall bladder requested my feedback on his Web site design. Physician groups ask patients to email them questions. In some rural and more remote areas, our mobile units, like the Joseph F. Sullivan Center's mobile at Clemson University, practice telemedicine. A nurse practitioner can put a probe in a patient's ear and send that picture to a doctor in Greenville, who can diagnose and suggest treatment immediately. Most mobile units, including ambulances and EMT or paramedic vehicles, are now equipped with telemetry devices that send the EKG tracing over phone lines to ER doctors or cardiologists for instant readings. Our electronic capabilities have indeed changed the face of medicine. With those changes come some challenges for writers in the health professions. We, too, have to constantly update our technical and software capabilities as well as understand and then master the changes in composition that new technological genres and formats demand. Especially when the technology changes so rapidly, it's difficult to quickly understand and articulate the new approaches and conceptualizations we have to use. Sometimes it feels as if just as soon as we get one format under our belts, the world shifts and new skills are demanded. When teaching writing for medicine courses, I find that it is better to teach students *how* to learn than it is to teach them individual software programs. In the following, you will learn about several of the more widely used electronic practices and their implications for writers in the health professions.

How Are Web Sites Used in Medicine?

In Chapter 1 you learned some of the ways Web sites are used in medicine. You now can analyze Web sites to understand an organization's or culture's members and what their priorities are. You know that Web sites also archive important information and documents when you need to quickly research history and context. In addition you learned about the various databases on the Web. Barbara Gastel's *Health Writer's Handbook* has a whole chapter (Chapter 5) devoted to on-line resources.

In the chapter on ethics, we highlighted the importance of making Web sites accessible to people with disabilities. In addition, the ethics of designing Web sites and using others' materials was also discussed. You learned that Web sites, besides supplying patient information, also provide access to support groups where people with the same illnesses can talk to and support each other. We discussed the ethics of quoting a person from a chat room or using people and their identities for your own ends, stressing the need for ethics in this situation. In that chapter, you also found extensive details regarding copyright and Web sites.[2]

As health writers you need to know how to design Web sites that achieve their purpose and reach their audiences. You'll need to know either how to write the code for the Web site (html = hypertextual markup language), be able to use a more simple or conventional text editor (such as Netscape Composer), or know the latest programs: Macromedia's Dreamweaver, Flash, and Fireworks.[3]

If your hospital system does not have good Web design software available, your first priority should be to get what you need to implement your designs. It's also possible to storyboard your ideas and hand them off to someone else to design or just give the person the content. You will need to supervise carefully at first, to make sure the design and navigation are appropriate. You should also give your designer the information you start any genre with: your audience's makeup and the purpose of your site. Articulate what the function of the Web site will be. Also, make sure there is audience feedback during the developmental process so that you can test the navigation and general usability.

If you are the designer/writer of the site, you might want to think of yourself as a landscaper. Unlike the way we approach written texts, your space is fairly limited. Many students think of the Internet as a sort of infinity in which they can launch all kinds of designs and/or data. In many ways this concept is true, but Web sites themselves are actually limited in form and format. Few think about the constraints of this medium.

As the designer/writer, information architect, webmaster, or whatever you choose to call yourself, you will want to create a Web site that fulfills the function for which it was intended.[4] Here is the global view of the steps in creating and launching a Web site:

- Analyze your audience and purpose
- Choose the software program for your work
- Create a preliminary design storyboard
- Design the site and its pages (this often means creating a template)
- Organize the content
- Storyboard the content onto your preliminary storyboard
- Create the pages within the site
- Test the site on audience members
- Revise using audience feedback
- Launch and test the site again (Make sure that you test the site in both major browsers, Internet Explorer and Netscape Navigator, because often these two browsers do not display sites in quite the same way.)
- Register the site with a good search engine
- Maintain or arrange for steady maintenance of the site

You learned about storyboarding in Chapter 3. If you have forgotten what that process involves, return to that chapter for a review.

A Web site can be deep or broad (shallow), depending on how you structure it. A shallow site (see Exhibit 11.1) will feature the home page with many pages linked on the second level only (Markel 2001, 613).

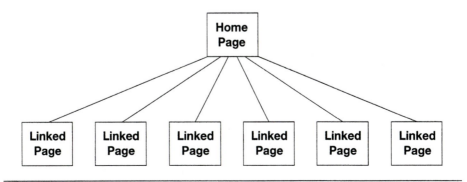

Exhibit 11.1 A Shallow Web Site

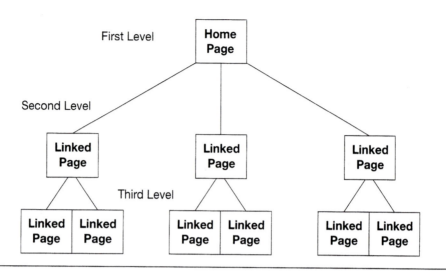

Exhibit 11.2 A Deeper Web Site

Be careful not to create a Web site that is frustrating for users to navigate. Your audience is accessing your site to search for important information, not to be entertained. Therefore, don't bury important information deep in the site. Research has confirmed that today's Internet users are not willing to click numerous times to find the information they need. Instead, three clicks are the maximum most of your audience will tolerate (see Exhibit 11.2). Therefore, you might not want to go deeper than three levels on your Web site. After you have a working draft of the site, get feedback on it from your audience.

Now that you have the global view of your medical Web site, you need to realize that your front page or home page should include the following data:

- Heading information
- Content area (or what we call the body in a written text)
- Comment/feedback and email address of contact person
- Copyright information
- Navigational bar with links to the pages within the site.[5] (Hawkes 1999, 118)

Remember that Web sites change rapidly and are a very fluid interface (meaning they shift forms as the new genre stabilizes, only to change again when the technology advances and gives designers even more options). None of these suggestions are set in stone. If you work for a large health system, you may want to study their live Web sites to see what styles they use. They may include certain guidelines that you must follow. Or your organization may ask you to contribute Web pages to their site, in which case they will most likely provide a template for you to use to make your pages consistent with the larger site.

One of the biggest difficulties in designing for the Web is the limited space you actually have. Although you can always add scroll bars to increase your Web page capability, users do not like having to scroll though lines and lines of text. Most users prefer a one-screen approach with limited text. If you must use scroll bars, unless it's an academic research site with lengthy texts, test audience reaction to the scrolling to make sure it's not frustrating for them.

Depending on the organization or individual you work for, you will have to decide whether to do the "latest, greatest" design work or do what is most practical. Your computer capabilities and audience will dictate what you are really able to produce. If you don't have the software programs (such as Macromedia Dreamweaver, Flash, or Fireworks) and your facility can't provide them, your design capability will be more limited. Your audience also provides constraints. You need to determine if the audience has the newest browsers or older ones that are not able to view later iterations of software. Or, if you have a senior citizen or geriatric audience, because of the vision problems of aging, cataracts, etc., you will need to keep text large and to a minimum and the site as simple to navigate as possible. At the same time, do not underestimate the senior populations. Many do try to keep up with the technology and succeed.

If a home page (also called a splash page) is successful, it sets a particular theme and initiates a tone that should be persuasive and credible to the viewer/user. This is why rhetoric is a good background for information designers, because it provides a basis for understanding how language and designs are persuasive and powerful. Ideally, your opening page should:

- Provide a theme in the heading information
- Give a good overview in the content section (e.g. http://www.asnc.org/)
- Identify the author, owner, or webmaster
- Establish restrictions or guidelines on use by others
- Provide a date of creation or update (may feature a hit counter)
- Supply links to other pages within the site on the second or third level (a mini site map) or to other Web sites.

One of the most popular voices in Web design is Jakob Nielsen. The 2002 release of his *Homepage Usability* text features 50 Web sites and his analysis. Most are corporate and well-known Web sites; one of the drawbacks of this book is that it doesn't address home pages in other areas, such as health, healthcare, medicine, or medical research. However, his advice on good home page design is still applicable here. The following represents his advice, filtered through a medical lens. Keep these tenets in mind as you design hospital or other health facility Web sites, patient self-help sites, and pharmaceutical sites.

- Show the health facility, pharmaceutical, or nonprofit logo in a reasonable size and noticeable location (e.g. http://www.asnc.org/termsofuse.cfm)
- Include a tag line that explicitly summarizes what the medical site or company does. (For example, Logo: Pacifica Support Association; Tagline: Resources and Support Groups for Dialysis Patients)[6]
- Emphasize what your site does that's most valuable to its target audience. (Sometimes you can do this in the tag line if you do it concisely.)
- Place the most important information, graphic, or path to pages in an upper middle location. Deemphasize less important information (can use color, graphics, etc., strategically to do this).
- Only have one official home page per medical site

- Use the term *web site* in the site only to refer to the entire Web presence. Don't refer to various sections as *sites*
- Have a recognizable icon or style on the home page so that it helps navigation (users can find their way back home from other pages) (Nielsen and Tahir 2002, 10–11).

What Kinds of Designs Are Best?

One caveat in Web design, especially for those of us who regard ourselves as creative, is to make sure any icons or logos we design from scratch are recognizable to our audience members. Some of my most creative students have designed interesting and unique logos and icons, only to discover that they were unusable because their audiences couldn't recognize or relate to them. If you are unsure about recognizability and don't have time to conduct formal usability tests, do some informal field-testing to see if people recognize the artwork you want to use. Especially in medical and health design, you can't assume "universal" symbols for all health issues and body parts. Not all cultures will recognize what a pair of lungs looks like, for example.

One mistake new Web site designers make is putting too much text on a site, regarding it as the same as the page of a book (which it is not). Web format (unless you are presenting medical archives, forms, articles, etc., in ".pdf" format deeper in the site) calls for minimum use of text and more graphical interfaces. Even though the amount of text is limited, as a medical writer, it's important that you craft the content well for maximum information (in minimal space), so that readers can scan the text and comprehend it quickly.

Here are guidelines for the content, links, navigation, and graphics on the Web site:

Content

- Make sure the sections, headers, and labels on your pages are divided and worded according to the needs of your audience member (the viewer), not from your standpoint (reader-based as opposed to writer-based language)
- Avoid repetitious content
- Don't use regional or generational idioms that many people will not understand
- Use consistent font formats, styles, and capitalizations (e.g., www.cfids.com)
- Don't label content that needs no labels
- Don't put a single item in a list or bullet (e.g. http://www.cardiosource.com/study)
- Don't give directives unless the action is mandatory, because users follow directions literally
- In the first instance of a name, spell out the abbreviation or acronym and follow it by the abbreviation
- Avoid exclamation marks

- Use all capitalizations (USE ALL CAPS) sparingly. Many think of it as shouting or emphasizing phrases
- Don't get too creative with spacing or punctuation. It confuses search engines and browsers and creates difficulties for visually impaired readers.

Links

- Use examples rather than description
- For each example, provide a link to the page with only that example on it
- Provide archives for past site features or information
- Don't say "more" or "click here;" be specific and give link name.

Navigation

- Put the main navigation in a prominent place
- Group items by similarity
- Don't provide multiple navigational devices for the same areas
- Don't make *Home* clickable on the Home page, only on other pages
- Make navigation categories clear and recognizable
- Use icons only if they are clear and recognizable
- Include a search box on the home page; don't bury it in the site
- Input boxes should be wide enough for readers to see and edit their text
- Don't put a Web search on your site, just a search for within your site if it is extensive (e.g. http://www.medscape.com/px/urlinfo)
- Don't put tools on the site unless they are really needed and appropriate
- Any high-priority tasks (and their tools) should be accessible on the home page.

Graphics

- Don't use gratuitous graphics (for decoration); use graphics keyed to content
- Label graphics and photos if they are not immediately recognizable
- Don't use (or limit) animation on the home page; it draws attention away from other page content
- Don't animate logos, tag lines, or headlines on pages (too difficult to read)
- If text elements have too much graphic and stylistic emphasis, the users think they are ads and avoid them (e.g. wwww.webmd.com)
- Don't overuse logos, especially if they are not your site logo (Nielsen and Tahir 14-23) [e.g. http://www.diabetes.org/homepage.jsp].

How Do I Write Software Documentation?

More and more information architects are expected to be able to compose software documentation for medical information products or medical technologies. While this is not a requirement for every job, it's a possibility you might want to know something about. As a cardiopulmonary technologist, it also fell to me in my hospital system to learn all the new cardiopulmonary equipment coming into our facility and rewrite the instructions for the health professionals who would be expected to put them into future use. It was only a matter of time before that skill would soon be stretched to include the software and help screens that needed to be added to present systems of health information and medical technologies. Two formats now need to be mastered to document new software: on-line and print-based.

RoboHelp is one of several software systems that helps you write on-line (Help) documentation. Except for the knowledge required to load a new documentation system, the writing of software documentation is not as complicated as one might think. It does require a new way of thinking, and a very simplified and brief form of writing. One's goal is to say the most with the fewest words possible. Clarity is of course also key here. One way to prime yourself for writing in this particular genre is to pull up any help document on your word processing program on your computer. Here is the Help screen text for converting files from another format to Word format:

1. Open the document you want to set compatibility options for.
2. On the **Tools** menu, click **Options,** and then click the **Compatibility** tab. Show Me
3. In the **Recommended options for** box, click the file format of the document you are working with.
4. Select **Custom** to specify individual options.
5. In the **Options** box, select the options you want.

   Note: Changing the options affects only the display of a document while you're working with it in Word. It does not permanently change any formatting in the document. If you later convert the document back to its original file format, the formatting appears as it did before you converted it to Word.

   Tip: To use the same options every time you convert a document you can *customize the way a file converter works.*

Once you get used to writing in this way, you adopt a style that becomes consistent and easier to write than your initial attempts will be.

Print-based documentation is similar to the on-line format, only the writer has more space in which to work. You still need to be brief and to the point. Basically, writing both forms of software documentation can be divided into three categories:

- Writing to teach: Tutorials
- Writing to guide: Procedures
- Writing to support: References (Barker 1998, 394–461).

How Do I Write Tutorials?

You will have to be able to sort out your purpose for the documentation project you do, and in the health professions, depending on your specific job and skill sets, you could be called upon to compose in any of the categories above. An example of tutorials in medical software is the documentation for a new electronic charting system or billing system in a rural doctor's practice. The staff is going to need to be able to understand the new system with a minimum of disruption and confusion. Their days will already be busy with patients and the present system that they use, so a new system will not always be greeted with wild enthusiasm. Ease of use becomes one of the guiding purposes for conversions from paper to digital systems. You may need to design in both formats for some audiences, as some users will be more comfortable with on-line documentation, while others want a print manual.

Don't make the mistake some tutorial writers do when they assume their audience is only made up of novice users. Many times tutorials are used to teach intermediate and advanced users who have used earlier iterations of the system (Barker 1998, 394). One of the purposes of a tutorial is for a "user to internalize certain skills or concepts about a program" (408). Users not only have to get familiar with a skill, but they also need to be able to perform it by memory at a later time. Here's where audience analysis is again important. You need to know for whom you are designing/composing and select the tasks that require documentation. You can't teach everything, so selectivity here is the key. Figure out what is the "typical-use scenario(s)" for your audience, and design around those tasks (409). Stand-alone designs, Judith Crandall suggests, must be composed when your audience members have only your documentation to rely on (no other teachers, textbooks, training sessions, Q & As, etc.). So the tutorial you compose needs to become the whole teaching environment (Crandall 1987, xvi).

Tutorials can stylistically be divided into two types: elaborate and minimal.[7] These categories are useful to guide your style choices. Research does support the idea of elaboration enabling improved retention of software skills. Davida Charney, Lynn Redder, and Gail Wells found that "elaborative elements helped them *apply* these tasks to real-world situations" (1988, 47–72). This style includes articulations of goals, summaries, examples, explanations, descriptions of scenarios, elements of good storytelling, and effective pacing.

Barker suggests that elaboration follows traditional patterns of lesson design:

- Instruction results in concrete skills
- Skills transfer to the real world
- Steps should be part of a logical, cumulative structure
- Specific instructions work best
- Give practice and feedback at each level
- Master one skill first, and then move to the next. (Barker 1998, 410–11)

The elaborative style does work best for novice users, people who are new to the system about which you are writing. Biomedical engineers and other highly trained users will respond better to the minimalist approach.

In the minimal approach, less equals more. No more introductions and reviews. This style is also supported by research that finds many users, especially experienced ones, approach software from different standpoints. John Carroll et al. suggest that users "jump the gun, skip information and like to lead" (1988, 123–53). Their research says that readers don't want to read introductions, but want to get started immediately in the new program. Often they won't read the manual, but they will just start using the program. In addition to these habits, they tend to skip information, not necessarily reading anything until they get stuck on a particular function. Then they look for help. Many users also don't like tutorials that over control and manipulate them, giving them few choices along the way (Barker 1998, 411).

For the minimal style of software documentation, again whether on-line or print-based, do the following:

- Focus on real tasks and activities
- Slash the verbiage (KISS: keep it short and simple)
- Encourage exploration
- Support error recovery (Barker 1998, 412).

Users who prefer the minimal style want to be able to do something right away with the software by way of instruction rather than just read about doing something. Cut to the chase quickly, in other words, and provide your minimalist users with a task to complete. Rather than elaborate, cut all the introductions; try to get chapters down to as few pages as is humanly possible and still instruct. Don't give lots of examples or practice sessions unless they are "real" and can satisfy the impatient user. As the documentation writer who must both keep the style minimal *and* encourage exploration, you are in a bit of a bind here. You have to find ways to encourage the user to forge ahead in different directions, but all the directions need to point the user toward learning the instruction and ending up in the same place as other users. Also, for the minimalist user, make it easy for him or her to recover from errors he or she makes and move on. Figure out in advance what mistakes users may make and write in the corrections for those mistakes. However, be careful not to discourage exploration.

How Do I Write a Procedural Tutorial?

In a new system of gathering epidemiological data, for example, you will need to make the procedure understandable to the technicians who will be using the new technology. You'll want to tap into the prior knowledge of the users, based on their present types of operating systems on the older technologies. Procedures include step-by-step instructions and how-to guides. Like the tutorial, you expect your users to memorize and internalize many of the instructions so that they can perform the tasks the next time without consulting your guidelines. However, you won't be encouraging your audience member to explore. There's only one way to complete the procedure (usually), and you have more control over your user than in a tutorial (Barker 1998, 421). Your steps may include some how-it-works and why-it-works data, too. This type of software documentation is actually the most common form you may be called upon to write, if you are expected to write software documentation.

Just because this form of documentation is quite concrete, it doesn't mean you won't be using a possible opening scenario for each segment. Do consider that you are a documentation writer guiding your readers step-by-step. If, for example, you were writing a set of procedures for a new software upgrade for ICU nurses on central venous pressure (CVP) line monitoring, you would want to paint a scenario of a doctor calling up for a CVP reading on his or her patient. The nurse at the ICU nursing station does not need to find that patient's nurse, but can simply press one button and relay the pressure to the physician. Based on the system now in place in ICU, you would want to immediately describe the scenario that would showcase what new or improved capabilities the user has and contextualize it within real day-to-day occurrences in the workplace.

Given the scenario above, then, you need to be able to guide the nurse or technical support person in installing the new software program. You may be called upon to design several procedural guides; such as one for installation and one for use (which needs to be defined specifically in terms of function, e.g. "How to Access Central Venous Pressure [CVP]"). Your audience may be different for varying functions.

The most difficult part of writing procedures is determining how much information to include. Part of this problem can be solved if you have a chance to test your documentation on real users. Usability testing is difficult in a healthcare setting because of the work pace and demands on nurses. Sometimes having access to a Master's in Nursing program and student nurses can be helpful as they can function as possible testing subjects. If you are centered at a healthcare facility, you should also work to build relationships with the practitioners. If you are more involved and have made an effort to work with them, then you can ask to interview them briefly. The night shift, when there is less demand on nursing time (unless patients are not stable), is often a good time to ask questions and test some of your steps.

The steps for writing a procedural tutorial include the following:

- Determine how much information your user needs
- Choose the appropriate instructional format
  Standard
  Prose
  Parallel
  Context-sensitive
- Follow an action-response pattern
- Test all procedures for accuracy (Barker 1998, 423–432).

In determining how much information your users need, you can look at the situation in which they must function. Do they need to get through the procedures as quickly as possible? In order to remember the instructions, do you need to build in a bit more rich detail? You will find, too, that if you are frequently writing in this genre, you will develop your own style; once that happens, you only need to develop some variations on it to make your documentation effective. Besides analyzing purpose and audience, your audience analysis should include the context in which your user functions. By situating the writer, you will more quickly be able to give audience members exactly what they need.

How do you determine which format is needed once you have scoped out the situation? The **standard format** includes "steps, notes, screens, and other elements aligned on the left margin and continuing in either one or two columns, in a numbered sequence from first step to last" (Barker 1998, 425). The standard format is the one you are probably the most familiar with in software manuals, help screens, and other instructional writing. The advantages of this standard format are that it:

- Is recognizable by users
- Flows from one page to next
- Is numbered for testing
- Is quickly visible via hanging indents

Some of the disadvantages include the amount of space required for even simple procedures and the cognitive difficulty of having to learn complex steps that seem to be given as much space and importance as simple steps. Also, in a complex interface, users can get mixed up about where they are in the order of the steps (Barker 1998, 426–27).

The **prose format** surrounds the steps in prose paragraphs. This format functions more effectively with procedures that require few steps and a simple interface. Prose format also works well in reference sections; it's a real space-saver. You can also mix styles (that is use the prose format for a cramped error message box). Prose format does not mean that the sentences are long and the paragraphs unending. Limit sentences to 20 words or less and paragraphs to 5 sentences or less. The advantages of the prose format are that it:

- Is good for basic, simple steps
- Saves space
- Works best for experienced users (Barker 1998, 428)

The disadvantages are that it is difficult to see steps within the paragraphs; graphics can't be included for individual steps; and it doesn't work well for inexperienced users.

Parallel format is best for a complicated procedural program, such as databases, invoices, order entries, and address-types. The steps and heading fields parallel each other, so that users can fill in blanks as they go along. See Exhibit 11.3 for an example. This format allows for illustrations to be embedded along the way. It is only problematic if the explanation goes so long as to move onto another page, and the reader has to flip back and forth from screen to screen (or page to page if it's a print manual). To make this format most effective, keep the following in mind:

1. Explain how the parallel format works in the beginning of the instruction set.
2. Give examples for each step.
3. Keep the terminology consistent. If you use the term *patient number*, refer every time to the same term (not *patient I.D.* or *patient identification* or even *patient no.*).
4. Keep the terms the same as the actual screen terms.
5. Keep the same font, style, and size as the screen.
6. Discuss one screen at a time.

Query on Patient Accounts Form

The Query on Patient Accounts Form allows you to systematically recall information about patient accounts.

To get to the Query on Patient Accounts Form follow these steps:

1. Press F2 from the Medical Menu
2. Press F1 from the Query and/or Recall Menu

The program will respond by displaying the following form.

Query on Patient Account Form

Query and/or Recall Fields on Patients

Patient no.: (...............) to (Patient no.): (...............)
Last Name: (...............)   First Name: (...............)
Balance Due: (...............) to (value): (...............)
Birth Date: (...............)   SS#: (...............)

Tab (to move among fields). F10 (done). Esc (exit).

Description of the fields in the Query on Patient Accounts Form.

Patient No.   Query on this field displays the patient number, home phone, family
(8 digits)   balance, last visit date, and check and credit status. You can enter a single
                   patient number, or a range of patient numbers.

Last Name: Query on this field displays the same information as the patient number
(up to 18)   query field. Enter the name, and press the Enter key. The program will
                   display the record of the first person with the name you entered.

First Name: Query on this field displays the same information as query on the patient
(up to 18)   number field. Enter the first name, and press the Enter key. The program
                   will display the record of the first person with the name you entered.

Balance Due: Query on this field displays a list of patients whose balance falls under the
                   specified range. For example, here you can get answers to questions such
                   as "Show me all patients whose balance due is between $100 and $250."
                   Enter $100 in the first space of this field and $250 in the second space in
                   this field.

                   *NOTE: You can get any range from $1 up by inquiring on this field. You
                   must enter something, so if you don't want to use this field, enter $9999 to
                   get all the patients.*

Exhibit 11.3 Query on Patient Accounts Form: A Procedural Tutorial

7. In illustrations, reproduce the screen format exactly in your instructions.
8. If you need to complete one step before another one, let your users know that.

The advantages of parallel format are that it:

- Is good for complicated screens and dialogue boxes
- Is best for short procedures
- Helps users stay on track (Barker 1998, 430).

Disadvantages include that this format is somewhat specialized and won't fit all procedures. Users can get mixed up if there are many similar screens and repetitive steps, and if it is keyed to the form and not necessarily to "logical" or intuitive steps.

Context-sensitive procedures can be used if you have the ability or technical capacity in your on-line documentation to embed tags in the instructions so that the program recognizes where the user is in the process. You know these types of documentation from the Windows and word processing programs you use. If you are working with a programmer, you can request that he or she put the tags in if you need help. Link the help screen to the various steps. You can also program the screen so that the help message will pop up when the user makes a mistake. Either way, you will have a sophisticated procedural tutorial that keys directly into each step or instruction.

Now that you understand what formats are available for procedural tutorial writing, we can discuss the action-response pattern. This concept is simply the pattern that when a user takes an action in the program procedure, the computer responds as it should. This pattern is repeated over and over, so that the audience member works through the procedure very systematically from beginning to end. The last step in the procedure writing process is to test each step to make sure it is accurate and does what it is supposed to do. Besides double-checking your work, we strongly advise that you also have members of the target audience test it (ideally people who really don't know how to use the software yet). Then, revise based on their feedback, and you're ready to release your documentation.

How Do I Write a Reference Tutorial?

Support and reference material includes all the help sections and manual references that you consult (often found in the back of the books as appendices or as chapter endings/troubleshooting guides). You can organize the content alphabetically, by menu, or by context sensitivity. Barker has divided reference into two forms: usual and special (1998, 444). Your reference users already have some knowledge, at least enough to know what they are looking for. The usual forms are very adaptable to putting on on-line and include appendices, update info/readme files, interface elements (explanations of menus, definitions of keys, labels of screen regions, and explanations of rulers), and glossaries.

The steps for designing support documentation are to:

- Choose the right form of reference

- Decide content to include
- Organize the info
- Tell the user how to use the reference
- Establish a pattern (Barker 1998, 444).

Creating documentation for the proliferation of software programs will probably emerge as one of the most important electronic writing tasks (in addition to WEB site and portal design) for those who specialize in medical writing. For more information on this important genre, consult the *Writing Software Documentation* text in this series (Thomas T. Barker).

How Are Audiotapes Used?

Audiotapes are still used most often by doctors to dictate notes following patient interviews (histories and physicals), patient followup visits, and procedures. Those of us working in the health professions look forward to the day when the voice recognition software and PDAs (personal organizers) will be able to hook up and enable flawless transcription. Because of the difficulty of voice recognition and the importance of total accuracy, that technology may still be a few years away. Medical transcriptionists in some offices still work with the standard equipment of rendering the tapes into text: a dictation machine that you can control via foot pedal or hand press that stops, starts, and rewinds the tape easily and quickly. Good medical transcribers are worth twice their weight in gold, because the combination of medical language and a contextualized voice is a difficult transcription task. There is no room for error in these transcriptions. Some of the important data is written on the chart via other means (EKG tracings, lab test results, pharmacy orders, etc.); those measures can be accessed to clarify certain audio quandaries. Once a transcriber has worked in a certain area, such as cardiology, with the same physicians, he or she becomes used to their patterns and diction, and the task becomes easier. Transcriptionists who have to rotate from department to department are in the most difficult transcribing situation.

Audiotapes are also used in other areas, including speech pathology. Because that's a highly specialized area, it is not covered here. As designers of health messages, you may also use audiotapes in public service announcements (PSAs) prepared for radio broadcasting. You can also, as a multimedia designer, embed digitized audio in multimedia presentations such as Macromedia Authorware, Macromedia Flash, Macromedia Fireworks, Macromedia Dreamweaver, Director, and even Microsoft PowerPoint and others. This also applies to video capabilities. Many Web sites and other health information programs run digitized video in their presentations. At this point in the technology, older video equipment does not compress videos well, but streaming video and other newer technologies will become more and more widespread and enable better, longer, and more usable video capture. Such capabilities require resources for the state-of-the-art equipment, well-trained individuals, and the attractive salaries to pay them. Smaller hospitals and health facilities will have difficulties affording such high-end work, but as more freelancers flood the market, perhaps their prices will come down. As a medical writer with these skills, you would be a very sought-after employee for the large health systems. Since the digitizing of both video and audio is fairly simple once one has the equipment, it's not necessary to go into detail here. However, Adobe Premier, Sound Forge, and other software programs make these tasks relatively simple and will get more affordable and easier over time, as the technology leaps forward.

How Do I Design Multimedia?

What used to be the realm of the professionals in Hollywood and New York is now possible, on a scaled-down version, for those of us who work in electronic formats. We talked above about audios, videos, and Web design. A multimedia presentation typically includes these forms, but can be defined as any presentation using more than one medium. Some regard Microsoft's PowerPoint as a multimedia presentation, which is a valid term, but that medium has its own constraints. So, how does one storyboard and design for a format that has multiple media and so many possibilities? Again you start with the basics: audience and purpose. Be careful not to be caught up with throwing everything into the production just because you can. If you are so enamored of the technological possibilities that you lose your audience, you haven't been an effective health communicator.

If you are asked to design a multimedia production for a health facility or a public health campaign, you will need to find out what your client means by his or her use of the term *multimedia,* since different people define it in different ways. What are the expectations? The more you can ascertain up-front, the more successful you will be in your design choices. Also, you will need to examine the budget somewhat closely and make sure your client has a firm idea of what resources are available. Videotaping and animation, for example, take time and money. In addition, make sure you know what resources you have to offer, both in equipment and personnel. Make sure you include enough time in your estimates.

Once you have gotten this far in the planning process, you'll want to storyboard and/or script the presentation. Make this graphic or text part of your proposal so that the client can see what you have in mind and what your expectations are. Before you sign any contracts (or commit to this project if you are in-house), meet with the clients after they have reviewed your storyboards and scripts to see whether you now have shared expectations or whether further description and dialogue are necessary. Again, the more of this type of preparatory work you do up-front, the better off you will be. Don't forget to factor in what capabilities the audience has. If you are doing a multimedia presentation and burning it onto a DVD, for example, make sure that your audience members have DVD players! You might assume that everyone has a particular form of technology that you have, but such an assumption would not be true for all segments of the audience. This advice is doubly true in cross-cultural or international work. The less you assume and the more you research, the better off you will be.

Sometimes a multimedia presentation can come across as fragmented, chaotic, and/or playful. For the most part in health and medicine, none of these approaches would work. If you stretch your imagination, perhaps in pediatric medicine and health, you could create a playful cartoon about a small child who has to undergo a procedure. Otherwise, you are usually expected to create a serious tone. Do consider the context in designing your multimedia presentation. Set up the situation or medical problem in the beginning of the multimedia design. Again, the scenario principle comes into play here, as it is particularly helpful in visual and video presentations. Create a small scene as your context: show a person who has the disease (or have an actor play the part). Don't just list the symptoms in an abstract context; it's not as persuasive. In addition, you do not need to avoid emotion or *pathos* in your multimedia presentation. Emotions can be powerfully persuasive in health campaigns and other forums. Just don't overdo it.

Learn to use the best medium for the content you wish to get across. An audio selection can be affective if we don't need a visual to give us more information. Make sure you give introductory material before each change of scene, introduction of audiotape segment or video shots, etc. Don't make your audience work hard or guess what it is you're showing and why. Do include smooth transitions. Just because you are using different media, you still must develop content logically, coherently, and intuitively. If you just throw pieces together in a presentation, the audience members will be jarred and dissatisfied with the presentation. If you use too many rapidly changing formats, your viewers will become frustrated or distracted by the poor quality of the multimedia presentation and not gain the information and knowledge for which the presentation was designed. Because a narrative structure is recognized by all cultures, you could organize the presentation around a narrative or a narrative theme, or you could organize it around a metaphor (extended metaphor), too. If you provide some unifying theme or construct to your presentation, your viewer's comprehension will be greatly enhanced. Also, make sure you have been given a time limit for your presentation (length of time it takes to view the presentation). Then rehearse your presentation several times to make sure it is not over or under the time mandated. If you are preparing a public service announcement (PSA), you only have 30 seconds, which is an extraordinarily short time to get content across clearly. The shorter the presentation is supposed to be, unless you can assume a well-informed audience base, the harder it is to design.

Although it was mentioned earlier, it's worth repeating: do be careful of the urge to push the envelope with the greatest and latest designs if it does not enhance the project for this particular audience. Instead, talk to audience members about their preferences and design according to those. I had one graduate student who insisted on a certain rather outlandish design using animation that did not prepare the viewer for what was coming (it had no "predictive" power). I flagged this student's opening splash screen and flash animation, telling him it was not well designed (nor was it navigable). He insisted that such Flash animation was intuitive. I argued that his audience was an older audience and included an international segment; I thought the animation was not at all intuitive unless you were a 24-year-old white American male who spent a lot of time with the latest high-tech equipment. He said he could win his committee over, although as his chair, I had very grave doubts. My fears were well founded, and after his project review before the whole committee, he had to redesign his splash screen and navigation.

One recent, very effective television ad for a local hospital features what looks very much like a Web-page format. Boxes with a particular health symbol or photograph of what the script is talking about float in and out of the television screen, much as animated graphics and texts fly in on Microsoft PowerPoint. Different voice-overs give another component of the definition for good health care as the box flies in. The spoken word or words coordinate exactly and persuasively with the symbol or photograph in the box. Because the design if visually powerful and the words in different voices punctuate each animation, the overall effect is cumulative, with the entire picture of this healthcare facility well cemented in the audience's mind by the end of the ad. In addition to their good graphics and effective design, their timing of all components was excellent. The voice-overs came in regular intervals so that a rhythm was established. Researchers have proven and rhetoricians have known for centuries that cadenced speech with its rhythmic intervals is a very powerful and effective memory technique for all human beings. It's much like our remembering words to many songs, even if we haven't heard those songs for a long time. When language is associated with a stylized, rhythmic beat, it's easier for us to remember and recall.

Since many, but not all, multimedia presentations do include videos, we should address some of the video formats that are now used in medicine.

How Are Videos Used in Healthcare?

Videotaping is used more and more in medicine for a number of reasons. One of the reasons is the increase in malpractice litigation. Today many operative procedures are routinely videotaped in case a lawsuit is filed. The videotape can be used as evidence in court, usually to prove that no unusual situation occurred in the operating room during the procedure. Color photographs are also used with more and more frequency, because camera lenses can now be made small enough to be embedded in small-bore catheters and regular-size scopes inserted into arteries, nerve channels, bronchial tubes, Eustachian tubes, and colons.

Videography is also used more and more to bring access to expertise not always available in other countries or in rural areas of the United States. I recently reviewed an article for a technical communication journal that described research on several rural healthcare sites in which trauma cases were connected via television cameras to trauma centers in large cities. With the television cameras and monitors at both sites, the urban trauma-center specialist was able to view and guide the rural providers in trauma procedures in which advanced expertise was needed. This system saved time and lives, because trauma patients' care was not delayed since they no longer had to be physically transported to the regional trauma center. Although the trauma center doctor was watching and giving advice on the rural procedure, if the rural doctors needed a specific demonstration, a switch was thrown so that the rural doctors could view the urban trauma specialist demonstrating a particular tie-off, scalpel technique, etc. The studies found decreased numbers of mortalities, complications, and impairments as a result of the video systems.

Although the televised system that was being reported on was in the Northeast United States, other regions also have similar systems. The University of Arkansas for Medical Sciences (UAMS) in Little Rock, Arkansas, for example, serves some 40 rural healthcare sites.

Although you may not be involved in setting up or monitoring the videotaping of such procedures, you may be called upon to develop brochures to advertise it. You may complete some of the paperwork or design programs for further funding, support, or uses of the system. Telemedicine is not only used for trauma procedures; UAMS also uses its system for medical consultations, followup care, and educational services (as do some other systems throughout the country). "Telehealth uses special high-speed phone lines and the most advanced computer and visual communications equipment to instantly connect specialists with patients in rural communities throughout Arkansas. . . . Two-way 'live' verbal and visual communication allows a patient to ask and answer questions, and explain symptoms to a physician" (UAMS). Of course such a system saves time and the inconvenience of traveling for consultations with specialists. It also makes it possible for your primary care physician and even family members to be present for the consult, to ask questions, and to receive instant feedback.

Because the equipment is not familiar to many people yet (just those who have to write about them or set them up), here are a few terms you need to know as a medical writer in today's video- and telehealth world:

- *Computer Connected Stethoscope:* Uses multiple frequencies to allow doctors to listen to heart and lung sounds as with a regular stethoscope.
- *Handheld Zoom Cameras:* The assistants or nurses use this equipment to zoom in on small details. Doctors can evaluate rashes, lesions, wounds, cuts, eyes, etc.
- *High-Resolution Cameras:* Enables physicians to observe patients' motor skills, flexibility, and mobility.
- *Telemedicine:* A computer-based telecommunications system that enables real time interactive audio-visual links between patients and doctors located at sites distant from each other.
- *X ray/Lightboard Camera:* Enables specialists to read X rays and EKG tracings.

Videotapes in VHS and now DVD format have long been used in patient education. Most of us can remember back to the films we viewed in health education in high school. Videotapes have come a long way now in education; health-related videos are sophisticated avenues for delivery of health information. Physicians also use videos and DVDs to highlight their own new surgical and other techniques for international and national conferences. They often want this form of documentation, as it is much faster and easier to understand new procedures without reading through pages of text. Later on, they may have professional illustrators draw the steps of the procedure for use in textbooks and journal articles. Clinicians can take distance learning courses or order tapes and CDs for classes (or individuals) in many areas of medical practice.

For example, some catalogues feature videotapes and CD-ROMs, such as the one on *Clinical Skills in Speech Pathology and Audiology.* This small catalogue features some 45 videotapes in VHS format and CDs ranging from 26 minutes to three hours on a variety of topics within these specialties. Some of the titles include: "The Work of the Speech and Language Therapist," "Speaking After Laryngectomy: Voices of Experience," "Assessing Craniofacial Speech Disorders," and "Programming for Children with Attention Deficit Disorder: The Role of the Speech-Language Pathologist." *Speech Science and Communication Disorders* from the same company features another 40-some videos and CDs within that area.

In medical education, although videotapes and DVDs are sometimes used, much of the education is still hands-on, as medical and healthcare students complete rotation cycles in various specialties. What has changed is computer animation and virtual environments for medical schools. Students can complete dissections, practice surgery, build 3-D models of organs or body systems for research purposes, etc., on computers. Human models also come with some technology so that nurses can practice taking vitals, listening to respiratory systems, bowel sounds, and cardiac rhythms without practicing on real patients yet.[8]

One of the many purposes of this text, and I repeat it here, is to remind us all of the many research areas open to us for exploration. An entire text could be and should be written on videotaping in medicine, its ethics, its visual appeal, its design, scripting, and digitizing. While it is only briefly addressed here, I would hope others would use the suggestion here for inspiration to follow through with many articles and texts in this and other areas.

What About PowerPoint Presentations?

Microsoft PowerPoint presentations have become the most popular new form of slide presentation since the old Kodak Carousel system. It seems that every discipline now uses PowerPoint, although few have written guidelines for its use. The practice has certainly outstripped the theory and design. The PowerPoint system is both flexible in terms of design possibilities and constrained within its own limits, much like the Web site format. Background designs within the various releases of the software have become very familiar to all of us, some to the point of ad nauseum. In medicine, however, an undistinguished background and simple graphics and text are probably the best rule of thumb for medical writers unless you are doing health education. In health education, tailor the PowerPoint to your audience. If you have children as your target audience, you could import a more interesting, child-centered design, employ fun graphics, use splashes of primary colors, surprising audio sounds, and animation. In the latest iterations of PowerPoint, we can now import videos and audios, making the productions more and more "multimedia" and less like the more traditional, now somewhat boring, usual slide presentations.

Still some useful standards and guidelines are pertinent to this genre. Some of the suggestions are common sense, yet they are violated every day in some venue. Keep your backgrounds the same, unless you have important reasons to change them, throughout the presentation (most iterations demand this, but it is possible to change them). Don't make the mistake of using too much text. Often new users of PowerPoint tend to see it as another blackboard or book page, covering the rectangular space with too much text that is impossible to read from a distance. (New Web site designers are also guilty of putting too much text on Web pages.) Because text is limited on the software, use sans serif font such as Arial or Helvetica. Make sure you use good figure-ground contrast and don't have the background so bright, busy, or clashing with the text that the viewer can't read the slide. Remember to put dark text on light backgrounds and light text on dark backgrounds. Make sure you don't include images and sounds (or videos) just for show and just because you can. They need to have a strong and necessary connection between what you are saying and showing. Don't irritate your audience members by reading every word of the slide for them. Instead, use the entire presentation as a structure or prompt for your talk that you then deliver without notes or memorization. Let the PowerPoint keep you on topic. Make sure what you are saying and doing is culture-, race- and gender-sensitive. Don't make the application itself so memory-intensive that you can't transfer it. In other words, if it is a large file (over 1 MB), burn it onto a CD or transfer it to the hard drive of the presentation computer. When doing the presentation itself, transfer it from any peripheral (CD, disk, etc.) onto the hard drive or desktop of the computer you are using. The presentation will load more quickly from the hard drive or desktop than from a peripheral. Also, do not send your PowerPoint over phone modem to someone unless there is no other alternative, especially if it is a large file. It will take too long to download. Some programs/connections time out before large files load. You can change your settings, too, if this becomes the only way of getting your file. Also, take a backup file (in case something happens to yours) and backup overheads with the PowerPoint printed on them (color transparencies) so that you still have a good presentation for your audience even if the computers on site don't function.

Summary

In this chapter you have learned that electronic or digital medical writing comes in many forms. You first reviewed the various electronic genres you had already come into contact with in previous chapters. Then you learned about resources for more information on medical technologies and their documentation. Computers have been used in health facilities for at least 30 years, and I worked with one in 1973 in a clinical laboratory.

You learned more elements of designing when writing for the Web. First we looked at the total project and listed the various steps in creating a Web site. There are both shallow and deep Web sites, and you need to be aware that users, especially those who are looking for health information, do not want to spend long periods of time ferreting out information that could be made readily accessible. You examined both home page design and site design. You looked at home pages from a structural perspective and then learned the elements of detailed design for the home page, especially what to avoid as you design. You learned what content needs to be on the entire Web site and what should and should not receive emphasis. Many of the design issues are based on usability research done by Nielsen and others on content, links, navigation, and graphics.

We spent much time discussing software documentation, because it's probable that this will become an important area in which medical writers work. Few sources have gone into depth on software documentation for medical purposes, so that section was more detailed than some. You learned that software documentation could be categorized as tutorial, procedural or research, and support. Within those categories are various formats that lend themselves to certain kinds of tasks and contents.

Audiotapes are designed for physician use and for use in educational materials. You learned that health professionals need to be aware of some of the writing and transcription issues. Audiotapes are also used in speech pathology and in other specialties involving speech (like laryngectomies). Transcriptions are difficult, but experienced transcribers are an important asset to any health facility. Audiotapes, like videotapes, can now be easily digitized using such software as SoundForge, Adobe Premier, and other programs.

Multimedia writing can include audios and/or videos as well as other media such as PowerPoint, interactive CD-ROMs, etc. Multimedia writing calls for careful scripting and design in order for health messages not to get lost in technologies.

Medical videos are designed in many large hospital systems for use with patients. They are also used in public health campaigns and used in telemedicine. You learned that there are several sites that have brought medical specialists' expertise to rural areas via the television. These broadcasts are two-way, so that the specialist can observe the health care the patient is receiving, and the rural health professionals can see techniques and other information that the urban specialist is demonstrating. Cameras are also embedded in tiny catheters that go into the body and capture photographs of internal organs and systems. You were introduced to the video terms that are most important when writing in telemedicine.

Because PowerPoints are now so popular, we talked about the "do's" and "don'ts" of composing PowerPoint presentations. This section outlined certain practices that are sure to lose or irritate your audience in such presentations; it also outlined what works well for PowerPoint presentations, such as backup systems, careful attention to background design, limited amount of text, clear font styles, good figure-ground contrast and use of the slide show to help you present clearly and naturally.

Discussion Questions

1. Discuss any experiences you've had as a health provider, health professional, or as a patient with a computerized system within a hospital or healthcare facility (including student health centers). Do you feel that the electronic capabilities at a healthcare facility enhance or cause problems in the healthcare

experience of patients and their families? What systems do you regard as particularly vital and which are less necessary?

2. What options do you feel a small doctor's practice has when it comes to employing a computer system? What suggestions would you make if systems were too expensive and elaborate for smaller organizations to use?

3. Why and how are Web sites/splash pages persuasive? What can you do increase the "power" of the Web site through design and language (information design or information architecture)? Give concrete examples to back up your answers. If you are in a computer class or have a laptop, you might look at and show the class your own Web page. If you are planning to go on the job market or are already a health professional, what does your Web site say about you? How could you redesign it to increase the persuasion for potential hirers or your superiors who may promote you? Have the class give you suggestions for revision.

4. Notice the three different kinds of software documentation genres. Can you articulate the differences? Discuss the various features, purposes, and audiences that distinguish one genre of documentation from another.

5. Look at various on-line tutorials as a class and discuss which ones you think are "elaborate" and which ones you think are "minimal." Is there a majority of one style choice over another? Which ones do you think would be easier to compose and why?

6. What ethical issues might arise in videotaping of patient histories and physicals or surgical procedures? What are the pros and cons of the practice? How would you feel about being videotaped in a doctor's office? How would your family members respond to videotaping? Do you feel guidelines are mandatory for videotaping? If so, what would they be?

7. What PowerPoint presentations have you seen? How were they designed? What were some of the elements of good presentations? What were some of the elements of the less effective presentations, and what could have been done to make them better?

8. How will you choose what medium to use for delivering health content? Brainstorm some examples of content and discuss why you would or would not use certain media for delivery.

Exercises

1. Research various computer systems used in medicine. In the class decide which of you is going to look at which area in the many systems used in healthcare and medical research. Conduct your research via the Internet and/or interviews with health professionals in hospital settings.[9]

2. Choose several health or medical Web sites to review. Check out the navigation. Is information easy to find? Why or why not? How many clicks does it take to find the information you are seeking? Did you immediately identify which links to other pages you needed, or did you have to make a few false

starts to get there? Write this information into a memo and be ready to deliver it verbally to your classmates with the Web sites you've identified.

3. Choose several health or medical Web sites to review. Analyze their front pages (or home or splash pages) using the criteria we have established above in the section on Web pages in this chapter. Write up your analysis on whether or not the pages establish the credibility and information they should provide. Then conduct a design or spatial analysis. How are the components arranged on the page? Ideally, choose several pages that have different spatial arrangements to review.

4. Find several examples of software programs in healthcare or medicine. Analyze them for type (tutorial, procedure, or reference). Are they well written or are there some problems? Why do you think so? Support your claims with evidence from your examples.

5. With a team of your classmates (three to five students), choose a software program that you already know, and without looking at the tutorial already designed as a part of the program, write your own tutorial. First go through the steps of deciding who your users will be and what you need to teach them about the software program. If you have access to medical software, use one of those programs. After spending at least two to three class periods analyzing the user and writing parts of the tutorial, as a team write a collaborative reflection about your experience of composing in this genre.

6. Follow the same instructions as in Exercise #5, but this time write procedures for a new medical software program. Again, write a reflection afterwards. Write your reflections individually and then combine your insights in a collaborative report.

7. Create a PowerPoint presentation using some health message or content in which you have an interest. Craft the presentation using the advice in this chapter on what elements to use in effective design and what elements to eschew in poorly designed presentations. Present your slide show to your class and have them critique it. Then gather up their comments and revise your PowerPoint presentation. Write a brief one- or two-pager about your insights based on the critique and on your revision.

Works Cited

American Diabetes Association. Updated Dec. 11, 2003. Accessed December 16, 2003. http://www.diabetes.org/homepage.jsp

American Society of Nuclear Cardiology. © 2003 American Society of Nuclear Cardiology. Accessed December 16, 2003. http://www.asnc.org/

Barker, Thomas T. 1998. *Writing Software Documentation: A Task-Oriented Approach.* Boston: Allyn and Bacon.

Cardiosource. A Collaboration of American College of Cardiology Foundation and Elsevier. © 2003, Elsevier, Inc. Accessed December 16, 2003. http://www.cardiosource.com/study

Carroll, John M., Penny L. Smith-Kerker, James R. Ford, and Sandra A. Aazur-Rimetz. 1998. "The Minimal Manual." *Human-Computer Interaction* 3: 123–53.

CFIDS Association of America. Chronic Fatigue and Immune Dysfunction Syndrome Web site. © 2003 The CFIDS Association of America, Inc. Accessed December 16, 2003. http://www.cfids.org

Charney, Davida, Lynn Redder, and Gail Wells. 1988. "Studies of Elaboration in Instructional Texts." *Effective Documentation: What We Have Learned from the Research.* Ed. Stephen Doheny-Farina. Cambridge, MA: MIT Press.

Cochrane Collaboration. 4/1/2003. July 8, 2003. http://hiru.mcmaster.ca/cochrane/default.htm

Crandall, Judith A. 1987. *How to Write Tutorial Documentation.* Englewood Cliffs, NJ: Prentice-Hall.

ECRI. 6/10/2003. July 9, 2003. http://www.hslc.org/emb

Food and Drug Administration (FDA) 7/1/2003. July 9, 2003. http://www.fda.gov

Gastel, Barbara. 1998. *Health Writer's Handbook.* Ames, IA: Iowa State University Press.

Hawkes, Lory. 1999. *A Guide to the World Wide Web.* Upper Saddle River, NJ: Prentice-Hall.

Insight Media. 2001. *Clinical Skills in Speech Pathology and Audiology.* Video and CD-ROM Catalogue (Fall). New York: Insight Media.

———2001. *Speech Science and Communication Disorders.* Video and CD-ROM Catalogue (Fall). New York: Insight Media.

Macromedia Web site. © 1995-2003 Macromedia, Inc. Accessed December 16, 2003. http://www.macromedia.com/software/fireworks/productinfo/upgrade/

Markel, Mike. 2001. *Technical Communication,* 6th ed. New York: Bedford/St. Martins.

Medscape from WebMD. © 1994-2003 Medscape. Accessed December 16, 2003. http://www.medscape.com/px/urlinfo

National Library of Medicine, Health Services/Technology. 6/30/2003. July 9, 2003. http://text.nlm.hig.gov/ftrs-v3/gateway

Nielsen, Jakob, and Marie Tahir. 2002. *Homepage Usability: 50 Websites Deconstructed.* Indianapolis: New Riders Publishing.

Office of Medical Applications of Research (OMAR). 6/15/2003. July 9, 2003. http://text.nlm.nih.gov/nih/uploadv3/About/OMAR/OMAR.html

Price, Jonathan, and Lisa Price. 2002. *Hot Text: Web Writing That Works.* New Riders.

Roche Pharmaceuticals. 7/10/2003. July 10, 2003. http:www.roche.com/home.html

UAMS (University of Arkansas for Medical Services). n.d. *Telehealth* brochure. Little Rock, AK: UAMS.

WebMD. © 1996-2003 WebMD, Inc. Accessed December 16, 2003. http://www.webmd.com

Williams, Robin. 2004. *The Non-Designer's Design Book,* 2nd ed. Berkeley, CA: Peachpit Press.

Endnotes

1. See the earlier reference to the National Library of Medicine in this text in the chapter on audience analysis. Also, the guidelines listed here may be invaluable if you are asked to prepare a boilerplate for some researcher's experiment or send out descriptions for an Institutional Review Board or to grantors. You'll want to make sure the research components fit and satisfy the guidelines. Researchers and physicians are unfortunately not always aware of them. Medical schools have enough to teach and cannot cover every aspect. You may be able to work with the legal and regulatory officer for your health facility on these issues.

2. For useful texts on Web design, consult Robin Williams, *The Non-Designer's Design Book,* 2nd ed.; Jakob Nielsen and Marie Tahir, *Homepage Usability: 50 Websites Deconstructed;* Jonathan Price and Lisa Price, *Hot Text: Web Writing That Works.*

3. Macromedia Dreamweaver MX is an html editor and Web site management system with templates for Web page consistency and easy and quick update and maintenance capability. The templates also help maintain consistency between the browsers (Internet Explorer and Netscape Navigator). Macromedia Flash MX movies can be imported into Dreamweaver with relative ease, and Flash is used to enable interactivity on Web sites. Macromedia Fireworks supplies a "suite of user interface enhancements designed to speed common tasks, collaborate[s] easily with consistent version control capabilities and built-in FTP . . . and adds realism and depth to illustrations with new drawing tools including contour gradients, and dashed strokes" (Macromedia).

4. For a good example of a Web site fulfilling its purpose, see: http://www.roche.com/home.html.

5. The navigation bar used to appear always on the left side in the early Web sites that used the inverted L design. Now it can appear in a number of places. Usually, though, the bar is located at the bottom of the site in a fairly small font. Often the links are repeated in a larger menu/navigation bar somewhere else on the site.

6. Some logos are self-explanatory and don't need tag lines.

7. Minimal style refers to on-line documentation that has basic instructions: Do this; do that . . . in a briefly stated list. The elaborate style, whose criteria is listed in the text, includes scenarios with more examples, putting a person in a scene in which he or she can apply the particular task being taught.

8. Such a system of vitals monitoring on models was recently put into place in Clemson University's Department of Nursing.

9. Hint: Often in the nursing department, nursing professors also work some shifts or summers in a hospital or other medical facility. Much biomedical research is carried out on campuses. You might ask those professors which databases they use and how they prepare their graphs, tables, and line drawings. See which computer programs they use for that work.

APPENDIXES

Common Greek and Latin Roots, Prefixes, and Suffixes

Most medical terminology can be broken down into one or more word parts. For example, any medical term may contain one, some, or all of these parts:

- roots
- prefixes
- suffixes
- linking or combining vowels

An example of medical terminology with three of the above parts is the medical term *pericarditis*, which means inflammation of the outer layer of the heart. Pericarditis can be divided into three parts:

- the prefix, *peri*, translates to surrounding
- the root, *card*, translates to heart
- the suffix, *itis*, translates to inflammation

Hence, pericarditis is an inflammation of the layer surrounding the heart, anatomically known as the pericardium.

Medical terminology always consists of at least one root, although it may contain more. A prefix or suffix only adds to the specificity of the medical term and is not required. An example of this medical terminology is *sternocleidomastoid*, which is a muscle that has attachments at the sternum, the clavicle, and the mastoid. The term *sternocleidomastoid* can be divided into three parts: stern - o - cleid - o - mastoid. Notice that there are vowels between the three roots. These are linking or combining vowels, which serve to make some medical terminology easier to pronounce.

Some Latin and Greek roots, prefixes, and suffixes:

a-, an-	not, negative
ab-	away from

-ac	used instead of -ic if an -i immediately precedes. Example: *cardiac*—having to do with the heart.
-acea, -aceae	having the quality of, having the nature of, resembling. Example: *rosaceae*.
ad-	to, toward, near
acro-	extremity. *Acrophobia* is a fear of heights.
adeno-	gland. *Adenoid* is a lymph gland found in the nasopharynx.
alba-	white. *Albinism* is the white appearance of skin lacking melanin.
algia-	pain. *Neuralgia* is a pain following the course of a nerve.
allo-, alieni-	another's
ambi-	both. *Ambidextrous* means both-handed.
ana-	up, back again, throughout, against
angi-	vessel. *Angioplasty* is the repair of a blood vessel.
-ant, -ent, -ance, -ence, -ancy, -ency	person who, that which, state of being, condition. Example: *residency*.
ante-	before, in front of
anti-	against
apo-	away from
-ary, -arium	place for. Example: *sanitarium*.
arthro-	joint. *Arthritis* is the inflammation of skeletal joints.
-asis, -ase	a state or condition of
-ation, -ure	action or condition resulting from an action. Example: *ligation*.
auto-	self. *Autolysis* is the destruction of body cells by bodily enzymes.
bary-, gravi-	heavy
-be	life
bi-, bis-	twice, double
bio-	living. *Biology* is the study of living organisms.
blast-	germ, bud. *Osteoblast* is the germ of a bone cell.
-ble, -cle	instrument, means of an action, place. Example: *receptacle*.
blephar-	eyelid. A *blepharoplasty* is eyelid surgery.
brachi-	arm. The *brachialis* muscle moves the arm.
brachy-, brevi-	short
broncho-	trachea, windpipe. *Bronchitis* is the inflammation of the respiratory system.
bucc-	cheek. The *buccinator* muscle is in the cheek.
capit-	head. *Decapitate* means "Off With the Head!"
carcin-	cancer. A *carcinogen* is a substance that triggers cancer formation.
cardia-	heart. A *cardiologist* is a heart specialist.
cata-	down, across, under
-ce	state of. Example: *jaundice*.
-cele	tumor. Example: *hydrocele*.
ceno-, coelo-, vacuo-	empty

cephal-	head. *Cephalon* is another term for the brain. (See also capit-)
cerebro-	brain. *Cerebrospinal* fluid (CSF) is fluid circulating within the brain and spinal cord.
chloro-, prasini-, viridi-	green
chole-	bile, gall. *Cholecestectomy* is removal of the gallbladder.
chondro-	cartilage. A *chondrocyte* is a cartilage cell.
chroma-	color. *Chromosomes* are so named because they take color easily when dye is added to a cell.
chryso-, cirrho-, aureo-, flavo-, fulvi-	orange
cili-	eyelash. *Supercilia* are eyebrows—the hairs above the eyelashes.
circum-	around
coccino-, erythto-, rhodo-, eo-, purpureo-, rubri-, rufi-, rutuli-, rossi-, roseo-, flammeo-	reds of various shades
con-	with
contra-	against
corpus-	body. *Corpus* albicans is the white body inside an ovary.
corona-	crown. *Coronary* arteries supply blood to the heart muscle run along the heart, encircling it like a crown. The *coronary* ligaments of the liver encircle the liver like a crown.
cost-	rib. *Costal* cartilages attach ribs to the sternum.
cryo-, psychro-, frigidi-	cold
crypto-, calypto-, operti-	hidden
cyano-, iodo-, ceruleo-, violaceo-	blue
cut-	skin. *Cutaneous* tissue is skin tissue.
cyclo-, gyro-, circuli-	round
cysti-	sac, bladder
cyto-	cell. *Cytology* is the study of cells.
dactyl-	digits. *Polydactylism* is the presence of more fingers than is normal.
de-	down, from, away from
derma-	skin. *Dermatologists* are skin specialists. (See also cut-)
dexio-, dextri-	right
di-	two, twice, double
dia-	through
dis-	apart, removed
dyo-, duo-	two

dys-, caco-, mali-	hard, difficult, bad
dura-	tough, hard. *Dura* mater is the tough covering around the brain and spinal cord.
e-, ex- [Latin], e-, ec- [Greek]-	out of
ecto-	outside
-ectomy	cut. Example: *appendectomy*.
-eme, -em, -emesis, -emea	vomit, regurgitate. Example: *emesis* basin.
-emia, -amia	blood. Example: *anemia*.
en-	inside
endo-	within
ennea-, novem-	nine
entero-	intestine. *Enteritis* is inflammation of the intestines.
epi-	upon, on
erythro-	red. *Erythrocytes* are red blood cells.
-escent, -escence, -sc	beginning to be, becoming, to be somewhat. Example: *excrescence*.
eu-, kalo-, kallo-, boni-	well, good, easy. *Euphonia* means harmony or good sounds.
eury-, platy-, lati-	wide
exo-	outside, outward
extra-	outside, beyond, in addition to
-fact, -facture, -faction	make, do, build, produce; forming, shaping. Example: *putrefaction*.
-fer, -fera, -ferous	to bear, carry, produce. Example: *odiferous*.
-fy, -fic, -fictation	make, do, cause, produce; having the state of. Example: *putrefy*.
galacto-	milk; *galactose-* milk sugar. *Galactosemia* is the lack of ability to digest one of milk's sugars.
gastro-	stomach, belly. *Gastric* juices are produced in the stomach.
-gen, -genous, -genic	descent, origin, birth, creation, beginning. Example: *eugenic*.
glosso-	tongue. *Hypoglossal* means "below the tongue."
glyco-, dulci-	sugar, sweet. *Glycosuria* is sugar in the urine.
-gram, -graph, -graphy	to write. Example: *electrocardiagram*.
hema-	blood. *Hemaglobin* is a large molecule of the red blood cell.
hemi-	half
hen-, uni-	one
hepato-	liver. The *hepatic* vein drains blood away from the liver.
hepta-, septem-	seven
hetero-, allo-, vario-	different
hex, hexa-, sex-	six
holo-, toti-	entirely
homo-, homeo-, simili-	alike
hygro-, hydro-, humidi-	wet
hyper-	over, above. *Hyperglycemia* is blood sugar above normal.
hypo-, imi-, intimi-	under, below. *Hypoglycemia* is blood sugar below normal.
hyster-	uterus. *Hysterectomy* is the removal of the uterus.

-ia, -y	act, state of. Example: *polyuria*.
-iasis, -iosis	a morbid condition, a process. Example: *spondyliosis*.
-iatry, -iatric, -iatrician	treatment, practice of, one who, connected with. Example: *pediatrician*.
-ic, -tic	pertaining to, having to do with Example: *emetic*.
-ical -ic + -al,	pertaining to, having the nature of. Example: *pharmaceutical*.
-ics	things having to do with art, science, study of [usually used with a singular verb]. Example: *optics*.
-ida, -idae, -id	descended from, related to. Example: *spina bifida*.
idio-, proprio-, sui-	one's own
ileo-	ileum (part of the small intestine)
ilio-	ilium (part of the hip bone)
in-	in, into, on. You often see this prefix as *im;* used with verbal roots.
in-	not, occasionally, beyond belief
infra-	below
inter-	between
intro-	within
intus-	within
-ion, -idion, -arion, -isk	small, diminutive. Example: *meniscus*.
iso-, equi-	equal
-ist, -est, -ast, -ician	someone who believes in, professes, or practices. Example: *bacteriologist*.
-ite	one connected with
-itis	inflammation of. Example: *tendonitis*.
-ium, -eum	part, lining or enveloping tissue. Example: *ileum*.
-ize	to make, to do. Example: *cauterize*.
lachry-	tears. *Lacrimal* glands secrete tears.
lepto-	thin
leuko-, albo-, argenti-	white. *Leukocytes* are white cells of the blood. (See also alba-)
lingua-	tongue. *Sublingual* glands are beneath the tongue. (See also glosso-)
lipo-	fat. *Liposuction* is the removal of fat by suction tube.
lith-	stone. Shock wave *lithotripsy* is a treatment for breaking up kidney stones.
-logy	study. Example: *cardiology*.
lumbo-	lower back. *Lumbar* vertebrae are located in the lower back.
-lysis, -lytic, -lyst, -lyte	loosening, dissolving. Example: *catalyst*.
-m, -me, -ma [base, -mat-]	result of. Example: *edema*.
malaco-, molli-	soft
macro-, dolicho, longi-	long
macul-	spot, blotch. The *macula* lutea is a spot on the retina of an eyeball.
mamm-	breast. A *mammogram* is an X ray of the breast.
mast-	breast. *Mastectomy* is the removal of a breast.
mega-, megalo-, makro-, magni-, grandi-	large

melano-, nigri-	black
-men, -ment, -mentum	action or condition resulting from action. Example: *specimen*.
meningo-	membrane. *Meninges* are the coverings of the brain and spinal cord.
meso-, medio-	middle
meta-	with, after, beyond. The cancer has *metastasized* (gone beyond the original site).
-meter, -metry, -metro, -metrics, -metrist, -meters	to measure. Example: *telemetry*.
metro-	uterus. *Endometrium* is the inner lining of the uterus.(See also hystero-)
micro-, parvi-	little. A *microscope* looks at small things.
morpho-	shape. *Endomorphs* are people whose physical shape extends to the limits of human dimension.
myelo-	spinal cord. *Poliomyelitis* is inflammation of the greymatter of the spinal cord.
myo-	muscle. A *myocardial* infarction is a problem with the heart muscle.
myri-, myriad-	large or countless number
necro-	death. *Necrosis* is the death of cell tissue.
neo-, ceno-, novi-	new
nephro-	kidney. *Nephrons* are the functional units of a kidney.
neuro-	nerve. *Neurons* are individual nerve cells.
non-	not. The tumor is *nonmalignant*.
octo-	eight
oculo-	eye. An *oculist* supplies eyeglasses.
odont-	tooth. *Orthodontics* refers to the repair of teeth.
-oid	resembling, like, shaped. Example: *fibroid*.
oligo-, pauci-	few
-oma, -ome, -omatoid	tumor, morbid growth; to swell, to bulge, to mass. Example: *hematoma*.
-on, -tron, -tronic	used in reference to an apparatus. Example: *electronic*.
onco-	tumor. *Oncology* is not doctors wearing pagers; it is the study of cancerous tumors.
ophthalm-	eye. *Ophthalmology* is the study of eye and its diseases.
opistho-	beyond
-or	that which. Example: *flexor*.
-orium, -ory	place which. Example: *sanitorium*.
oro-	mouth. The *oral* cavity is the other name for the mouth.
orchido-	testicle. *Orchidectomy* is removal of a testicle.
osse-, osteo-	bone. *Osteoporosis* is porosity of bone.
-osis, -sis, -sia, -sy, -se	actor, process, condition or state of, result of.
oto-	ear. *Otosclerosis* is the formation of bone in the ear. *Otomycosis* is a fungal infection in the ear.
oxy-; acri-	sharp
pachy-, pycno-, steato-, crassi-	thick
paleo-, archeo-, veteri-, seni-	old
palin-	again
para-	alongside of, beside

patho-	disease. *Pathogens* are agents that cause disease.
penta-, quinque	five
peps-	digestion. *Pepsin* is an enzyme found in the digestive system.
per-	through, thorough, complete
peri-	around, near. *Pericarditis* is an inflammation around the heart.
phago-	eat. *Phagocytes* are cells (cyto-) that eat foreign material.
philo-, -phily, -phil	love, to have an affinity for. *Hydrophilic* molecules are attracted to water (hydro-).
phleb-	vein. *Phlebitis* is inflammation of the veins.
-phobe, -phobia	to fear. Example: *claustrophobia*.
-phor, -phoresis, -phori, -phoro, -phoria, -phore	to bear, to carry, turning, directing, producing.Example: *leukophoresis*.
phren-	diaphragm. *Phrenic* refers to a diaphragm.
picro-, amari-	bitter
pneumo-	lung. *Pneumonia* is a disease of the lungs.
-poesis	make
polio-, glauco-, amauro-, cani-, cinereo-, atri-	gray
poly-, multi-	many
porphyro-, puniceo-, purpureo-	purple
post-	after, behind. A *postoperative* therapy regime is one after an operation.
pre-	in front of, before. A *preop* medication is one givenbefore an operation.
pro-	before, in front of
proso-, proto-; frontali-	onwards, in front
pulmo-	lung. *Pulmonary* functions are tests to determine lung functioning.
pyo-	pus. *Pyruria* is pus in the urine.
quadrati- rectanguli-	square
re-	back, again
ren-	kidney. The *renal* artery supplies blood to the kidney.
retro-	backward
rhin-	nose. *Rhinoplasty* is a nose job.
scaio-, scaevo- levi, sinistri-	left
scler-	hard. *Atherosclerosis* is hardening of the arteries. (See also dura-)
-scope	see into. Example: *microscope*.
semi-	half
sesqui-	one and one-half
-sis	act, state, condition of. Example: *hemolysis*.
stasis-	stand still. *Homeostasis* is the process of maintaining constant conditions within the body.
-stomy	mouth, opening. A *colostomy* is the surgical formation of an opening to the anus.
steno-, angusti-	narrow
stheno-, validi-, potenti-	strong

sub-	under, below. A *subdural* hematoma is below the skull (dura).
super-, supra-	above, upper
syn-	with
tauto-, identi-	same
-tes, -ter, -tor, -te, -t	the agent, he who, that which, means of, place for. Example: *gamete*.
tetra-, tessaro-, quadri-	four
thermo-, calidi-	hot
thromb-	clot, lump. *Thrombosis* refers to a clot in the heart or blood vessel.
-tia	quality, condition, state
trans-	across
tri-	three
trich-	hair. *Trichosis* is a disease of the hair.
-tude	condition, state, quality
ultra-	beyond
-ulum, culum, trum	instrument, means of an action, place. Example: *speculum*.
vas-	vessel, duct. *Vas* deferens is the vessel that carries* sperm from the epididymus.
viscer-	organ. *Visceral* refers to organs.
xantho-, ochreo-, fusci-, luteo-	yellow
xero-, sicci-	dry
-y	condition, state
zoo-	animal. *Zoology* refers to the study of animals.

Works Cited

Medical Supplies Web site. Accessed Dec. 17, 2003. http://www.medical-supplies-4u.net/medical-terminology.html

Nguyen Hy Hau Scientific Terminology Web site. Accessed Dec. 17, 18, 2003. http://www.lpmc.ens.fr/~nguyen/SuffList.html, http://www.lpmc.ens.fr/~nguyen/SGFrame.html, http://www.lpmc.ens.fr/~nguyen/SLFrame.html

Technion Web site. Accessed Dec. 17, 2003. http://www.technion.ac.il/medicine/Students/latin&Greekprefixes.html

University of Idaho Web site. Accessed Dec. 17, 2003. http://www.class.uidaho.edu/luschnig//EWO/18.htm

Wordsources Web site. Accessed Dec. 17, 18, 2003. http://www.wordsources.com/cgi/ice2-for.cgi

Frequently Used Visual Design Terms

Arrangement is one of the five modes of rhetoric. In the past *arrangement* referred to the words within a text or a speech. For our purposes here, we include *arrangement* of text and visuals on pages, screens, and a variety of media. How one arranges information is key in medical texts. We *arrange* information both logically in terms of content and visually in terms of design. The *arrangement* of information must make sense intuitively, too, and be appropriate for your users. For example, you would not suggest that paramedics begin to resuscitate someone by blowing two breaths into the mouth before clearing the airway.

Chartjunk is a term coined by Edward Tufte, a well-known visual communicator, referring to the overloading of graphs, charts, and other visual representation with too much data or too many visuals. (See Works Cited in Chapter 10.) (1997, 65).

For example, if there is so much information or too many graphics to get a sense of what the important information is, then you are guilty of the *chartjunk* phenomenon that Tufte describes. Especially in medicine, it's key to have information presented in the most effective way, since patients' lives and well-being are at stake. Extraneous materials can distract the user from the most important information at best and confuse them at worst.

Chunking is a concept well known to designers and educators. One *chunks* information about the same topic, not just in paragraph form, but also in smaller segments in documents and on screens so that the reader can quickly see that these pieces of information relate to each other.

For example, if a health practitioner needs to fill out a procedure form that includes both a cardiac section and a pulmonary section on one page, the user/reader would expect to have all the cardiac information *chunked* together and all the pulmonary information *chunked* together on a different part of the page.

Color cuing refers to "using colors to focus attention, simplify information, group elements, and create separate layers of information" (Hilligoss 147).

For example, for the situation above, you might want to have the font color of the emergency information in red, especially if your audience is a Western (Euro-American) one.

Color schemes or *palettes* focus on using colors that fit both the rhetorical situation as well as fit well together. For example, you would not produce flyers about LaLeche classes for nursing mothers at your clinic in blacks, grays, and whites, unless the only printer available was black and white. Instead, you would use colors associated with babies or a warm and nurturing female audience. If you have a large document or a group of different documents associated with a program, choose colors and logos by creating a color palette that you use for all the publications, whether print-based or electronic. Be aware that print-based documents use one set of colors (CYMK), while electronic documents use another (RGB). CYMK refers to cyan (blue), yellow, magenta, and black; RGB refers to red, green, and blue. However, even if you design across numerous platforms, you can match colors closely enough for your various projects. Also, drawing software with its tool kit can help you duplicate colors (showing both the value numbers and allowing you to "eyedropper" colors from one document to another within the same or similar software programs). An "eyedropper" is one visual tool used by programs such as Adobe PhotoShop to duplicate exactly the color from one site to another site; the tool is an icon of an eyedropper.

When one of my graduate classes designed an 87-page bilingual health diagnostic booklet, *Diagnosticando la Salud del Trabajador,* we gained permission from the National Center for Farm Worker Health to use artist Tony Ortega's *La Reunion* on the front cover of our publication (Heifferon et al.). (See Works Cited in Chapter 2.) Because of the rich and varied tones, using Adobe Illustrator, we were able to use the colors from the painting to create our own customized color palette for the publication. The section dividers, title bars, etc., were all colors within that palette, which gave the large document both consistency and a good, aesthetic visual appeal. Although you cannot see the colors in this text, Chapter 2 Exhibit 2.2 illustrates the principle of using a color palette to help keep colors from clashing and distracting the users. If you outsource large printing jobs, you may need to send the palette with your other electronic files so printers can reproduce the colors that you desire.

Contrast means that within documents and screens, information and/or visuals need to stand out and be distinguishable from one another. *Contrast* can be accomplished through use of color, patterns, fonts, italics, point-size, and numerous other means.

For example, you might want to have emergency information in a booklet designed for patients in a home healthcare situation set in *contrast* to routine care information so that patients and caregivers can immediately find the emergency care information when needed.

Conventions refer to reader or user expectations. Many genres are so familiar to your audience that they will already have expectations about certain medical forms, documents, health Web sites, or charts that you will want to honor. When you begin to consider your design, you will want to decide early in the process how much change you want to introduce in a genre you are using and how much you want to meet reader expectations. Sometimes breaking *conventions* works well, but often you will get resistance or confusion from users if you do this without rhyme or reason.

For example, many doctors' offices meet some resistance when they move from traditional print-based charting and medical records to electronic record keeping. Good managers know, as do good document designers, that when one is introducing some changes, it is often better not to change everything at once.

Earlier in Chapter 3, we mentioned paying attention to *conventions* in regard to style of font, size of font, headings, case choices, margins, and justifications. Style sheets are the *Bibles* of such generic conventions. If your organization has one, you will be obligated to use it.

Figure-ground separation is a concept familiar to visual and graphic designers, architects, and artists. Because more and more written communication tends to be visual, however, we include it here as part of good document design. This concept is similar to contrast, in that we, as visually capable humans, depend on being able to distinguish what is closer to us and what is further away. We need to be able to organize visuals into what is background and what is foreground, or what is less important and what figure or image is being emphasized.

For example, we have all visited amateurs' Web sites in which the text and images get lost in a busy background, or colors are too similar in intensity or saturation. As users, we are frustrated if we cannot distinguish one from the other and have to work to read text or view graphics.

Focal points are an obvious reference to the main places in a document that draw readers' eyes. As a good document designer, if some key text or an image is what you want your reader to see first, do not then create another *focal point* of less important information. Western readers' eyes tend to move from left to right and top to bottom in a document; therefore don't embed the focal point at bottom right on the page or screen.

For example, an intriguing border on a flyer may distract from the key focal point: an announcement of training sessions on new ER equipment. Or putting a picture of the equipment outside of the natural path of readers' eyes, which includes the middle of the page as an appropriate site, would possibly draw attention away from the image you want readers to notice first. This distracting focus problem also can relate to Tufte's term *chartjunk*, since too many different focal points on a document can make it difficult for readers to determine what is the most important information.

Graphics, like images and visuals, include charts, graphs, photos, line drawings, pictures, icons, and usually anything that is not text. We know that to be able to reach many learning styles, we often need to include graphics to reach visual learners. Graphics do need to match text or illustrate something within the text and need not be used randomly, at least in the Western Euro-American cultures. (See the chapter on multicultural design for other perspectives.)

For example, do include graphics for health education materials and adapt them to the particular audience you are attempting to reach. Few audiences within North American culture respond well to text-heavy documents particularly on the Web, unless their profession requires such documents. Even professional journal articles are beginning to include more graphics.

When you do include graphics, make sure they illustrate what it is you intend to convey. Sometimes users are confused when graphics don't seem appropriate to the text. Make sure, too, if a graphic is difficult to decipher (such as the picture of a new piece of equipment) that the text carefully explains the parts of the machine. Good labeling and use of diagonal lines drawn between labels and components can be very helpful to your user.

Hues refer to colors themselves, whether primary (red, yellow, blue in pigment; or red, green, blue in light) or secondary colors (orange, green, violet in pigment; or cyan, magenta, yellow in light) [http://www.artlex.com]. Document designers use RGB (red, green, blue) for on-line work and CMYK (cyan, magenta, yellow, black) for offset printing, a.k.a. four-color printing [http://www.webopedia.com]. Computer software design packages include choices when you package a design. If the design is for Web work and other electronic media, you can choose to save it as RGB color choice. If your design is to be printed, either in-house or sent out, save it as CMYK so that the colors will print up true to the ones you intend. It's also a good idea to double-check your color choices for print documents by asking to see a copy before the printer does the complete run of your print job. Many printers will volunteer to do this check.

Icons are small, stylized images often used to represent areas or categories on Web sites or to serve as buttons to click to open up other Web pages.

For example, if you use icons in a health Web site, you must make sure they are recognizable to your audience and used consistently on the site to denote the same meaning throughout. Don't use a red heart-shaped button for cardiac information early in the site and then switch to another symbol later in the site.

Orientation should not be confused with other forms of orientation; for our purposes, the term refers to directionality in document design. How one orients a page or a graphic or text within a document is somewhat related to arrangement, but *orientation* is a term that printers and designers will also recognize and refers to physical location of material.

For example, if you use Microsoft Word to word process, you are familiar with setting up the printer to print either landscape (horizontal) or portrait (vertical). We also need to include legal landscape or portrait size, brochure sizes, poster sizes (11 X 17 inches and larger), and computer landscapes and portrait sizes, depending on the most common size of the screen currently in use (Kostelnick and Roberts 1998, 184. See Chapter 10 for Works Cited.) This information is important to include as you describe your design project. In some document designs, either orientation could work, so it's important to specify to others how you want the work displayed.

Production refers to the process that includes all the steps involved in developing your design into a deliverable product. Whether you are part of the production process or the design process or both will depend on where you work and how you function within that unit.

Portrait Orientation   **Landscape Orientation**

Exhibit B.1 Orientations

In communication, design, and educational departments within large health systems, you will produce deliverables (print, electronic, and visual products) for various health audiences. To manage larger projects well, you will need to plan your production schedule carefully. At the end of Chapter 3, you'll find a section to help you understand the planning process.

Saturation refers to the intensity of a color. Like hue, this measure is important to use consistently and determines how the color looks in your health communication project.

For example, you will need different saturations depending on what medium you use. Color saturations differ whether they are on print-based materials or on the Web, so you will want to test-run certain colors in different media. Colors also differ on the Web from browser to browser, so you may want to test them in the two main browsers, Netscape Navigator and Internet Explorer.

Scale is the size and perspective of images, graphs, and charts. You will need to think of *scale* as you choose fonts and graphics for documents. Scale means how components appear in perspective to others and should reflect both the message you want to deliver and the logic and sense of size you portray to your audience. In order for your user to correctly process visual information, you need to keep elements of your design within an appropriate scale. Don't include, for example, a large foot and a very small child on the same poster to illustrate a pediatric foot specialist. Make the foot fit the same scale as the child (make the child larger and circle the foot to help users logically process the information).

Similarity, like chunking, is a concept that groups like with like not on the basis of proximity to each other or content, but instead on the basis of form, shape, texture, color, or direction. If you think back to early childhood classes in which you had to circle what two things were like each other, you see that in this culture we are trained to process information this way. In other words, we have certain *schema* or slots already set up for processing information. Good health communicators will design materials to fit the schemas in whatever culture in which they work.

Value is part of our color terminology and refers to shades of color, particularly the amount of black or white added to pure colors that make various *values* of the original color.

For example, if you work for a large health corporation, you will have a brand or logo that goes out on the materials you develop as a health communicator. You will need to use the exact shade or value of the color that is standard for the corporation. You must have permission to vary it. In Web-site design you can use the identical code or use one of the tools in the newer software programs to replicate color.

APPENDIX **C**

Frequently Used Clinical Abbreviations

AARP	American Association of Retired Persons
ABGs	arterial blood gases
AC	acromioclavicular
AC>BC	air conduction greater than bone conduction
ADHD	attention deficit hyperactivity disorder
ADL	activities of daily living
AFP	alph-fetoprotein
ALT	alanine transaminase
AP	anterior posterior
AROM	active range of motion
ASA	aspirin
ASAP	as soon as possible
AST	aspartate transaminase
AUA	American Urology Association
AV	anteverted
AV nicking	arteriovenus nicking
Bands	banded neutrophils
BASO	basophil
BCP	birth control pill
BID	two times a day
BM	bowel movement
BP	blood pressure
BPH	benign prostatic hypertrophy
BPPV	benign paroxysmal positional vertigo
BS	bowel sounds, breath sounds
BSE	breast self exam
BUN	blood urea nitrogen

BUS	Bartholin's, urethral, and Skene's glands
C/O	complains of
CBC	complete blood count
C & S	culture and sensitivity
CA	cancer
CABG	coronary arterial bypass graft
CAD	coronary artery disease
CBE	clinical breast exam
CDC	Centers for Disease Control
CHF	congestive heart failure
CMT	cervical motion tenderness
CN	cranial nerve
COPD	chronic obstructive pulmonary disease
CPR	cardiopulmonary resuscitation
CSF	cerebrospinal fluid
CTA	clear to auscultation
CV	cardiovascular
CVA	cerebrovascular accident
CVAT	costovertebral angle tenderness
CXR	chest X ray film
D & C	dilation and curretage
d/c	discharge
DDST	Denver developmental screening test
D & E	dilation and evacuation
DIP	distal interphalangeal
DM	diabetes mellitus
DMPA	Depo-Provera
DOE	dyspnea on exertion
DTAP	diptheria, tetanus, acellular pertussis
DTR	deep tendon reflexes
DUI	driving under the influence
DVT	deep venous thrombosis
ELISA	enzyme-linked immunosorbent assay
ED	emergency department
EDC	estimated date of confinement
EMG	electromyogram
ENT	ear, nose, throat
EOMI	extraocular movements intact
ESR	erythrocyte sedimentation rate
ETOH	ethyl alcohol
FB	foreign body
FBS	fasting blood sugar
FEV1	forced expiratory volume in 1 second
FH	family history
FROM	full range of motion
FSH	follicle-stimulating hormone

FTT	failure to thrive
FVC	forced vital capacity
FX	fracture
GAS	general anxiety syndrome
GBS	group B streptococcus
GC	gonococcus
GERD	gastroesophageal reflux disease
GGT	gamma-glutamyl transpeptidase
GI	gastrointestinal
GPA	gravida para abortions
Gtt	drop
GU	genitourinary
Gyn	gynecological
H/A	headache
HCG	human chorionic gonadotropin
HCT	hematocrit
HCTZ	hydrochlorothiazide
HDL	high density lipid
HEENT	head, ears, eyes, nose, throat
Hep B	hepatitis B vaccine
Hgb	hemoglobin
Hib	hemophilus B influenza
HIV	human immunodeficiency virus
HJR	hepatojugular reflex
HPV	human papilloma virus
HR	heart rate
HRT	hormone replacement therapy
HS	hour of sleep
HSM	hepatosplenomegaly
Ht	height
HTN	hypertension
HZ	herpes zoster
IADL	independent activities of daily living
IBS	irritable bowel syndrome
IPV	intermuscular polio vaccine
IVDA	IV drug abuser
JVD	jugular vein distention
L & D	labor and delivery
LA	lymphadenopathy
LCTA	lungs clear to auscultation
LD	learning disabled
LDL	low density lipid
LH	luteinizing hormone
LLQ	left lower quadrant
LNMP	last normal monthly period
LOC	loss of consciousness

LUQ	left upper quadrant
LVH	left ventricular hypertrophy
Lymph	lymphocytes
mammo	mammogram
MCH	mean corpuscular hemoglobin
MCHC	mean corpuscular hemoglobin concentration
MCL	midclavicular line
MCP	metacarpal phalanx
MCV	mean corpuscular volume
MDI	multi-dose inhaler
MGF	maternal grandfather
MGM	maternal grandmother
MI	myocardial infarction
ML	midline
MM	mucous membranes
MMR	measles, mumps, rubella
Monos	monocytes
MRG	murmur, rub, gallop
MS	multiple sclerosis
MSG	monosodium glutamate
MTP	metatarsal phalanx
MVA	motor vehicle accident
MVI	multi-vitamin
N/V	nausea and vomiting
NAD	no acute distress
neuro	neurologic
NIDDM	non-insulin dependent diabetes mellitus
NKA	no known allergies
NKDA	no known drug allergies
NSAIDs	nonsteroidal anti-inflammatory drugs
NSR	normal sinus rhythm
NVD	nausea/vomiting/diarrhea
OA	osteoarthritis
OBG	obstetrical/gynecological
OC	oral contraception
OCP	oral contraceptive pills
OD	right eye
ODD	oppositional defiant disorder
OM	otitis media
OPV	oral polio vaccine
OS	left eye
OTC	over the counter
OU	both eyes
PCN	penicillin
PE	physical examination
PEFR	peak expiratory flow rate

PERRL	pupils equal round reactive to light
PERRLA	pupils equal round reactive light & accommodation
PFT	pulmonary function test
PGF	paternal grandfather
PGM	paternal grandmother
PID	pelvic inflammatory disease
PIP	proximal interphalangeal
PMI	point of maximal impulse
PND	paroxysmal nocturnal dyspnea
PPD	purified protein derivative
PRN	as needed
PSA	prostate specific antigen
PTT	partial thromboplastin time
PUD	peptic ulcer disease
QD	every day
QID	four times a day
QOD	every other day
QS	quantity sufficient
R	respirations
RA	right arm
RAM	rapid alternating movements
RBC	red blood cell (count)
RCM	recommended daily
RDA	recommended daily allowance
RDW	red (cell) division width
RLQ	right lower quadrant
ROM	range of motion
ROS	review of symptoms
RR	respiration rate
RRR	regular rhythm, rate
RTC	return to clinic
RUQ	right upper quadrant
S/P	status post
SEG	segmented neutrophils
SEM	systolic ejection murmur
SLR	straight leg raise
SOB	shortness of breath
SPF	sun protection factor
STD	sexually transmitted disease
T	temperature
T&A	tonsillectomy and adenoidectomy
TID	three times a day
TM	tympanic membrane
trich	trichomoniasis
TSH	thyroid-stimulating hormone
U/A	urinalysis

UGI	upper gastrointestinal
URI	upper respiratory infection
UTI	urinary tract infection
VA	visual acuity
VS	vital signs
WBC	white blood count
w/o	without
WNL	within normal limits
Wt	weight
y/o	years old

CREDITS

Pages 37–38, Exhibit 2.1. Reprint of the American Medical Writers Association Code of Ethics. Used with the permission of the American Medical Writers Association.

Page 46, Exhibit 2.3. Reprint of cover of *Diagnosticando la Salud del Trabajador* with permission of the authors, Jacob Barker, Melissa Tidwell, et al. Clemson, S.C.: Clemson University, 2002. Cover art used with permission of the National Farmworkers Association.

Page 74, Exhibit 3.5. Communicare Booklet. Reprinted with the permission of the Northeast Georgia Medical Center Public Relations Department.

Page 77, Exhibit 3.7. Hackos' Publications Development Life Cycle used with permission of John Wiley and Sons.

Page 95, Exhibit 4.1. Schematic of Clinical Decision Making. Printed with permission of Lippincott Publishers in Denise L. Robinson's *Clinical Decision Making,* 2nd ed., 2002.

Page 117, Exhibit 5.1. Sullivan Center Encounter Form Part 1. Reprinted with permission of Joseph F. Sullivan Center, Clemson University, Clemson, S.C.

Page 117, Exhibit 5.2. Review of Systems. Reprinted with permission of Joseph F. Sullivan Center, Clemson University, Clemson, S.C.

Page 118, Exhibit 5.3. Sullivan Center Physical Exam Section. Reprinted with permission of Joseph F. Sullivan Center, Clemson University, Clemson, S.C.

Page 119, Exhibit 5.4. Encounter Form—Description of Activities. Reprinted with permission of Joseph F. Sullivan Center, Clemson University, Clemson, S.C.

Page 120, Exhibit 5.5. Social History and Language. Reprinted with permission of Joseph F. Sullivan Center, Clemson University, Clemson, S.C.

Page 121, Exhibit 5.6. Sullivan Center—Problem List. Reprinted with permission of Joseph F. Sullivan Center, Clemson University, Clemson, S.C.

Page 122, Exhibit 5.7. Sullivan Center—Review of History. Reprinted with permission of Joseph F. Sullivan Center, Clemson University, Clemson, S.C.

Page 137, Exhibit 6.1. Cover of *Healthwise Handbook.* Reprinted with permission of Partners for a Healthy Community. *HealthwiseHandbook.* Boise, ID: Healthwise, Inc. 1997.

Page 139, Exhibit 6.2. Ear Infections (*Healthwise Handbook,* page 170). Reprinted with permission of Partners for a Healthy Community. *Healthwise Handbook.* Boise, ID: Healthwise, Inc. 1997.

Page 140, Exhibit 6.3. Ear Infections continued (*Healthwise Handbook,* page 171). Reprinted with permission of Partners for a Healthy Community. *Healthwise Handbook.* Boise, ID: Healthwise, Inc. 1997.

Page 141, Exhibit 6.4. Inside back cover (*Healthwise Handbook,* 1999). Reprinted with permission of Partners for a Healthy Community. *Healthwise Handbook.* Boise, ID: Healthwise, Inc. 1999.

Page 152, Exhibit 6.7. Mammogram bookmark (U.S. Department of Health and Human Services. Pub. 2/4/98).

Page 217, Exhibit 9.2. Flow Chart for Translation Document (Surgical Manual)—Manufacturer. Based on information from Bruce Maylath's "Translating User Manuals: A Surgical Equipment Company's 'Quick Cut.'" *Global Contexts: Case Studies in International Communication.* Deborah S. Bosley, Ed. Boston: Allyn & Bacon, 2001. Used with permission.

Page 218, Exhibit 9.3. Flow Chart for Translation Document (Surgical Manual)—Translation Co. Based on information from Bruce Maylath's "Translating User Manuals: A Surgical Equipment Company's 'Quick Cut.'" *Global Contexts: Case Studies in International Communication.* Deborah S. Bosley, Ed. Boston: Allyn & Bacon, 2001. Used with permission.

Pages 241–245, Exhibit 10.1. Facilities Guidelines Table 7.2. Used with permission of the Facilities Guidelines Institute.

Page 273, Exhibit 11.3. Query on Patient Accounts Form: A Procedural Tutorial. Used with permission of Allyn & Bacon/Pearson.

INDEX

Aazur-Rimetz, Sandra A., 269
A.D.A.M., Inc., 155
Aetna IntelliHealth, 156
Albers, Josef, 253
American Diabetes Association, 259
American Society of Nuclear Cardiology, 265
American Health Lawyers Association (AHLA), 57
American Medical Association Manual of Style, 40, 45, 47
American Medical Writers Association (AMWA), 32, 36–38, 59–61, 112
American Specialty Health Networks (ASH Networks), 155
American Specialty Health Plans of California, Inc. (ASH Plans), 155
Archives of Dermatology, 238, 246
Arrangement
defined, 297
Artlex Art Dictionary, 300
Asher, J. William, 83, 228
Audience
multicultural, 152–153
segmentation, 20
usability testing and, 26–27
writing for differently abled, 51–52
Audience analysis
assumptions and attitudes, 15, 22
class, 8–12
determining audience needs, 6, 9–10, 21

education and, 8, 13
focus groups, 23–25
gender, 8, 11
geography of, 2–3, 7
global, 10
history of, 2–3, 7
race, 8, 11
Audiotapes, 275–276

Barker, Jacob, 46
Barker, Thomas T., 268–272, 274–275
Barnett, Rebecca, 208
Barrells, Howard S., 92
Bates, Barbara, 93
Benton, David C., 42
Berger, John, 66, 233–234
Bergstrom, Nancy, 114
Bernhardt, Stephen A., 196–197
BOBBY, 51, 61
Bonk, Robert, 197
Bosley, Deborah S., 208
Braden Risk Assessment Scale, 106, 114
Braden, Barbara, 114
Braun, K. L., 208
Brewer, Douglas J., 230
Briscoe, M. H., 253
Brownson, Ross C., 205, 208
Brunard, Philip, 48
Budget. *See also* Grants.
grants, 191
red flags in, 191
Bulechek, G., 108

Campbell, C. P., 205, 212–214, 220–221
Capra, Donald J., 154
Cardiosource, 266

Carroll, John M., 269
CDM (Clinical Decision Making)
defined, 93–96
heuristic, 96
online charting, 104–109
patient history, 97–99
physical exam, 99–102
tentative diagnosis, 96
CFIDS Association of America, 266
Charney, Davida, 269
Chartjunk, 297, 301
Chin, Arthur E., 59
The Cholera Inquiry Committee, 232
Chou, C., 22
Clemson University Institutional Review Board, 58, 176
CMS (Centers for Medicare and Medicaid Services), 199
Cochrane Collaboration, 259
Color cuing, 298
Color palettes, 298
Color schemes, 298
Context sensitivity, 274
Contrast, 298
Conventions, 298
Cooper-Patrick, L., 208
Copyright
Act, 42
guidelines, 38–39
illustrations, 44
law, 42
digital copyright, 47
duplicate publication, 40–41
ethics and, 32
freeware, 48
graphics, music and video, 49

DATE			